STABLE ISOTOPES

CONFERENCE ORGANISING COMMITTEE

T. A. Baillie
D. S. Davies
C. T. Dollery
G. H. Draffan
A. M. Lawson

STABLE ISOTOPES

Applications in Pharmacology, Toxicology and Clinical Research

Proceedings of an International Symposium on Stable Isotopes held at the Royal Postgraduate Medical School, London, U.K., on January 3rd–4th, 1977, and sponsored by the British Pharmacological Society

Edited by

T. A. BAILLIE

Department of Clinical Pharmacology, Royal Postgraduate Medical School, Hammersmith Hospital, London, U.K.

First published 1978 by
THE MACMILLAN PRESS LTD
London and Basingstoke
Associated companies in Delhi Dublin
Hong Kong Johannesburg Lagos Melbourne
New York Singapore and Tokyo

Printed in Great Britain by
UNWIN BROTHERS LIMITED
The Gresham Press, Old Woking, Surrey
A member of the Staples Printing Group

British Library Cataloguing in Publication Data

International Symposium on Stable
Isotopes, *London, 1977*
Stable isotopes.
1. Isotopes – Congresses 2. Pharmacology – Congresses 3. Medicine, Experimental – Congresses
I. Title II. Series III. Baillie, T. A.
616′.007′2 R853.I/

ISBN 0-333-21747-0

Contents

Bodies Providing Support to the Symposium

The Royal Society
Abbott Laboratories Ltd
Allen and Hanburys Ltd
Beecham Pharmaceuticals
Boots Company Ltd
Finnigan Instruments Ltd
Geigy Pharmaceuticals
Glaxo Research Ltd
Hoechst UK Ltd
Imperial Chemical Industries Ltd
Jeol (UK) Ltd
LKB Instruments Ltd
Lilly Research Centre Ltd
May and Baker Ltd
Merck, Sharp and Dohme Ltd
Nicholas International Ltd
Organon Laboratories Ltd
Parke-Davis
Pfizer Ltd
Prochem
Roche Products Ltd
Sandoz Products Ltd
G. D. Searle and Company Ltd
Syntex Pharmaceuticals Ltd
Upjohn Ltd
Varian MAT GmbH
VG-Organic Ltd
The Wellcome Research Laboratories
Sterling-Winthrop Group Ltd

The Contributors

M. M. Ames, Department of Pharmaceutical Chemistry and Division of Clinical Pharmacology, University of California, San Francisco, California 94143, USA

R. A. Anderson, Department of Forensic Medicine, University of Glasgow, Glasgow G12 8QQ, UK

E. Änggård, Department of Pharmacology, Karolinska Institutet, 104 01 Stockholm, Sweden

M. Appler, Department of Pharmacology and Experimental Therapeutics and of Medicine, The Johns Hopkins University School of Medicine, Baltimore, Maryland 21205, USA

M. Axelson, Department of Chemistry, Karolinska Institute, Stockholm, Sweden

T. A. Baillie, Department of Clinical Pharmacology, Royal Postgraduate Medical School, Ducane Road, London W12 0HS, UK

J. D. Baty, Department of Biochemical Medicine, Ninewells Hospital, Dundee DD2 1UD, UK

H. R. Black, Lilly Laboratory for Clinical Research, Wishard Memorial Hospital, Indianapolis, Indiana 46202, USA

C. Bogentoft, Department of Organic Pharmaceutical Chemistry, Biomedical Center, Box 574, S-751 23 Uppsala, Sweden

A. R. Brash, Department of Clinical Pharmacology, Royal Postgraduate Medical School, London W12 0HS, UK

K. Brendel, University of Arizona, Tucson, Arizona, USA

R. B. Brundrett, Department of Pharmacology and Experimental Therapeutics and of Medicine, The Johns Hopkins University School of Medicine, Baltimore, Maryland 21205, USA

A. R. Buckpitt, NIH, Bethesda, Maryland, USA

A. L. Burlingame, Space Sciences Laboratory, University of California, Berkeley, California, USA

N. Castagnoli, Department of Pharmaceutical Chemistry and Division of Clinical Pharmacology, University of California, San Francisco, California 94143, USA

M. Colvin, Department of Pharmacology and Experimental Therapeutics and of Medicine, The Johns Hopkins University School of Medicine, Baltimore, Maryland 21205, USA

M. E. Conolly, Department of Clinical Pharmacology, Royal Postgraduate Medical School, London, W12 0HS, UK

P. J. Cox, Chester Beatty Research Institute, Institute of Cancer Research, Royal Cancer Hospital, Fulham Road, London SW3 6JB, UK

T. Cronholm, Department of Chemistry, Karolinska Institutet, Stockholm, Sweden

T. Curstedt, Department of Chemistry, Karolinska Institutet, Stockholm, Sweden

B. Dahlström, Department of Pharmaceutical Pharmacology, Biomedical Center, Box 573, S-751 23 Uppsala, Sweden

D. S. Davies, Department of Clinical Pharmacology, Royal Postgraduate Medical School, Ducane Road, London W12 0HS, UK

G. H. Draffan, Inveresk Research International, Inveresk Gate, Musselburgh, Midlothian, EH21 7UB, UK

H.-G. Eibs, Pharmakologisches Institut der Freien Universität Berlin, Abteilung Embryonalpharmakologie, West Germany

S. Evans, AEI Scientific Apparatus Ltd, Manchester, UK

P. B. Farmer, Chester Beatty Research Institute, Institute of Cancer Research, Royal Cancer Hospital, Fulham Road, London SW3 6JB, UK
C. Fenselau, Department of Pharmacology and Experimental Therapeutics, The Johns Hopkins University School of Medicine, Baltimore, Maryland 21205, USA
G. C. Ford, G. D. Searle and Co. Ltd, High Wycombe, Bucks, HP12 4HL, UK
A. B. Foster, Chester Beatty Research Institute, Institute of Cancer Research, Royal Cancer Hospital, Fulham Road, London SW3 6JB, UK
C. R. Freed, Department of Pharmaceutical Chemistry and Division of Clinical Pharmacology, University of California, San Francisco, California 94143, USA
J. C. Frölich, Departments of Medicine and Pharmacology, The Vanderbilt University School of Medicine, Nashville, Tennessee 37232, USA
W. A. Garland, Hoffmann-La Roche Inc., Nutley, New Jersey, USA
B. Gregg, Kinderklinik, Freie Universität und Pharmakologisches Institut, Freie Universität, Berlin, West Germany
C. T. Gregg, Los Alamos Scientific Laboratory, University of California, Los Alamos, New Mexico, USA
S. J. W. Grigson, G. D. Searle and Co. Ltd, High Wycombe, Bucks, HP12 4HL, UK
C. M. Gruber, Jr., Lilly Laboratory for Clinical Research, Wishard Memorial Hospital, Indianapolis, Indiana 46202, USA
D. L. Hachey, Section of Medical Development, Division of Biological and Medical Research, Argonne National Laboratory, Argonne, Ill. 60439, USA
K. D. Haegele, University of Texas Health Science Centre, San Antonio, Texas, USA
N. J. Haskins, G. D. Searle and Co. Ltd, High Wycombe, Bucks, HP12 4HL, UK
H. Helge, Kinderklinik, Freie Universität und Pharmakologisches Institut, Freie Universität, Berlin, West Germany
A. F. Hofmann, Gastroenterology Research Unit, Mayo Clinic and Mayo Foundation, Rochester, Minnesota, USA
W. Hoppe, Institut für Chemie Weihenstephan der Technischen, Universität Munchen, D-8050 Freising-Weihenstephan, Federal Republic of Germany
M. G. Horning, Baylor College of Medicine, Houston, Texas, USA
J. A. Hoskins, MRC Unit for Metabolic Studies in Psychiatry, University Department of Psychiatry, Middlewood Hospital, PO Box 134, Sheffield, S6 1TP, UK
U. Jacob, Pharmakologisches Institut der Freien Universität Berlin, Abteilung Embryonalpharmakologie, West Germany
I. Jardine, Department of Pharmacology and Experimental Therapeutics, The Johns Hopkins University School of Medicine, Baltimore, Maryland 21205, USA
M. Jarman, Chester Beatty Research Institute, Institute of Cancer Research, Royal Cancer Hospital, Fulham Road, London, SW3 6JB, UK
R. A. D. Jones, MRC Unit for Metabolic Studies in Psychiatry, University Department of Psychiatry, Middlewood Hospital, P.O. Box 134, Sheffield S6 1TP, UK
A. Kalir, Department of Pharmaceutical Chemistry and Division of Clinical Pharmacology, University of California, San Francisco, California 94143, USA
M.-N. Kan, Department of Pharmacology and Experimental Therapeutics and of Medicine, The Johns Hopkins University School of Medicine, Baltimore, Maryland 21205, USA
P D. Klein, Section of Medical Development, Division of Biological and Medical Research, Argonne National Laboratory, Argonne, Ill 60439, USA
S. Knies, Kinderklinik, Freie Universität und Pharmakologisches Institut, Freie Universität, Berlin, West Germany
P. C. J. M. Koppens, Scientific Development Group, Organon International BV, Oss, The Netherlands
M. J. Kreek, Rockefeller University, New York, NY 10021, USA
C. R. Lee, MRC Unit for Metabolic Studies in Psychiatry, University Department of Psychiatry, Middlewood Hospital, PO Box 134, Sheffield, S6 1TP, UK
K. Lertratanangkoon, Baylor College of Medicine, Houston, Texas, USA
T. Lewander, Psychiatric Research Centre, Ulleråker Hospital, 750 17 Uppsala, Sweden
C. Lindberg, Department of Organic Pharmaceutical Cehmistry, Biomedical Center, Box 574, S.751 23 Uppsala, Sweden
R. Medina, Institut für Chemie Weihenstephan der Technischen, Universität München D-8050, Freising-Weihenstephan, Federal Republic of Germany

K. L. Melmon, Department of Pharmaceutical Chemistry and Division of Clinical Pharmacology, University of California, San Francisco, California 94143, USA

J R. Mitchell, NIH, Bethesda, Maryland, USA

D. Nagel, Pharmakologisches Institut der Freien Universität Berlin, Abteilung Embryonalpharmakologie, West Germany

E. Neill, Department of Clinical Pharmacology, Royal Postgraduate Medical School, Ducane Road, London W12 0HS, UK

S. D. Nelson, NIH, Bethesda, Maryland, USA

D. Neubert, Kinderklinik, Freie Universität and Pharmakologisches Institut, Freie Universität, Berlin, West Germany

I. Nötges-Borgwardt, Kinderklinik, Freie Universität und Pharmakologisches Institut, Freie Universität, Berlin, West Germany

J. A. Oates, Departments of Medicine and Pharmacology, The Vanderbilt University School of Medicine, Nashville, Tennessee 37232, USA

B. D. Obermeyer, Lilly Laboratory for Clinical Research, Wishard Memorial Hospital, Indianapolis, Indiana 46202, USA

L. Paalzow, Department of Pharmaceutical Pharmacology, Biomedical Center, Box 573, S-751 23 Uppsala, Sweden

R. F. Palmer, G. D. Searle and Co. Ltd, High Wycombe, Bucks, HP12 4HL, UK

R. J. Pollitt, MRC Unit for Metabolic Studies in Psychiatry, University Department of Psychiatry, Middlewood Hospital, PO Box 134, Sheffield, S6 1TP, UK

J. L. Reid, Department of Clinical Pharmacology, Royal Postgraduate Medical School, Ducane Road, London W12 0HS, UK

J. J. de Ridder, Scientific Development Group, Organon International BV, Oss, The Netherlands

P. R. Robinson, Sterling Winthrop Laboratories, Metabolic Studies Section, Fawdon, Newcastle-upon-Tyne, NE3 3TT, UK

H.-L. Schmidt, Institut für Chemie Weihenstephan der Technischen, Universität München D-8050 Freising-Weihenstephan, Federal Republic of Germany

D. A. Schoeller, Department of Medicine, Pritzker School of Medicine, The University of Chicago, Chicago, Ill. 60637, USA

H. W. Seyberth, Departments of Medicine and Pharmacology, The Vanderbilt University School of Medicine, Nashville, Tennessee 37232, USA

B. Sjöqvist, Department of Pharmacology, Karolinska Institutet, 104 01 Stockholm, Sweden

J. Sjövall, Kemiska Institutionen, Karolinska Institutet, Stockholm, Sweden

K. Sjövall, Department of Obstetrics and Gynaecology, Karolinska Hospital, Stockholm, Sweden

H. Spielmann, Pharmakologisches Institut der Freien Universität Berlin, Abteilung Embryonalpharmakologie, West Germany

C. N. Statham, NIH, Bethesda, Maryland, USA

B. J. Sweetman, Departments of Medicine and Pharmacology, The Vanderbilt University School of Medicine, Nashville, Tennessee 37232, USA

P. Tippett, Department of Clinical Pharmacology, Royal Postgraduate Medical School, Ducane Road, London, W12 0HS, UK

G. Traut, 11 Medizinische Klinik der Universität des Saarlandes, D-6650 Homburg/Saar, Federal Republic of Germany

Y. Vaishnev, Hoffmann-La Roche Inc., Nutley, New Jersey, USA

W. J. A. VandenHeuvel, Merck, Sharp and Dohme Research Laboratories, Rahway, New Jersey 07065, USA

J. T. Watson, Department of Pharmacology, Vanderbilt University, Nashville, Tennessee 37232, USA

B. Weber, Kinderklinik, Freie Universität und Pharmakologisches Institute, Freie Universität, Berlin, West Germany

R. Weinkam, Department of Pharmaceutical Chemistry and Division of Clinical Pharmacology, University of California, San Francisco, California 94143, USA

D. M. Wilson, Space Sciences Laboratory, University of California, Berkeley, California, USA

R. L. Wolen, Lilly Laboratory for Clinical Research, Wishard Memorial Hospital, Indianapolis, Indiana 46202, USA

E. A. Ziege, Lilly Laboratory for Clinical Research, Wishard Memorial Hospital, Indianapolis, Indiana 46202, USA

Preface

In recent years, the use in the biomedical sciences of compounds labelled with stable isotopes has undergone a dramatic expansion, partly as a direct consequence of the developments which have taken place on analytical instrumentation for the detection and accurate measurement of isotope enrichment. Combined gas chromatography–mass spectrometry (GC–MS), more than any other physicochemical technique, has played a key role in this field, and the continuously growing demand for stable isotope-labelled internal standards for use in GC–MS assay procedures has been, and is likely to remain, a major factor in stimulating the commercial production of the stable isotopes of carbon, hydrogen, nitrogen and oxygen.

The use of stable isotopes as tracers in metabolic studies continues to be a major area of application; the simultaneous administration of a radioactive and corresponding stable isotope-labelled compound is rapidly becoming the technique of choice for investigations of drug metabolism in animals. However, the growing concern in many countries over the health hazard associated with the use of radioactive tracers for metabolic studies in humans has led many investigators to rely solely on compounds labelled with stable isotopes for such purposes; this is particularly true, and indeed is often mandatory, for studies in young children and in pregnant women.

Novel applications of labelling with stable isotopes have been to studies on the metabolic fate of individual enantiomers of optically active drugs and to the investigation of *in vivo* isotope effects – the concept of 'metabolic switching', resulting from substitution of deuterium for specific hydrogen atoms, could have widespread implications in toxicology and future drug design. Another application of stable isotopes with considerable potential centres on the technique of *in vivo* labelling of coenzymes with deuterium; this approach has been used to obtain a wealth of information on the dynamics of biosynthetic pathways and on the compartmentalisation of enzymic reactions at the cellular level.

Examples of each of these areas of stable isotope usage are to be found in this volume, which represents the *Proceedings* of an International Symposium on Stable Isotopes, held in London on 3–4 January 1977, and devoted to applications in pharmacology, toxicology and clinical research. It is to be hoped that this book will provide a broad introduction to the great diversity of applications of stable isotopes in these sciences as well as serve as a valuable reference to investigators actively engaged in the field.

Finally, thanks are due to all those who took part in the organisation of the Symposium and, in particular, to Dr Harry Draffan, at whose suggestion this Meeting was originally conceived. Ms Dorothy Boland, Anne Davis and Helen Hughes are thanked for their invaluable assistance with conference management and the preparation of these proceedings, while the generous financial support of those bodies listed, without whose help the Symposium could not have taken place, is gratefully acknowledged.

London, 1978 T.A.B.

Section 1
Stable isotopes: scope of application and analytical techniques

1

Stable isotopes: essential tools in biological and medical research

P. D. Klein and D. L. Hachey (Section of Medical Development, Division of Biological and Medical Research, Argonne National Laboratory, Argonne, Ill. 60439, USA), M. J. Kreek (Rockefeller University, New York, N.Y. 10021, USA) and D. A. Schoeller (Department of Medicine, Pritzker School of Medicine, The University of Chicago, Chicago, Ill. 60637, USA)

The renaissance of interest in stable isotopes of hydrogen, carbon, nitrogen, oxygen and sulphur within the last ten years is based upon the development of new instrumentation, on the greater availability of enriched isotopes and on an espousal of non-radioactive techniques for human studies on ethical grounds. In order to put the present use of stable isotopes into perspective, we should examine both its antecedents and its expectations, so that we can judge the benefits to be expected.

If one pursues the trail of papers on stable isotopes backwards in time, it stretches back 50 years to 1927, when Aston first described a new mass spectrometer with a resolution of 600. This instrument showed the presence of three sulphur isotopes at masses 32, 33 and 34 (Aston, 1927) and prompted him to espouse the whole number rule for elements, as opposed to the fractional masses obtained by chemical analysis. There followed within quick succession the detection of ^{13}C in optical spectra from graphite arcs (King and Birge, 1929), ^{18}O and ^{17}O in atmospheric solar spectra (Babcock, 1929; Giaque and Johnston, 1929), ^{15}N in the absorption bands of NO (Naudé, 1930) and, finally, the discovery of deuterium (Urey, Brickwedde and Murphy, 1932). This sequence of discovery within a five-year period was not accidental; it hinged on the abundance of each isotope and the available means of detection, and thus went from 4.3 per cent for ^{34}S to 0.15 per cent for ^{2}H.

On the other hand, the production of enriched fractions of the stable isotopes was achieved on the basis of the magnitude of the isotope effect each exerted in a physical or chemical process, and it was deuterium, the last isotope to be discovered, that was first produced in highly enriched form in Urey's laboratory. According to Clarke (1948), the first toxicological studies on D_2O were conducted within a year after the first published report of the existence of deuterium and tested the ability of goldfish, earthworms and protozoa to survive in 92 per cent D_2O (Taylor *et al.*, 1933). Urey was fully aware of the tracer potential of deuterium and through a grant from the Rockefeller Foundation was able to subsidise the production of heavy water and to attract a young PhD, David Rittenberg, to whom he gave the commission to find applications for the isotope. The heritage of that grant, in the collaboration of Schoenheimer and Rittenberg, is difficult to overestimate, for it brought the concept of a dynamic equilibrium in metabolic processes to the then-static physiological chemistry of the day. Clarke's account of these fruitful years from 1933 through 1942 deserves periodic re-reading to keep in mind just how extensively the findings from the use of deuterium and, later, ^{13}C and ^{15}N defined the areas of modern biochemistry.

In September of 1947 a symposium was held at the University of Wisconsin entitled 'The Use of Isotopes in Biology and Medicine'. It marked the watershed in stable isotope usage, for in 1945 the first shipments of ^{14}C had been made by the newly established Atomic Energy Commission. In this historic meeting the two techniques for methods of preparation, synthetic procedures, instrumentation and quantitation were presented side by side; one for stable isotopes and the other for radioactive isotopes. The sense of excitement at the potential of radioactive tracers, then newly introduced, is keenly evident in the proceedings, and the literal explosion of radioactive tracer studies in the next decade completely overshadowed deuterium and ^{13}C applications.

The reasons for this explosion of radioactive tracer usage are quite clear: detection was simpler, it required far less sample preparation and, as we all know, the techniques were soon fully automated. Whereas the availability of stable isotopes depends upon the magnitude of the isotope effect exerted by the difference in mass, and upon the starting abundance of the desired isotope, no such limitations exist for radioactive isotopes. Stable isotope production is ultimately limited by the thermodynamic considerations, whereas radioactive isotopes depend only on the flux of neutrons available, and the higher the flux, the higher the specific activity of the isotope produced. Introduction of liquid scintillation counting in the late 1950s should perhaps be regarded as the great equaliser of tracer methodology, because this equipment made it possible for any research worker, regardless of previous experience, to enter the field.

All of this history transpired more than 25 years ago, and in our present experience we are faced with the search for significance in the use of stable isotopes. Clearly the halcyon days of 1940, teeming with ground-breaking opportunities in every tracer experiment, are gone, not likely to return. Nor are the majority of users of radioactive isotopes likely to foresake the ease and convenience of liquid scintillation counting for the headaches of gas chromatography–mass spectrometry that we all know and experience daily. It is true that when there are no alternatives, the slow, labour-intensive mass spectrometric measurement of stable isotope ratios will go on at any cost. To appreciate this fully, one need only to add up the number of papers dealing with ^{15}N applications in agricultural science journals. The absence of usable radioactive oxygen isotopes has created a demand for ^{18}O and ^{17}O that is not likely to be challenged. It is also true that both deuterium and ^{13}C are essential for the determination of fragmentation pathways, but, it should be emphasised, these types of studies are usually non-recurrent applications, just as the introduction of ^{13}C into an antibiotic for NMR studies is a unique event.

Where, then, are the areas of application that justify our convening in various spots around the globe at frequent intervals? It is clearly not simply the choice between using radioactive versus stable isotopes in human studies, because the use of 1–10 μCi of ^{14}C or 10–50 μCi of ^{3}H has not been banned in adults, nor is it likely to be, under the present risk/benefit ratio. We need to remind ourselves that if, in the attending physician's opinion, a procedure requiring the administration of radiation or radioactivity is clearly beneficial to the diagnosis or prognosis of the patient's condition, the procedure will be used. These procedures include the use of external emitters in scanning techniques in infants or women of child-bearing age, which involve higher doses of radiation than would be encountered in most tracer applications.

On the other hand, ethical considerations require that the physician avoid *any* procedure that is not both harmless and non-invasive in normal patients. Without a doubt, these limitations are most stringently imposed when the studies involve pre-term or term infants, children and women in potential or actual child-bearing status. Since pregnancy is a transitory state in health care delivery, there have been few studies reported in which tracer techniques were required to evaluate the state of the pregnancy or of the conceptus, and where stable isotopes have actually been used. On the other hand, stable isotopically labelled drugs have been used in studies of lactating mothers (Horning *et al.*, 1973); and in a number of infant studies including kinetic measurements of bile acid metabolism (Watkins *et al.*, 1973, 1975), alanine and glucose fluxes (Bier *et al.*, 1973) and the pathway of deuterated aromatic amino acid metabolism (Curtius, Völlmin and Baerlocher, 1972). There is a large body of information on adult metabolic processes that could be extrapolated to the child and infant; extrapolation of this information will, in many cases, extend the diagnostic and therapeutic benefits of adult experience to the child. Paediatric applications of stable isotopes thus seemed assured of considerable expansion.

It is in the area of pharmacology, however, that two types of applications appear to be particularly dependent upon the use of stable isotopes for recurrent measurements. (The use of stable isotopic labelling to provide ion pairs useful in drug metabolite identification that was developed by Hammar, Holmsted and Ryhage, 1968, is excluded in these considerations because of its non-repetitive application in the same molecule.) The first of these is in the measurement of drug or metabolite concentration in plasma, when the circulating level is at or below the microgram level and labelled forms of the drug are used as carriers and internal standards. The second is in the measurement of drug kinetics in plasma when these low levels are present, and when the total body pool is large. These two applications are best exemplified in our own experience in the study of methadone metabolism in humans, as conducted at Rockefeller University by Dr Mary Jeanne Kreek.

A penta-deuterium-labelled methadone has been synthesised by Dr Hachey according to the scheme shown in Figure 1.1. This material, which can be provided as the racemate or as individual optical isomers, is used either as the internal standard for quantitative measurements of plasma and urinary levels or as the pharmacological maintenance dose in kinetic measurements (Hachey, Mattson and Kreek, 1976). Similar studies have also been reported by Dr Änggård and his co-workers (Änggård *et al.*, 1976). Figures 1.2 and 1.3 show the

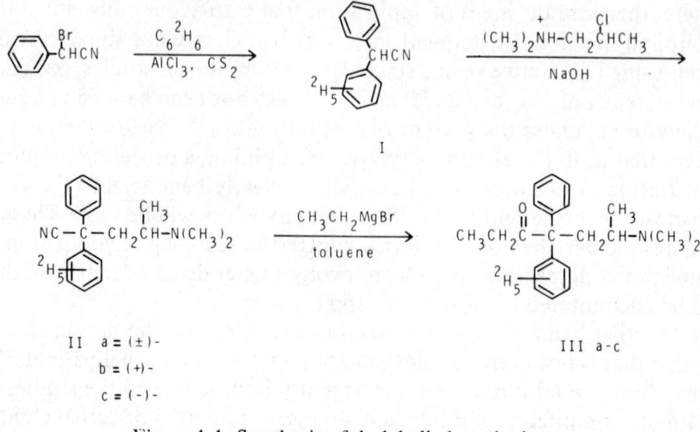

Figure 1.1 Synthesis of d_5-labelled methadone

quantitation of methadone in plasma and in urine, respectively, and illustrate the linearity over more than three orders of magnitude that can be achieved with this d_5 internal standard.

Methadone metabolism is unusual in that the plasma levels of the drug fluctuate within relatively narrow, but low, levels: typically 100–500 ng/ml, even following the daily oral dose of approximately 1 mg/kg. The excursions in a patient's plasma values on two successive days are shown in Figure 1.4, which

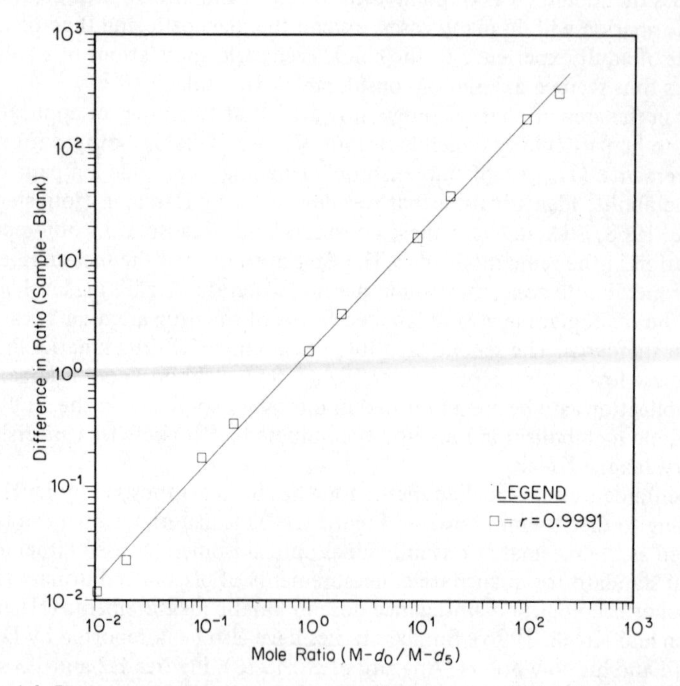

Figure 1.2 Plasma methadone inverse isotope dilution curve (540 ng internal standard)

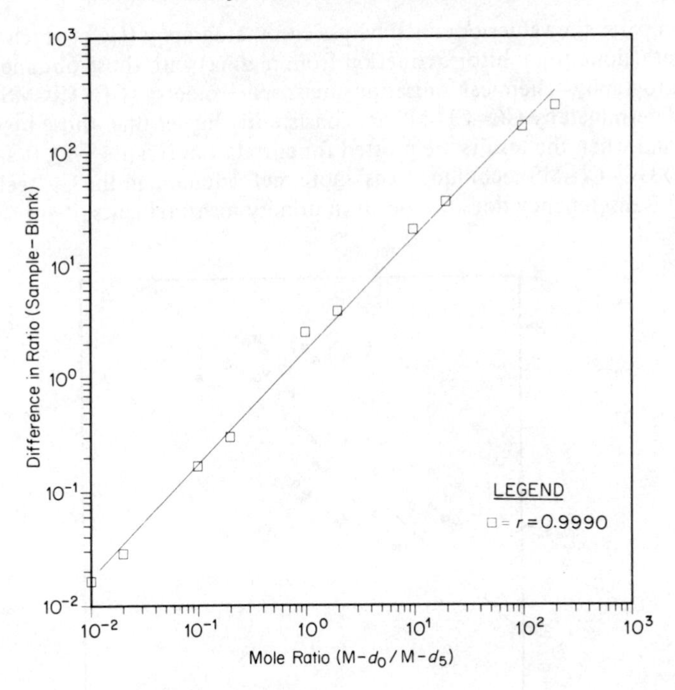

Figure 1.3 Urinary methadone inverse isotope dilution curve (540 ng internal standard)

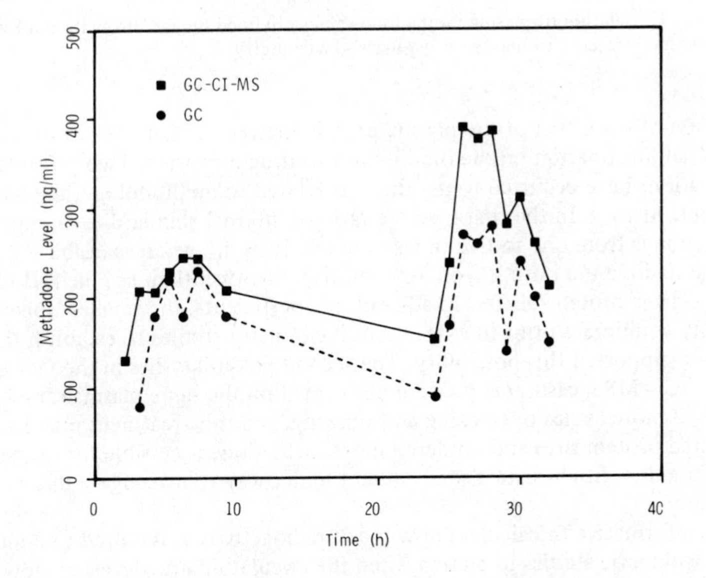

Figure 1.4 Plasma methadone levels in a patient over a 36 h period measured by gas chromatography and gas chromatography–mass spectrometry

also compares the values obtained by gas chromatography (GC) (which employs ^{14}C-methadone to monitor extraction from plasma) with those obtained by gas chromatography–chemical ionisation–mass spectrometry (GC–CI–MS). The levels determined by GC–CI–MS are consistently higher than those by GC alone, and when the results are plotted for correlation (Figure 1.5), it appears that the GC–CI–MS technique 'sees' more methadone than the GC technique. Since this discrepancy does not occur in urinary measurements, it appears to be a

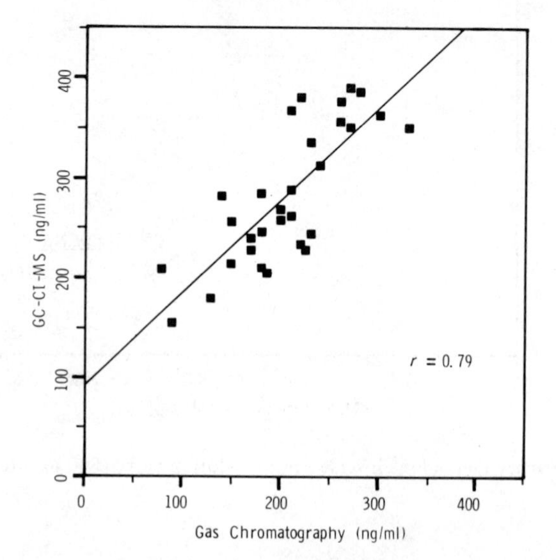

Figure 1.5 Correlation of plasma methadone values obtained by gas chromatography with those obtained by gas chromatography–mass spectrometry

reflection of the effect of the plasma protein matrix and may be a source of additional information on the binding of this drug in plasma. Two possible explanations have occurred to us: the first related to methodology, the second to sample history. In the first case the ratio of internal standard to endogenous methadone is from one to five times as much d_5 as d_0, whereas in the ^{14}C-methadone the ratio is from one-tenth to one-twentieth as much. If binding to the plasma proteins differs in adsorption coefficients, the smaller 'spike' may not fully equilibrate prior to extraction. Preliminary studies to establish this have not supported this possibility. The second possibility lies in the fact that the GC–CI–MS measurements were performed on the same plasma samples after 2–3 more cycles of freezing and thawing, and this treatment may have denatured protein sites and rendered more methadone accessible for exchange and extraction. Studies are also at present under way to investigate this possibility.

It is informative to calculate how much radioactivity is required to conduct pharmacokinetic studies in plasma when the circulating drug levels are low in comparison with the total body burden. Table 1.1 lists the assumptions that were used by Dr Schoeller in calculating the radioactivity required for

Table 1.1 Calculation of radioactive tracer dose required for kinetic measurements in plasma

Assumptions

Pool size: 1 g	Molecular weight: 400
Counting efficiency: 100%	Counting time: 10 min
Background rate: 10 d.p.m.	Labels/molecule: 1

Intermediate data

Counting accuracy: ±1% requires 1005 d.p.m. or 457 pCi
±5% requires 44 d.p.m. or 20 pCi

For 1 ml plasma samples, required dose administered is:

Concentration	1% accuracy	5% accuracy
1 mg/ml	0.46 μCi	0.02 μCi
1 μg/ml	457 μCi	20 μCi
1 ng/ml	exceeds carrier-free specific activity	20 000 μCi

measurements having ±5 per cent or ±1 per cent counting accuracy. These quantities of radioactivity are 20 pCi and 457 pCi, respectively. Depending upon the concentration of the drug in plasma, the amount of radioactivity required to produce these levels, when the body burden is 1 g, range from 0.2 μCi to amounts that exceed the levels of specific activity found in carrier-free material. It can be seen, for example, that even with counting accuracies of ±5 per cent it is difficult to follow the plasma kinetics of a metabolite in the 10–100 ng/ml range without exceeding the limits of safety. On the other hand, it is feasible to administer the entire dose in a stable isotopically labelled form, in either a single or a multiple dose, without hazard. Since the physical quantity of labelled material is not a limiting factor, one can achieve labelling concentrations in the range where stable isotopic ratio determinations can be carried out on the quantities available in 1 ml of plasma. This permits the labelled drug to be followed for a much longer period of time. Figure 1.6 illustrates the isotopic abundance in plasma methadone in a maintenance patient following the substitution of a single dose of methadone-d_5 for the regular daily dose. The existence of two pools is suggested by the computer fit to a double exponential function, one with a half-life of 1.9 h and a second with a half-life of 31.4 h. Note that isotopic ratio measurements could be carried out for more than a week before the sample size became limiting. In urine, where the concentration was higher, it was possible to follow the disappearance of label for more than ten days, as shown in Figure 1.7. These measurements also suggested two kinetic components with half-lives of 7.2 and 46.2 h, respectively.

A number of patients have been studied with the use of the d_5 methadone, and we are particularly interested in the opportunities to study drug–drug interactions in methadone maintenance patients, such as concurrent alcohol or barbiturate use, or therapy with rifampin (Kreek *et al.*, 1976). From the standpoint of stable isotope usage, however, several aspects of the studies of methadone kinetics deserve further comment as they bear upon the future of

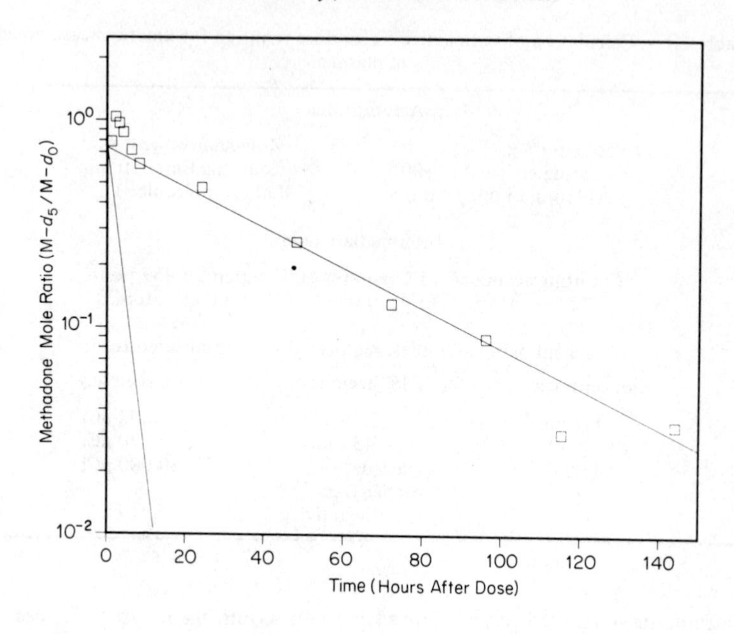

Figure 1.6 Plasma disappearance of methadone following a single 100 mg oral dose of methadone-d_5

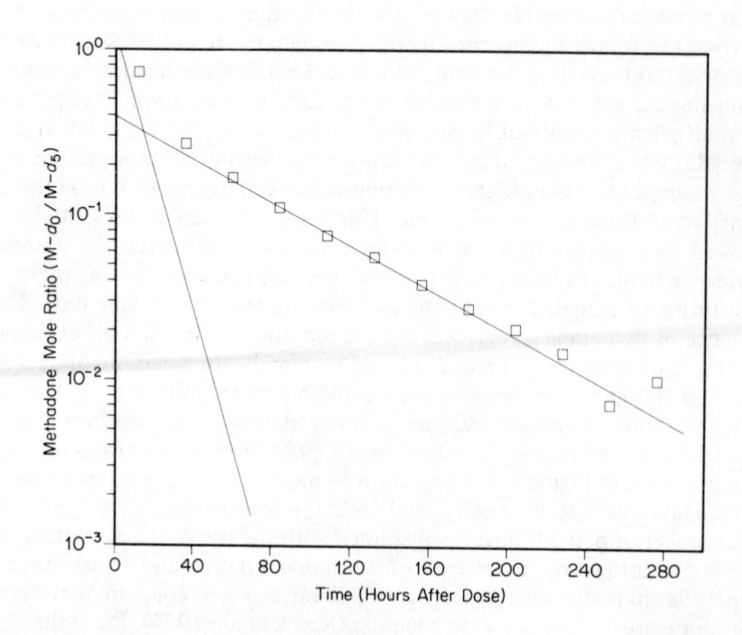

Figure 1.7 Urinary excretion of methadone following a single oral 100 mg dose of methadone-d_5

these techniques. First, even the relatively simple extraction procedures used in these studies and the short gas chromatographic retention times become limiting factors in the speed with which samples can be processed. When the total number of samples involved in plasma, urine and faecal kinetic measurements are considered, it is evident that these studies require far more labour than radioactive studies in animals. Moreover, because the instrumentation cost is high, it is not feasible in most laboratories to run such studies in parallel on two or three instruments at the same time. This emphasises the aspect of stable isotope applications that require GC–MS which must be kept uppermost: that the applications should be low sample number: high information gain problems where at all possible. The present effort required to measure isotope ratios in intact organic molecules means that problems involving large numbers of samples will have limited opportunities to be explored.

It was for reasons of labour cost that we began several years ago to look at stable isotope applications in which the end measurement of isotopic abundance could be simplified. In the area of clinical applications it seemed that the then-existent ^{14}C 'breath tests' were beginning to show significant diagnostic potential and that the substitution of ^{13}C-labelled substrates would permit these diagnostic procedures to be used in children. The analytical requirement was then to determine the abundance of $^{13}CO_2$ in respiratory CO_2, and since all of the tests had in common the evolution of labelled CO_2, a single procedure could accommodate a diversity of applications. A major contribution to defining the analytical system requirements arose from the work of Dr Schoeller and his associates (Schoeller *et al.*, 1976), who documented the limiting precision of respiratory $^{13}CO_2$ measurements. Of the total variance (0.518 per thousand) in replicate measurements over a 6 h period in fasting patients, 0.470 was due to metabolic variations in endogenous $^{13}CO_2$ abundance, even under fasting conditions (Table 1.2). This meant that the 95 per cent confidence limit for a

Table 1.2 Variance introduced in each step of $^{13}CO_2$ analysis

Source	Variance (‰)^2	Number of analyses
Total variance	0.518	39
Ratio determination	0.002	10
CO_2 release	0.017	9
CO_2 collection	0.029	14
Residual (due to patient)	0.470	

significant change in isotopic abundance of the respiratory CO_2 was 1.4 per thousand or 0.14 per cent. The significance of this finding is twofold: it defines the limiting instrumental sensitivity required (far less than originally suspected, and much more easily achieved) and it defines the minimum substrate oxidation rate that can be anticipated to be detectable against the background of endogenous $^{13}CO_2$ and its variation. This rate *must* be in excess of $0.14\ \mu mol\ kg^{-1}\ h^{-1}$, and should preferably be 5–20 times as large. This information enables us to rule out, on the basis of radioactive measurements, those substrates whose oxidation would be undetectable if they were labelled with ^{13}C instead of ^{14}C. For example, 2 mg/kg aminopyrine produces a significant increase in $^{13}CO_2$ abundance when the *N*-dimethyl ^{13}C form is administered, but the oxidation rates of ^{13}C-methyl labelled valium or methadone would not be detectable with physiological doses. Doses that *could* produce measurable levels of $^{13}CO_2$ would be pharmacologically unacceptable.

Much of the cost of stable isotopic ratio measurements is the labour involved in processing samples. It seemed to us that the automation of $^{13}CO_2$ isotope ratio measurements would effect a considerable economy, and we have been pursuing this objective. A key requirement is the ability to collect, ship and store individual samples of respiratory CO_2 in a manner that is simple, inexpensive and without effect on the isotope content. This requirement has been solved by using evacuated 50 ml containers, sealed with rubber septa and available commercially as Vacutainers. Figure 1.8 illustrates the stability of isotope

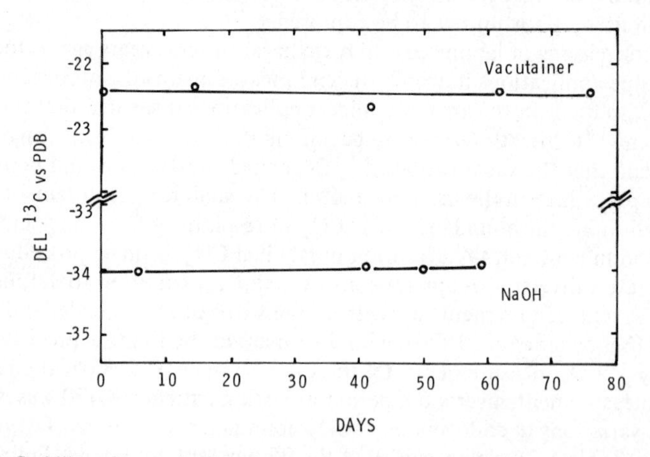

Figure 1.8 Stability of isotopic abundance values of respiratory CO_2 stored in Vacutainers or trapped as $NaHCO_3$ in NaOH. Abundance values are expressed as

$$\delta\, ^{13}C_{PDB} = \left[\frac{(^{13}C/^{12}C)_{unknown}}{(^{13}C/^{12}C)_{PDB}} - 1 \right] \times 10^3$$

where the isotopic abundance of the unknown is expressed relative to that of a known standard limestone sample (PDB)

abundance measurements on Vacutainer samples over a 7 week period and the comparison of these values with those obtained on CO_2 trapped in NaOH and released by acidification (Schoeller and Klein, 1976). An additional bonus is that the isotopic depletion that occurs when CO_2 is trapped in a dynamic flow system in NaOH is absent when Vacutainers are used and the absolute value of the $^{13}C/^{12}C$ abundance ratio is preserved.

At the present time we are constructing a prototype system for automatic $^{13}CO_2$ measurements, using Vacutainers as sample containers. The flow chart for the steps in the analysis is shown in Table 1.3, and these steps are being programmed for control by a dedicated microprocesser that also collects the ion-current ratio data. The ultimate objective, of course, is the ability to load a rack of 50 samples into the system at the end of the clinical working day and have answers printed out by the next morning. Furthermore it is our conjecture

Table 1.3 Automated breath $^{13}CO_2$ analyser operation cycle

Puncture Vacutainer septum and withdraw an aliquot
Pass gases through a dry-ice trap to remove water
Collect the CO_2 in a liquid nitrogen trap
Transfer the CO_2 to the mass spectrometer inlet
Adjust the volume to produce desired pressure
Measure the ratio of $^{13}CO_2$ to $^{12}CO_2$ in the sample
Measure the ratio of $^{13}CO_2$ to $^{12}CO_2$ in a reference
Calculate and print out the $^{13}CO_2$ enrichment
Advance to the next Vacutainer and begin again

that if discrete samples can be analysed in this manner, combustion processes for total label recovery, or selective degradation steps, can also be programmed for automatic analysis. Such a capability should increase the opportunities for the clinical applications of ^{13}C-labelled compounds by eliminating the analytical bottlenecks that exist at present.

As we begin the second 50 years of stable isotope applications, there are still many of the original limitations of cost and difficulty present in the beginning. Nevertheless we appear to be moving into a new era of human studies in which the answers we seek cannot be obtained in any other way, and, once more, stable isotopes have become essential tools in biological and medical research.

ACKNOWLEDGEMENTS

This work was supported by the US Energy Research and Development Administration and by the National Institutes of Health through AM 18741 and PHSDA 01138.

REFERENCES

Änggård, E., Holmstrand, J., Gunne, L.-M., Sullivan, H. R. and McMahon, R. E. (1976). *Proceedings of the Second International Conference on Stable Isotopes* (ed. E. R. Klein and P. D. Klein), US ERDA CONF-751027, National Technical Information Service, US Dept. of Commerce, Springfield, Va. 22161, p. 117

Aston, F. W. (1927). *Proc. Roy. Soc. (London)*, **A115**, 484

Babcock, H. D. (1929). *Proc. Nat. Acad. Sci. USA*, **15**, 471

Bier, D. M., Leake, R. D., Gruenke, L. D. and Sperlman, M. A. (1973). *First International Conference on Stable Isotopes in Chemistry, Biology and Medicine* (ed. P. D. Klein and S. V. Peterson), AEC CONF 730525, National Technical Information Service, US Dept. of Commerce, Springfield, Va. 22161, p. 397

Clarke, H. T. (1948). *A Symposium on the Use of Isotopes in Biology and Medicine*, University of Wisconsin Press, p. 3

Curtius, H. Ch., Völlmin, J. A. and Baerlocher, K. (1972). *Clin. Chim. Acta*, **37**, 277

Giaque, W. F. and Johnston, H. L. (1929). *Nature*, **123**, 831

Hachey, D. L., Mattson, D. H. and Kreek, M. J. (1976). *Proceedings of the Second International Conference on Stable Isotopes* (ed. E. R. Klein and P. D. Klein), US ERDA CONF 751027, National Technical Information Service, US Dept. of Commerce, Springfield, Va. 22161, p. 518

Hammar, C.-G., Holmstedt, B. and Ryhage, R. (1968). *Anal. Biochem.*, **25**, 532

Horning, M. G., Nowlin, J., Lertratanangkoon, K., Stillwell, R. N., Stillwell, W. G. and Hill, R. M. (1973). *Clin. Chem.*, **19**, 845

King, A. S. and Birge, R. T. (1929). *Nature,* **124**, 127
Kreek, M. J., Garfield, J. W., Gutjahr, C. L. and Giusti, L. M. (1976). *New Engl. J. Med.,* **299**, 1104
Naudé, S. M. (1930). *Phys. Rev.,* **36**, 333
Schoeller, D. A. and Klein, P. D. (1976). *Proceedings of the 24th Conference on Mass Spectrometry and Allied Topics*, p. 407
Schoeller, D. A., Klein, P. D., Schneider, J. F., Solomons, N. W. and Watkins, J. B. (1976). *Proceedings of the Second International Conference on Stable Isotopes* (ed. E. R. Klein and P. D. Klein), US ERDA CONF 751027, National Technical Information Service, US Dept. of Commerce, Springfield, Va. 22161, p. 246
Taylor, H. S., Swingle, W. W., Eyring, H. and Frost, A. A. (1933). *J. Chem. Phys.,* **1**, 751
Urey, H. C., Brickwedde, F. G. and Murphy, G. M. (1932). *Phys. Rev.,* **39**, 164
Watkins, J. B., Ingall, D., Szczepanik, P. A., Klein, P. D. and Lester, R. (1973). *New Engl. J. Med.,* **288**, 431
Watkins, J. B., Szczepanik, P. A., Gould, J., Klein, P. D. and Lester, R. (1975). *Gastroenterology,* **69**, 706

2
Mass spectrometric methods of isotope determination

J. Throck Watson (Department of Pharmacology, Vanderbilt University, Nashville, Tenn. 37232, USA)

INTRODUCTION

Although there are several analytical techniques for determining the isotopic composition of a sample, combined gas chromatography–mass spectrometry (GC–MS) is often the most suitable for analysis of samples of biological origin (Watson, 1976). For example, biological samples are rarely pure (even after extensive sample processing) and the biologically active component may represent only a trace constituent of the total sample. The combination of a separation technique with an identification technique in GC–MS permits a complex mixture to be separated in the time domain so that characteristic mass spectra can be obtained from picomole to nanomole quantities of sample depending on the mode of instrument operation.

Mass spectrometry can reveal specific, characteristic, structurally related information about a compound. During ionisation of the sample molecules (usually by a beam of low-energy electrons) molecular ions are produced. Depending on the excess energy level of the molecular ion, it decomposes into fragment ions. The molecular ion (the ionised, intact molecule) provides a direct indication of the molecular weight. The fragment ions represent subsets of atoms which may be related to functional groups or structural components of the original molecule. The array of molecular and fragment ions represented by peaks in a mass spectrum can be called a fragmentation pattern. In general, the fragmentation pattern for a given compound is sufficiently unique to be used as a means of identification.

TYPES OF MASS SPECTROMETERS

Magnetic sector

The mass analyser in this type of instrument is a wedge-shaped (or sector) magnetic field which disperses the total, unresolved ion beam from the ion source into discrete ion beams of individual *m/e* values (mass per unit charge) through a process of direction focusing (Farmer, 1963). As indicated in Figure 2.1, the instrumental design is symmetrical, the ion source and detector being equidistant from the magnet.

The assemblage of ions constituting the total ion beam from the ion source enters the sector magnetic field with a total kinetic energy (KE) approximately equal to the value of the accelerating potential (e.g. 4000 V). However, because the total ion beam consists of ions of various masses, these ions will have different velocities because KE = $\frac{1}{2}mv^2$, where m is the mass of the ion and v is its velocity. Therefore the total ion beam is dispersed into many individual ion

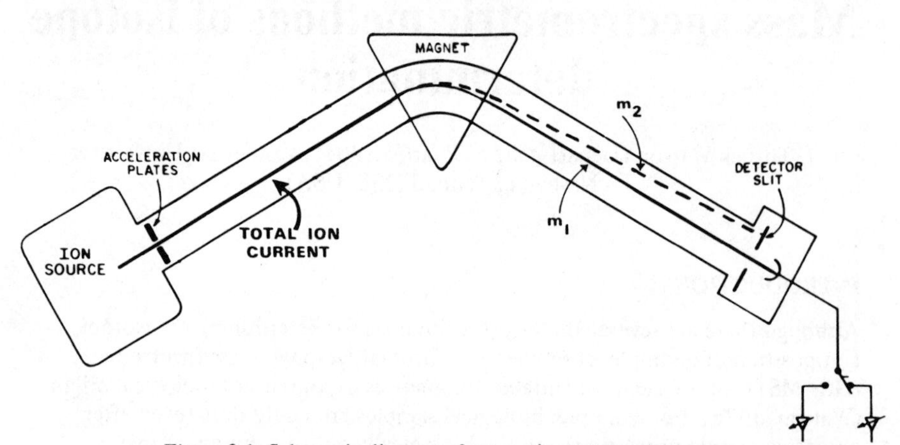

Figure 2.1 Schematic diagram of magnetic mass spectrometer

beams, each having its characteristic trajectory in the region of the magnetic field. As illustrated in Figure 2.1, the imposed accelerating potential (V) and magnetic field (H) permit those ions of mass m_1 to follow a trajectory to the detector, whereas the heavier ion m_2 follows a trajectory under these given conditions of V and H, which lead to collision with the wall of the analyser.

The mass per unit charge (m/e) of any ion can be related to the magnetic field and accelerating potential by the equation

$$m/e = R^2H^2/2V$$

in which previously unidentified terms are e, the electronic charge, and R, the radius of curvature (fixed for a given instrument). To obtain the complete mass spectrum either H or V is scanned while the other is held constant.

Quadrupole

This non-magnetic mass spectrometer employs a combination of d.c. and radiofrequency (r.f.) potentials as a mass 'filter'. Mechanically, the quadrupole consists of four parallel rods arranged symmetrically as indicated in Figure 2.2. Ideally, these four rods should have the shape of a hyperbola on the cross-section so that idealised hyperbolic fields could be produced according to quadrupole theory (Farmer, 1963); however, in practice, cylindrical rods are often used to approximate the hyperbolic field requirements. Opposite rods

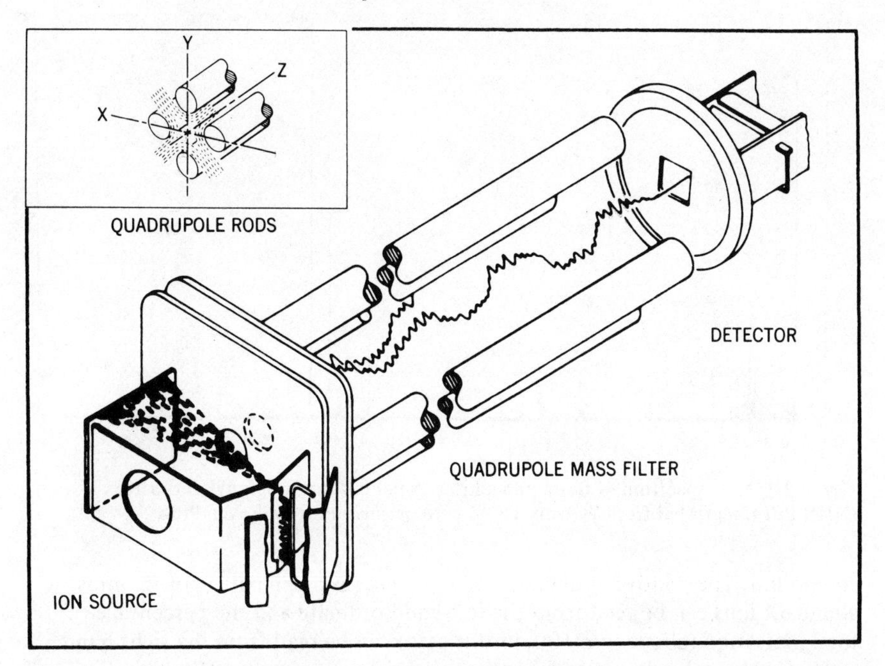

Figure 2.2 Schematic of quadrupole mass spectrometer illustrating irregular flight path of ions from source through central space between rods to the detector; reprinted from Waller, 1972, with permission from Wiley-Interscience

(i.e. those diagonally opposite) are connected electrically to r.f. and d.c.-voltage generators. Positive ions are extracted from the ion source and are accelerated (5–15 V) into the quadrupole region along the longitudinal axis of the four rods. The positively charged ions in the quadrupole region are influenced by the combined d.c. and oscillating r.f. fields, but they still have a momentum vector from the acceleration region towards the detector. Of course, to reach the detector, the ions must traverse the quadrupole filter without colliding with the metal rods (length of rods is typically 10–25 cm). As indicated in Figure 2.2, the ion trajectories through the central space are very complicated and for any given level of r.f./d.c. voltage, only ions of a specific m/e avoid collision with rods and successfully traverse the quadrupole filter to reach the detector; all other ions collide with the rods at some point and never reach the detector. The entire mass spectrum is obtained as the voltages are scanned from a pre-established minimum to a maximum value, but at a constant r.f./d.c. ratio.

APPLICATIONS

Conventional mass spectrometry

Regardless of the type of mass spectrometer, the conventional mass spectrum is presented in bar graph form as represented in Figure 2.3. In this format the investigator can readily examine or interpret the fragmentation pattern of a

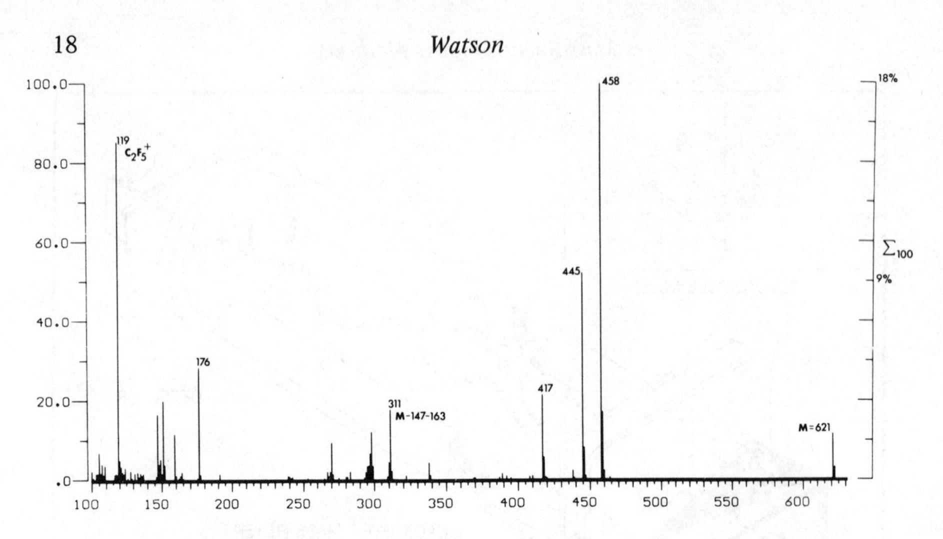

Figure 2.3 Mass spectrum of normetanephrine as pentafluoropropionyl derivative (NMN-PFP); reprinted from Watson, 1976, with permission from Raven Press

compound. The relative abundance of any peak (as a percentage of the most abundant ion) can be read from the left-hand ordinate and the percentage of total ionisation (above *m/e* 100, in this case) can be read from the right-hand ordinate. Manual reduction of mass spectral data for presentation as that in Figure 2.3 might easily require one hour. However, owing to the pioneering computer applications by Hites and Biemann (1967) and many others, including Sweeley *et al.* (1970) and Holmes, Holland and Parker (1971), this type of presentation is available within milliseconds for oscilloscope display or within minutes for hard-copy plots. Given the rapid scanning GC–MS instrumentation of today, it is not practical to cope with the huge amounts of data from a complex sample mixture without a computer. Hites and Biemann were the first to demonstrate the utility of the computer in this type of data processing using the mass chromatogram or reconstructed ion-current profile as a means of evaluating or classifying several hundred consecutively recorded mass spectra (Hites and Biemann, 1970). This technique has been refined to the point that only those *m/e* values at which the ion current has attained maximum abundance during a particular scan are plotted in the final data record; this procedure avoids having the mass spectrum of a minor component distorted by the mass spectrum of another component that emerges from the GC as an overlapping peak (Biller and Biemann, 1974).

As an adjunct to the use of this technology, stable isotopes can be used to give a distinctive appearance to the mass spectra of compounds of interest (Bush *et al*, 1972; Knapp *et al.*, 1972; Braselton *et al.*, 1973; Vore, Sweetman and Bush, 1974; and Pohl *et al.*, 1975). This 'twin peak' effect (Braselton, Orr and Engel, 1973) is especially useful in studying the biotransformation of a drug. For example, consider the following example describing a study of the metabolism of *N-n*-butyl barbital (Vore *et al.*, 1974). The ratio of $^{15}N : ^{14}N$ was 50 : 50 in the parent drug, a feature that would only be apparent by means of its mass spectrum. This stable-isotope-enriched drug together with a trace of radioactive drug was administered to experimental rats. The radioactivity (^{14}C) facilitated isolation of urinary metabolites on various chromatographic systems.

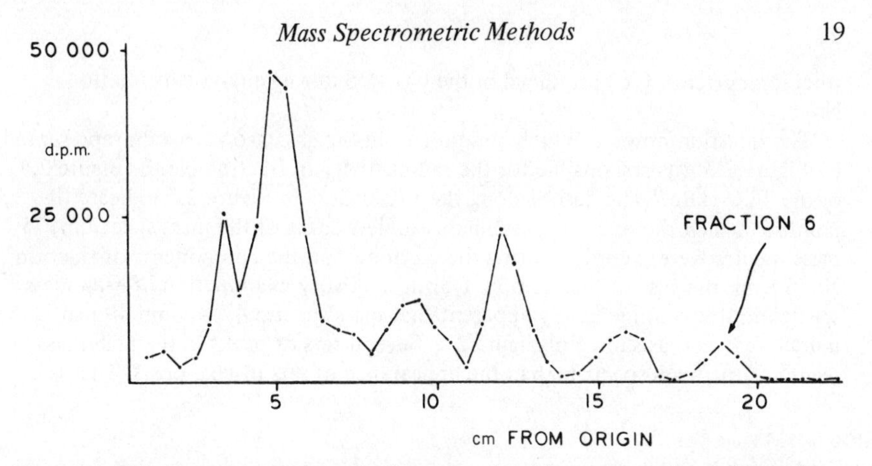

Figure 2.4 Thin-layer chromatogram (radioactivity eluted from consecutive sections of plate) of material extracted from rat urine by ether (Vore *et al.*, 1974); reprinted from Watson, 1976, with permission from Raven Press

Figure 2.4 shows the distribution of radioactivity as a thin-layer chromatogram (TLC) of the material extracted from rat urine with ether (Vore *et al.*, 1974).

For purposes of this example, the question to be pursued will be: 'What is the identity of the radioactive material in fraction No. 6?' The silica gel corresponding to fraction No. 6 on the TLC plate was scraped off and the material was eluted with acetone. An aliquot of this eluted material was then injected into the GC–MS. Although fraction No. 6 may have been radiochemically pure, it contained many non-radioactive compounds, as indicated by the gas chromatogram in Figure 2.5, which represents the profile of

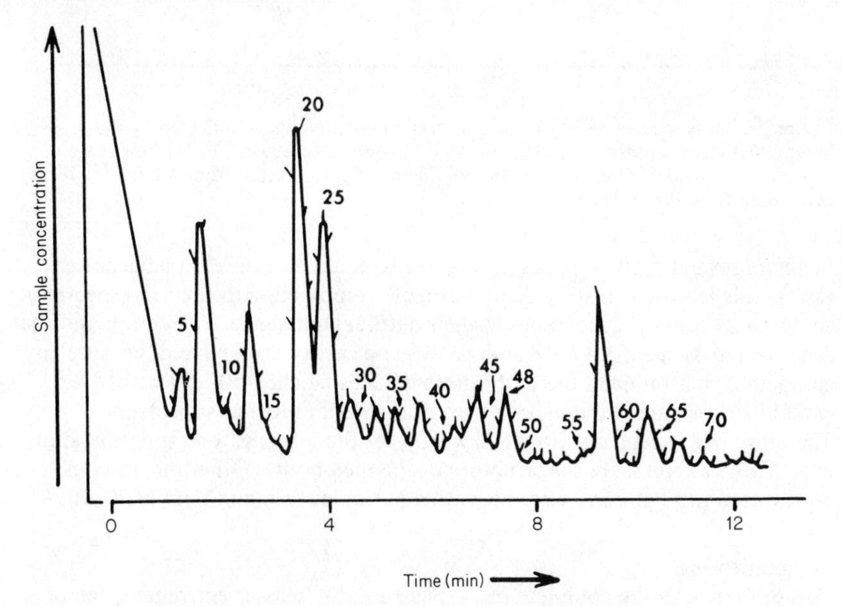

Figure 2.5 Total ion-current chromatogram obtained from the gas chromatograph–mass spectrometer during analysis of fraction No. 6 from TLC as indicated in Figure 2.4. The mass spectrum was scanned 75 times, as indicated by the 'tick' marks on the data record. Reprinted from Watson (1976), with permission from Raven Press

total ion current (TIC) produced in the GC–MS during analysis of fraction No. 6.

The question now is: 'Which component in the gas chromatogram represented by Figure 2.5 was responsible for the radioactivity in fraction No. 6 (Figure 2.4) of the TLC plate?' The dashes along the TIC profile in Figure 2.5 indicate the points at which the operator obtained complete scans of the mass spectrum; 75 mass spectra were recorded during the elution of all the components of fraction No. 6 from the gas chromatograph. During a cursory examination of each mass spectrum, it was immediately apparent that mass spectra 47, 48 and 49 had unique features, as shown in Figure 2.6. Several sets of peaks in the high-mass region of the mass spectrum had the appearance of sets of goal-posts. That is,

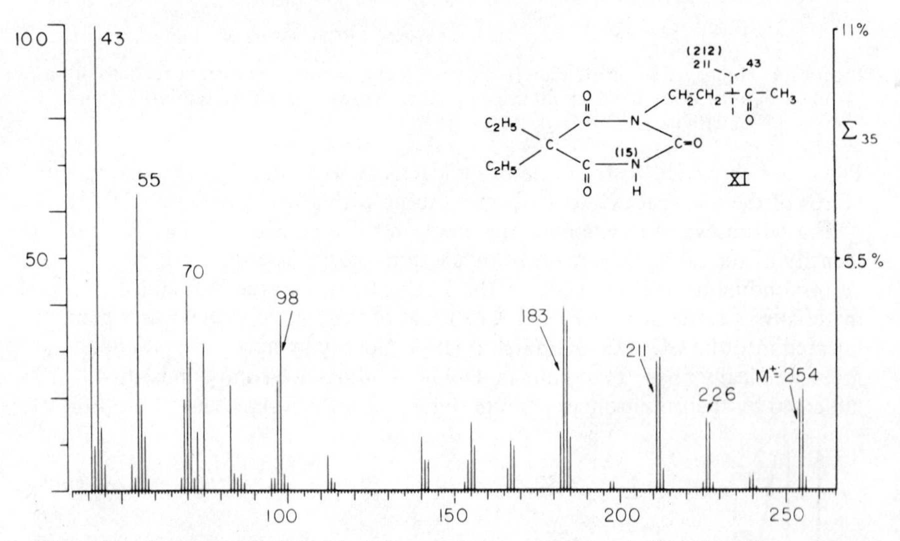

Figure 2.6 Mass spectrum (70 eV) of drug-related compounds as indicated by the ion-doublet effect at *m/e* 211, 212, *m/e* 183, 184, etc. (Vore *et al.*, 1974); this is the spectrum obtained during scan No. 48 (see Figure 2.5). Reprinted from Watson (1976), with permission from Raven Press

there are several pairs of peaks having nearly equal intensities at adjacent *m/e* values; this feature is readily distinguishable compared with the mass spectra of compounds containing isotopes in their natural abundance. The compound that produced mass spectra 47, 48 and 49 was apparently drug-related, because any compound that retained the barbiturate nucleus labelled 50 : 50 with ^{15}N : ^{14}N would have produced the unique twin peak effect in its mass spectrum. Therefore mass spectrum No. 48 is worthy of the investigator's time and study in an effort to elucidate the structure of the metabolite. Other metabolites represented in Figure 2.4 were identified in similar fashion (Vore *et al.*, 1974).

Ion monitoring

Rather than scan the complete mass spectrum, the mass spectrometer can be adjusted to monitor the ion current at selected *m/e* values during the analysis of a sample. High selectivity is achieved by this technique because only the ions that are characteristic of the compound of interest are selected for monitoring. A peak in the characteristic ion-current profiles at the expected retention time is good evidence for the presence of the compound of interest. The useful sensitivity of ion monitoring is approximately 1000 times greater than that for rapid scanning because of the improved signal-to-noise ratio obtained when only a few selected ions are monitored. The high sensitivity and selectivity of selected ion monitoring have made it an invaluable technique for analysing samples of biological origin (Falkner, Sweetman and Watson, 1975).

The teamwork of Sweeley *et al.* (1966) made it a practical technique for 'simultaneously' monitoring two or three ions with a magnetic mass spectrometer equipped with an accelerating voltage alternator. Holmstedt, Hammar and co-workers (Hammar, Holmstedt and Ryhage, 1968; Hammar *et al.*, 1971; Holmstedt and Palmér, 1973; Holmstedt and Lingren, 1974) refined the technique, which they prefer to call 'mass fragmentography', and played a major role in demonstrating its analytical potential to the biomedical community. The term 'selected ion monitoring' (Hoffenberg *et al.*, 1973) seems to be the most descriptive of the technique (Watson, Falkner and Sweetman, 1974).

Qualitative application of selected ion monitoring

The selectivity available with selected ion monitoring is illustrated in the following example. In the development of analytical methodology for some of the urinary metabolites of catecholamines (Heath *et al.*, 1974), electron-capture detection (ECD) was evaluated as a means of quantifying biological samples upon analysis by GC. One of the metabolites of interest was normetanephrine; a perfluoropropionyl derivative (PFP) of normetanephrine (NMN) was prepared (Änggård and Sedvall, 1969) as shown in Figure 2.3. Standard solutions of NMN-PFP were used to demonstrate that 50 pg of this derivative injected on-column could be readily detected by ECD. However, when an aliquot of derivatised biological extract was injected onto the column under the same GLC conditions, interfering materials precluded effective use of the ECD. A more selective, yet equally sensitive, detector (comparable to ECD) was needed to determine whether low levels of NMN-PFP were present in the biological extract.

Remembering that the mass spectrum of authentic NMN-PFP (Figure 2.3) indicated two characteristic peaks at m/e 445 and 458, the investigators adjusted the MS to monitor only the ion current at these two m/e values throughout the analysis of the biological sample by GC–MS. In a preliminary experiment by ion monitoring, analysis of authentic NMN-PFP gave simultaneous peaks in the ion-current profiles of m/e 445 and 458 at a retention time of 3.7 min.

The data records for analysis of the biological sample by the ion-monitoring technique and the non-selective total ion current (TIC) detector are compared in Figure 2.7. The upper panel in Figure 2.7 is a profile of the total ion current leaving the ion source and entering the magnetic field (see schematic of instrument in Figure 2.1). Clearly, the TIC detector is essentially useless as a selective detector; note that in the upper panel of Figure 2.7 there is no discernible peak at 3.7 min, the expected retention time of NMN-PFP. In contrast, the bottom panel of Figure 2.7 shows the output from the selected ion detector, which in this case only responds to selected ions of mass 445 or 458. Although there is a substantial peak with a retention time of approximately

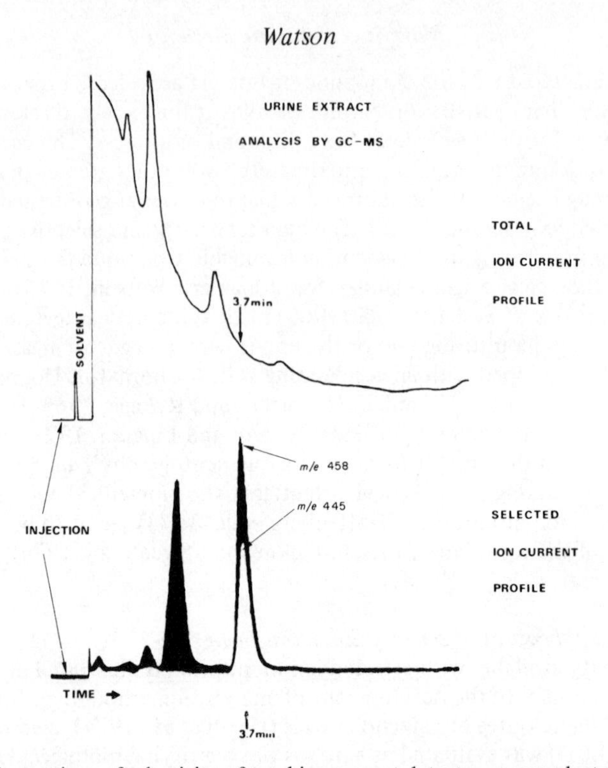

Figure 2.7 Comparison of selectivity of total ion-current detector versus selected ion-current detector during analysis of biological extract for the pentafluoropropionyl derivative of normetanephrine (Figure 2.3). Data records were recorded simultaneously from the two different detectors; see text for details. Reprinted from Watson (1976), with permission from Raven Press

2 min, this compound produces only ion current at *m/e* 458. However, at 3.7 min the ion currents at both *m/e* 445 and 458 rise and fall simultaneously, indicating that they are produced from the same source. Furthermore the ratio of the peaks in these ion-current profiles (*m/e* 458 : *m/e* 445) is approximately 2 : 1, as expected from the mass spectrum of authentic NMN-PFP (Figure 2.3).

Therefore observation of simultaneous peaks in the two selected ion-current profiles in the expected abundance ratio and at the expected retention time provides conclusive evidence that NMN-PFP was present in the derivatised urine extract.

Quantitative application of ion monitoring

This technique provides the opportunity to use a stable-isotope-labelled analogue of the compound of interest as an ideal internal standard for quantitative analysis. This application of selected ion monitoring (SIM) will be illustrated in the following example of prostaglandin methodology.

In the development of quantitative SIM for prostaglandin E_2 (PGE_2), the mass spectrum of a derivative of PGE_2 was examined for suitable ions to be monitored (Gréen, 1969). In Figure 2.8 the abbreviated bar graph spectrum of

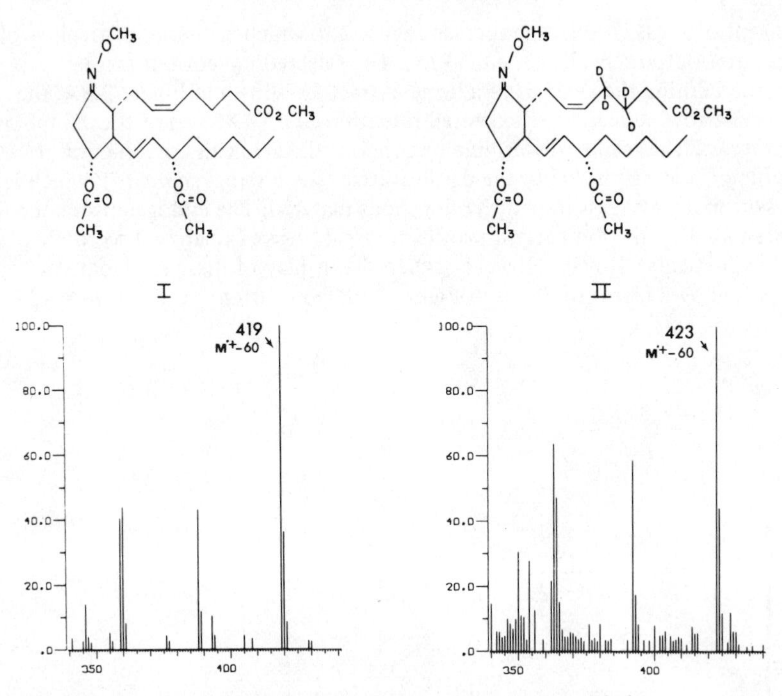

Figure 2.8 Left: Segment of mass spectrum of prostaglandin E_2-methyl ester-methoxime-bisacetate (I) indicating fragment ion of mass 419. Right: Comparable segment of mass spectrum of tetradeutero-PGE_2; note that the fragment ion has shifted four mass units, as indicated by peak at *m/e* 423. Reprinted from Watson (1976), with permission from Raven Press

the methyl ester (ME), methoxime (MO), bisacetate (Ac) derivative of PGE (I) shows a relatively prominent ion at *m/e* 419. This ion will be monitored to represent quantitatively the amount of endogenous (d_0) PGE_2 as the ME-MO-Ac derivative present in a biological extract. A deuterium-labelled analogue, 3,3,4,4-tetradeutero-PGE_2, available through the generosity of the Upjohn Company (Dr U. Axen), can be used as an internal standard. The abbreviated mass spectrum of the ME-MO-Ac derivative of this tetradeutero (d_4) analogue (II) is also shown in Figure 2.8. The molecular ion of the d_4 analogue also loses 60 mass units owing to elimination of the elements of acetic acid ($C_2H_4O_2$), to produce a prominent fragment ion at *m/e* 423 because this fragment retains all four of the deuterium atoms. Thus, by monitoring the ion current at *m/e* 419 and 423, the mass spectrometer can distinguish the endogenous and labelled compounds which co-chromatograph through the gas chromatograph.

In practice, a large excess (approximately 100–1000 times the amount of endogenous compound anticipated) of the d_4 analogue is added to the biological sample to function both as an internal standard and as a carrier to minimise adsorptive and other losses of the endogenous compound (Samuelsson, Hamberg and Sweeley, 1970; Axen *et al.*, 1971).

The analytical results are profiles of ion current at the selected values of *m/e* (419 and 423, in this case); the data record is similar in appearance to a conventional gas chromatogram (see Figure 2.9, which is a series of tracings of computer output: Frolich *et al.*, 1975). The selected ion-current profiles recorded during the analysis of a urine extract are shown in Figure 2.9A; the vertical arrow indicates the expected retention time (4.8 min) of the compounds of interest. The readily discernible peak above the arrow in the ion-current profile of *m/e* 423 is due to the d_4 derivative (the internal standard), which is present in large excess over the endogenous material. The endogenous material represented by the ion-current profile at *m/e* 419 gives a barely discernible peak in this particular display. However, when the display of these same data is amplified by a factor of 32, as in Figure 2.9B (ion-current profile at *m/e* 423

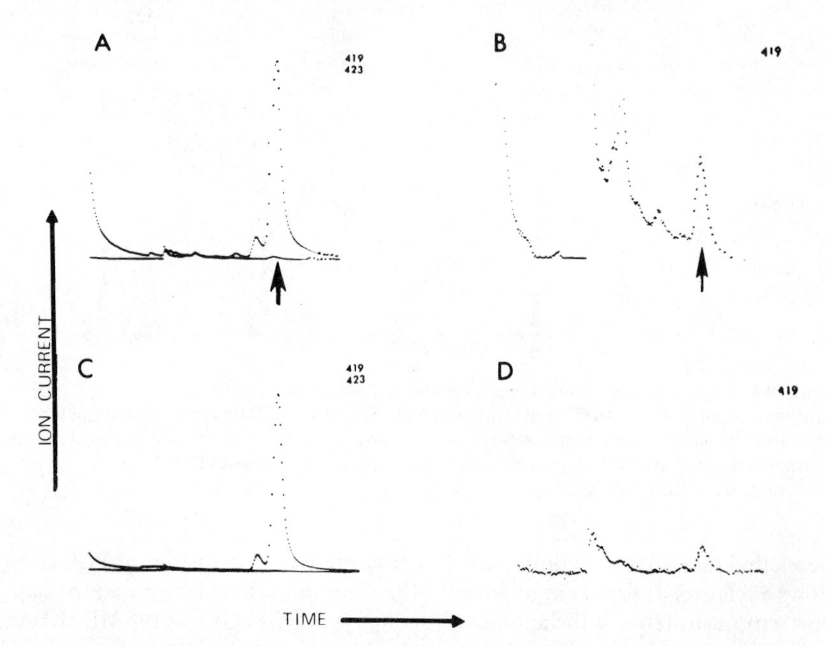
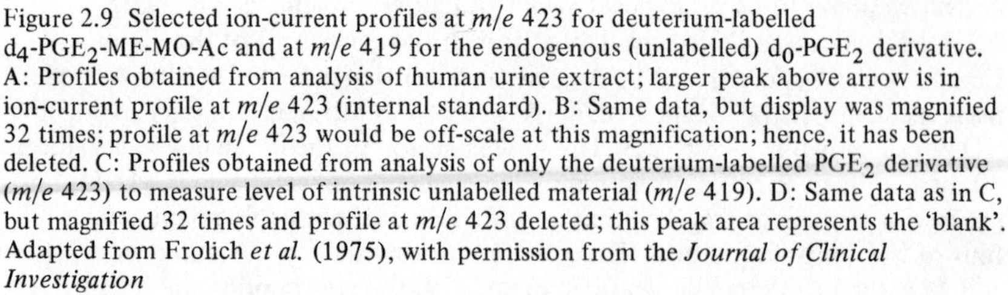

Figure 2.9 Selected ion-current profiles at *m/e* 423 for deuterium-labelled d_4-PGE_2-ME-MO-Ac and at *m/e* 419 for the endogenous (unlabelled) d_0-PGE_2 derivative. A: Profiles obtained from analysis of human urine extract; larger peak above arrow is in ion-current profile at *m/e* 423 (internal standard). B: Same data, but display was magnified 32 times; profile at *m/e* 423 would be off-scale at this magnification; hence, it has been deleted. C: Profiles obtained from analysis of only the deuterium-labelled PGE_2 derivative (*m/e* 423) to measure level of intrinsic unlabelled material (*m/e* 419). D: Same data as in C, but magnified 32 times and profile at *m/e* 423 deleted; this peak area represents the 'blank'. Adapted from Frolich *et al.* (1975), with permission from the *Journal of Clinical Investigation*

now deleted by computer: Watson *et al.*, 1973), there is a readily discernible peak in the ion-current profile at *m/e* 419 for the endogenous material at the expected retention time (vertical arrow).

Not all of the area under the indicated peak in Figure 2.9B is related to the endogenous material in the sample; some of this area is the result of 'blank' or the intrinsic amount (approximately 0.4 per cent) of protium (d_0), in the imperfectly labelled (d_4) analogue. The relative importance of the blank is illustrated by comparison of Figure 2.9C and D with Figure 2.9A and B. The combined display of ion-current profiles at *m/e* 419 and 423 in block C results from analysis of a standard aliquot of only the d_4 derivative. Again, these same data are amplified by a factor of 32 in the display in Figure 2.9D (the ion-current profile at *m/e* 423 deleted by computer). Thus the net peak areas (that in Figure 2.9B minus that in Figure 2.9D) can be related to the amount of endogenous PGE_2 in the original urine sample.

The procedure for quantifying the amount of endogenous PGE_2 in the urine sample involves calculating the net ratio of d_0/d_4 from the peak areas in the ion-current profiles, namely the ratio of the peak area in the ion-current profile at *m/e* 419 to that in the profile at 423 for both the urine extract (Figure 2.9A) and the blank (Figure 2.9C). In this case the ratio of peaks indicated that the gross ratio of d_0/d_4 in the biological extracts was 0.0133, whereas the ratio of d_0/d_4 in carrier-only represented in Figure 2.9C was 0.0042; thus the net ratio of d_0/d_4 was 0.0091. Because 2500 ng of d_4-PGE_2 was added to the original 200 ml sample of urine, the amount of endogenous PGE_2 in the sample was 0.0091×2500 ng = 22.8 ng or 114 pg PGE_2 per millilitre urine (Frolich *et al.*, 1975).

ACKNOWLEDGEMENTS

This work was possible through collaboration with Dr B. J. Sweetman, Dr W. C. Hubbard, Dr J. C. Frölich and Mr D. R. Pelster, as supported by the Research Center for Clinical Pharmacology and Drug Toxicology (NIH-GM-15431). JTW was supported by a Research Career Development Award (NIH-50290).

REFERENCES

Änggård, E. and Sedvall, G. (1969). *Anal. Chem.*, **41**, 1250

Axen, U., Gréen, K., Horlin, D. and Samuelsson, B. (1971). *Biochem. Biophys. Res. Commun.*, **45**, 519

Biller, J. W. and Biemann, K. (1974). *Anal. Lett.*, **7**, 515

Braselton, W. E., Orr, J. C. and Engel, L. L. (1973). *Anal. Biochem.*, **53**, 64

Bush, M. T., Sekerke, H. J. Jr., Vore, M., Sweetman, B. J. and Watson, J. T. (1972). *Proceedings of Seminar on the Use of Stable Isotopes in Clinical Pharmacology* (ed. P. D. Klein and L. J. Roth), Technical Information Center, CONF-711115, Springfield, Va., USA, p. 233

Falkner, F. C., Sweetman, B. J. and Watson, J. T. (1975). *Appl. Spectrosc. Rev.*, **10**, 51

Farmer, J. B. (1963). *Mass Spectrometry* (ed. C. A. McDowell), McGraw-Hill, New York, Chapter 2, p. 7

Frölich, J. C., Wilson, T. W., Sweetman, B. J., Nies, A. S., Carr, K., Watson, J. T. and Oates, J. A. (1975). *J. Clin. Invest.*, **55**, 763

Gréen, K. (1969). *Chem. Phys. Lipids*, **3**, 254

Hammar, C.-G., Alexanderson, B., Holmstedt, B. and Sjöqvist, F. (1971). *Clin. Pharmacol. Therap.*, **12**, 496

Hammar, C.-G., Holmstedt, B. and Ryhage, R. (1968). *Anal. Biochem.*, **25**, 532

Heath, E. C., Falkner, F. C., Hill, R. E. and Watson, J. T. (1974). *Proceedings of the 22nd Annual Conference on Mass Spectrometry and Allied Topics, Philadelphia*, American Society for Mass Spectrometry, p. 286

Hites, R. A. and Biemann, K. (1967). *Anal. Chem.*, **39**, 965

Hites, R. A. and Biemann, K. (1970). *Anal. Chem.*, **42**, 855

Hoffenberg, R., Lawson, A. M., Ramsden, D. R. and Raw, P. J. (1973). *Mass Spectrometry in Biochemistry and Medicine* (ed. A. Frigerio and N. Castagnoli), Raven Press, New York, p. 303

Holmes, W. F., Holland, W. H. and Parker, J. A. (1971). *Anal. Chem.*, **43**, 1806

Holmstedt, B. and Lindgren, J.-E. (1974). *The Poisoned Patient: The Role of the Laboratory,* Ciba Foundation Symposium 26, ASP (Elsevier–Exerpta Medica–North-Holland), Amsterdam, p. 106

Holmstedt, B. and Palmér, L. (1973). *Advan. Biochem. Psychopharmacol.*, **7**, 1

Knapp, D. R., Gaffney, T. W. and McMahon, R. E. (1972). *Biochem. Pharmacol.*, **21**, 425

Pohl, L. R., Nelson, S. D., Garland, W. A. and Trager, W. F. (1975). *Biomed. Mass Spectrom.*, **2**, 23

Samuelsson, B., Hamberg, M. and Sweeley, C. C. (1970). *Anal. Biochem.*, **33**, 301

Sweeley, C. C., Elliott, W. H., Fries, I. and Ryhage, R. (1966). *Anal. Chem.*, **38**, 1549

Sweeley, C. C., Ray, B. D., Wood, W. K., Holland, J. F. and Krichevsky, M. I. (1970). *Anal. Chem.*, **42**, 1505

Vore, M., Sweetman, B. J. and Bush, M. T. (1974). *J. Pharmacol. Exp. Ther.* **190**, 384

Waller, G. R. (1972). *Biochemical Applications of Mass Spectrometry*, Wiley-Interscience, New York

Watson, J. T. (1976). *Introduction to Mass Spectrometry for Biomedical, Environmental and Forensic Application*, Raven Press, New York

Watson, J. T., Falkner, F. C. and Sweetman, B. J. (1974). *Biomed. Mass Spectrom.*, **1**, 156

Watson, J. T., Pelster, D. R., Sweetman, B. J., Frölich, J. C. and Oates, J. A. (1973). *Anal. Chem.*, **45**, 2071

3
Stable isotopes in human drug metabolism studies

G. H. Draffan (Inveresk Research International, Edinburgh EH21 7UB, UK)

INTRODUCTION

The resurgence of interest in the stable isotopes of hydrogen, carbon, nitrogen and oxygen in biomedical research is mainly attributable to the development of increasingly refined technology for their determination. As a result, new areas of application have opened up, and this is particularly the case in pharmaceutical research. The purpose of this paper is to consider the rôles for stable isotope labelling in the conduct of drug metabolism studies in man, and Scheme 3.1 summarises a number of these applications. At present, the largest area of activity concerns the use of labelled drugs as standards in quantitative measurement following administration of the unlabelled substance. However, administration of the stable-isotope-labelled drug can also provide valuable information and, in other cases, complements the use of a radiotracer.

In considering the techniques available for the analysis of stable isotopes, mass spectrometry provides a unique combination of sensitivity with the provision of structural information. The applications cited below all involve mass spectrometry, but it should be noted that there are two fundamentally different approaches to the determination of stable isotopic abundance. The first is indirect analysis, in which the sample is converted to a gas (hydrogen, nitrogen, carbon dioxide), followed by analysis of isotopic enrichment in an isotope ratio mass spectrometer. This offers extremely high precision (0.01 per cent), but requires relatively large sample sizes (0.1–10 mg). It would be the technique employed in, for example, a ^{13}C balance study where the total isotope eliminated (drug plus metabolites) had to be determined. The second approach is direct analysis, in which the compound of interest is separated without degradation. The partial mass spectrum, possibly using ion-monitoring techniques, is obtained enabling the ratios of the labelled to the unlabelled species to be determined. Here the precision of ratio determination is much lower (1–10 per cent), but the technique is specific and offers high sensitivity. This approach, which utilises conventionally designed mass spectrometers, is the technique applied in most of the studies in the following discussion.

A. Stable isotope dilution

In which the unlabelled drug is administered and labelled drug or metabolites are subsequently added to biological fluids as internal standards/carriers in quantitative determinations.

B. Labelled drug administration

1. Metabolic fate of the label:

    (a) combustion techniques in which $^{13}CO_2$ excess is determined in 'total' extracts;
    (b) reverse dilution techniques in which individual metabolites are quantified;
    (c) metabolite recognition by the 'isotope cluster' technique in which mixtures of labelled and unlabelled forms of the drug are administered.

2. Isotope effects *in vivo* (deuterium):

    (a) mechanistic studies;
    (b) novel drug design in which deuterium is introduced to decrease metabolism and enhance bioavailability.

C. Pharmacokinetic studies following co-administration of labelled and unlabelled forms of a drug

1. Bioavailability comparison following two routes of administration.
2. Bioavailability comparison following the administration of two formulations by the same route.
3. 'Pulse-labelling' at steady state.

Scheme 3.1 Some roles for stable isotopes in drug metabolism studies in man

STABLE ISOTOPE DILUTION

The enthusiasm with which mass spectrometry has been adopted by many clinical pharmacologists is in large part attributable to its potential as a highly specific and sensitive measuring technique for use in pharmacokinetic studies. Inevitably there is another body of opinion which suggests that the technique is used excessively for this purpose and that in many cases other sensitive methods involving a fraction of the capital cost tend to be under-utilised. There is some justification for this, but, in general, it is always of value to have available a highly specific procedure against which other methods may be checked. This is particularly evident in the relationship between radioimmunoassay (RIA) and mass spectrometry to be found in the prostaglandin literature. RIA offers exceptional sensitivity and speed of analysis but is inherently suspect in terms of specificity. However, once proved against a mass spectrometric method, RIA would become the method of choice.

There are also many situations in which mass spectrometry provides the best solution. A particularly important possibility, when the mass spectrometer is the detector, is that the internal standard for quantitative analysis can be a stable-isotope-labelled form of the sample. This method, stable isotope dilution, normally ensures minimal distinction between sample and standard throughout the recovery sequence from biological fluid until final differentiation and ratio measurement is made by the mass spectrometer. The instrument is usually operated in the so-called 'selected ion monitoring' (SIM) or 'mass fragmentographic' mode (Hammar, Holmstedt and Ryhage, 1968), and output may be in the form of continuous traces recording the *m/e* values selected for the sample and the standard. As noted above, this is now virtually a routine technique and for a broader indication of its use in metabolic studies, reference may be made to reviews by, for example, Falkner, Sweetman and Watson, (1975), Lawson and Draffan, (1975) and Roncucci *et al.*, (1976). Some examples indicating the range of applications are cited below.

A typical problem in the standardisation of an extraction procedure arises when the compound of interest is labile and recovery from biological fluids tends to be variable. The detection and measurement of the minor metabolite of dopamine, salsolinol (Figure 3.1), provides a simple example of this situation (Draffan *et al.*, 1974). The substance has excellent gas chromatographic properties as its pentafluoropropionyl derivative and can be detected at trace levels in urine extracts using SIM methods. Figure 3.1 shows detection at *m/e* 617 (molecular ion), 602 and 603, and illustrates the sensitivity and specificity of the SIM technique. It was, however, found that recovery from replicate extracts could vary by up to tenfold. Standardisation of these losses was readily obtained by adding salsolinol-d_4 to the samples, and reliable quantitative measurements

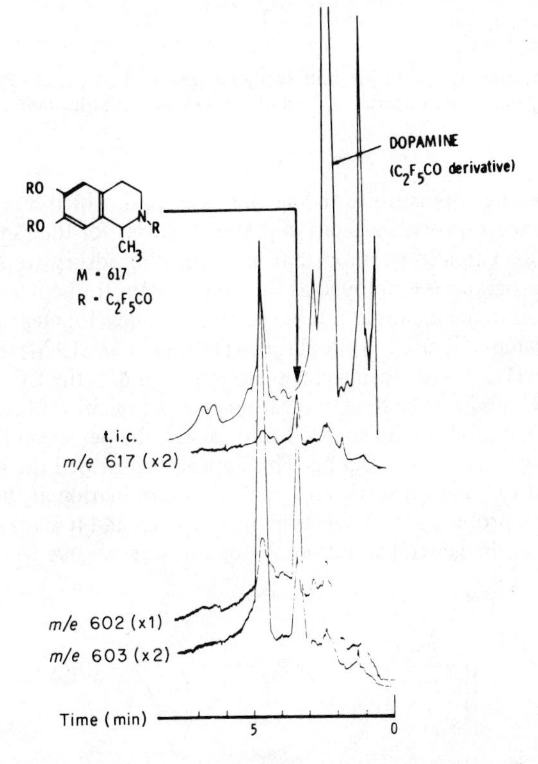

Figure 3.1 Detection of the dopamine metabolite salsolinol (as its pentafluoropropionyl derivative) in human urine following administration of L-dopa. Reproduced from Draffan *et al.* (1974)

could then be made in the range 5 ng–1 μg/ml. Problems may also be encountered in a multi-stage separation procedure requiring several chromatographic methods prior to final mass spectrometric analysis. Thus in the analysis of the major human urinary metabolite of prostaglandins $F_{1\alpha}$ and $F_{2\alpha}$ (Figure 3.2) nine stages are involved prior to mass spectrometry (Brash *et al.*, 1976). These include hydroxy-acid/lactone equilibration, selective ester hydrolysis, and thin-layer and gas chromatography. The prospect of standardising this sequence by any technique other than isotope dilution is extremely remote. It was found that with the trideutero metabolite as the internal standard, measurements could be made at concentrations of 1 ng/ml with a standard deviation of 10 per cent. At higher concentrations precision was typically ± 2 per cent.

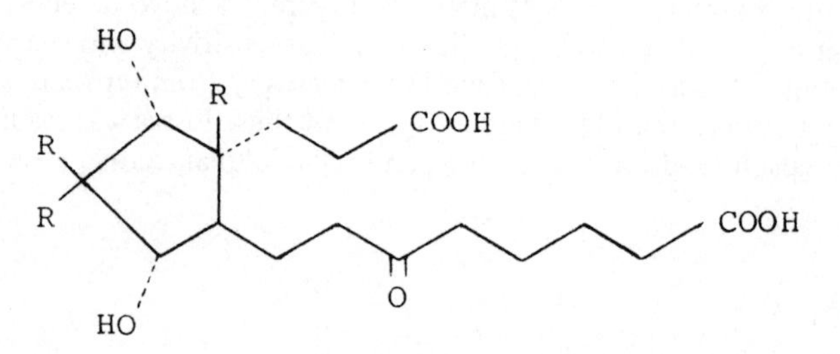

Figure 3.2 Structure of the major urinary metabolite of $PGF_{1\alpha}$ and $PGF_{2\alpha}$ in man, 5α, 7α-dihydroxy-11-ketotetranorprosta-1,16-dioic acid (R = H) and its deuterated analogue (R = ^{2}H)

In the preceding applications, the labelled standard, which was present in 10–100-fold excess, probably acted in part as a carrier for the sample. A carrying function can also be important in minimising adsorptive losses where the gas chromatograph is employed as the inlet system to the mass spectrometer. A particularly extreme example was encountered in development of an assay for the drug indoramin (Figure 3.3) in plasma (Draffan *et al.*, 1974). Because of the combined effects of losses during chromatography and in the GC–MS interface, the compound could not be detected carrier-free below 50–100 ng. However, by the use of indoramin-d_5 in up to 2000-fold excess, the detection limit for indoramin-d_0 was reduced to 0.5 ng. This approach provided the basis for an assay in plasma to 5 ng/ml, sufficient for drug determination at therapeutic dose levels. There are problems with this type of approach and it is scarcely to be recommended in the general situation. It does, however, serve to indicate that a

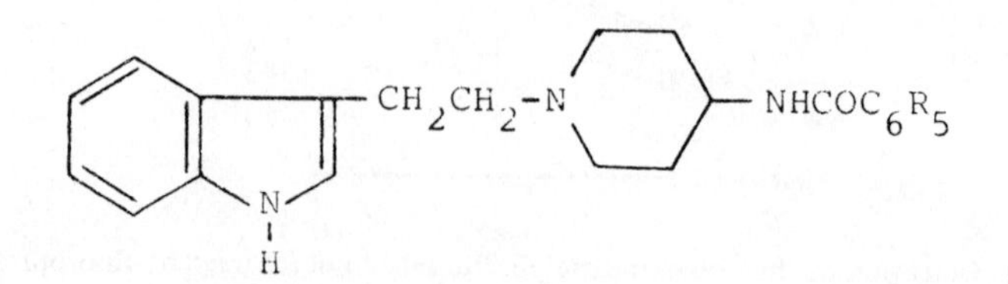

Figure 3.3 Structure of indoramin (R = H) and its deuterated analogue (R = ^{2}H)

labelled compound can be used to compensate for gross chromatographic deficiencies.

Methods developed for the determination of clonidine (Figure 3.4) in plasma provide a more typical example of the use of GC–MS and stable isotope dilution. Methods of the highest sensitivity were required, since in the original work employing ^{14}C-labelled clonidine (Rehbinder, 1970) peak plasma concentrations in the 1–2 ng/ml range following a 5 μg/kg oral dose were recorded. The availability of clonidine-d_4 with a residual d_0 contribution of approximately 0.5 per cent provided the basis for the required assay (Draffan *et al.*, 1976). The drug was extracted from plasma following addition of clonidine-d_4 at a level of 100 ng/ml. The mixture was then methylated for SIM GC–MS analysis recording $m/e = 257$ and $m/e = 261$, the two molecular ions.

Measurements could be made to a limit of 0.2 ng/ml and with a precision of ± 11 per cent at 0.25 ng/ml. The use of this method in concentration-effect and pharmacokinetic studies is discussed by Davies *et al.* (1977) and in the accompanying chapter by Davies *et al.* Further work with the drug involving the administration to man of the deuterated compound and the quantification of its hydroxy metabolite are noted in the following section.

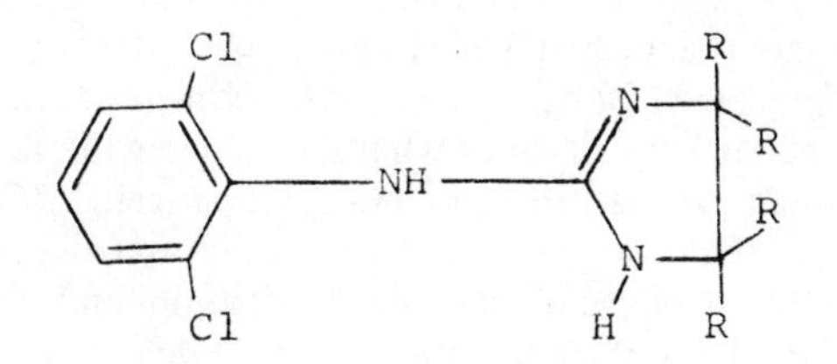

Figure 3.4 Structure of clonidine (R = H) and its deuterated analogue (R = ^{2}H)

LABELLED DRUG ADMINISTRATION

The use of dilution techniques employing stable-isotope-labelled drugs in quantitative analysis was briefly described in the preceding section, and this is currently the single largest area of activity. There are, however, many other rather more interesting possibilities in which the labelled form, either alone or in combination with the unlabelled drug, is administered to man. Several of these are outlined in this and in the following section.

Metabolic fate of the label

The ideal in a human drug metabolism study would be to dispense with radiotracers and conduct all aspects of the work with a stable isotopic label. While much can be achieved, the absence of a simple means of quantifying the total fate of the dose tends to restrict the range of applications. Thus, with a radiotracer, the conduct of an excretion/retention 'balance' study is the simplest of tasks. An equivalent experiment with, for example, ^{13}C calls for highly sophisticated equipment and considerable expertise in sample preparation. The fundamental problem is the natural abundance of ^{13}C (approximately 1.1 per cent) and the need to determine the slight changes in this background brought about by excess ^{13}C in the excreta. This problem is compounded by the differences in the $^{12}C:^{13}C$ ratio which will be encountered as a result of carbon source dietary variation (see, for example, von Unruh *et al.*, 1974). In a true *de novo* 'balance study' with ^{13}C, no assumptions may be made on the route of excretion or on the nature of metabolites and their extraction properties. Combustion of excreta to CO_2 is required, followed by gas analysis for isotopic enrichment using an isotope ratio mass spectrometer.

Techniques for the determination of carbon isotope ratios are highly developed partly as a result of their use in geology and the physical sciences. Thus precision in the order of 0.001 per cent is attainable (Beckinsale *et al.*, 1973) and the specification for a double collector isotope ratio mass spectrometer will normally be of the order of 0.01 per cent (0.1‰, i.e. parts per mil). The precision available is more than adequate for the task. However, for biological material the accuracy attainable may not be so high as a result of the technical problems involved in sample preparation, combustion and gas introduction to the mass spectrometer. Nevertheless, with care in sample preparation and taking account of natural variation in carbon isotopic enrichment, it is considered that ^{13}C tracer studies are practical at a level down to 5×10^{-5} g excess ^{13}C/g of carbon sample combusted (50 pg excess/μg) (von Unruh *et al.*, 1974). These authors, in an excellent illustration of what is attainable in practice, conducted an experiment in man using aspirin containing 92 per cent ^{13}C in the carboxyl group. They showed that even with the high dilution of 0.1 per cent of this labelled form in a 320 mg dose of the unlabelled compound, drug and metabolites could be determined as total ^{13}C in crude extracts of 0–24 h urine. Strictly, to fulfil the requirements of a balance study, extractability would not be assumed, and freeze-dried urine and faeces would be examined. Such studies appear practical. It should also be noted that the instrumentation used in this work was not of the conventional type. Ion counting was employed in obtaining good precision at sample sizes lower than those normally required for ratio determination in a double collector mass spectrometer (Schoeller and Hayes, 1975).

The next stage in a radiotracer study would be separation of the individual metabolites, their quantification as a percentage of the administered dose and, finally, structural identification. There is no ready means of distinguishing a metabolite from the endogenous background where the tracer is a stable isotope. One means of aiding detection is by the use of the 'twin isotope' or 'isotope cluster' technique used by, for example, Knapp, Gaffney and McMahon (1972) and McMahon *et al.* (1973). Here administration of a mixture of the labelled and unlabelled forms of the drug leads to the recognition of metabolites by the presence of characteristic doublet ions in their mass spectra. A more recent example is to be found in the identification of *N*-hydroxyamylobarbitone as a urinary metabolite of amylobarbitone in man. Here, the ^{15}N-labelled drug was dosed, mixed with unlabelled amylobarbitone (Inaba, Tang and Kalow, 1976). It is significant, however, that a ^{14}C tracer was also employed in monitoring the extent of excretion and directing attention to the fraction containing the novel metabolite.

An interesting alternative means of detection of ^{13}C-labelled metabolites has been described by Sano *et al.* (1976). This involves catalytic combustion of the effluent from a gas chromatographic column to CO_2 and passage to the mass spectrometer, where $m/e = 45$ ($^{13}CO_2$) : $m/e = 44$ ($^{12}CO_2$) ratios are determined. Peaks enriched in ^{13}C are detected as shown in Figure 3.5, in which a number of human urinary metabolites of ^{13}C-labelled aspirin may be distinguished. Peaks A, D, F and G were subsequently identified as salicylic acid, unchanged aspirin, gentisic acid and hydroxyhippuric acid.

In the foregoing studies some assumptions are made: for example, that metabolites are extractable or that they will be amenable to GC–MS after a particular derivatisation procedure. Stable isotope labelling as an aid in the detection of metabolites is, therefore, likely to remain in a supporting role to conventional radiotracer work. Since some measure of 'guess work' is clearly required, a logical extension is the prediction of metabolite structure, synthesis and the use of reverse dilution to confirm their presence and to enable quantification. A metabolic study in man will normally only be conducted after detailed animal studies are complete and ideas of the metabolic fate in man will already be formed. Studies of the extent of clonidine *para*-hydroxylation illustrate the use of reverse stable isotope dilution.

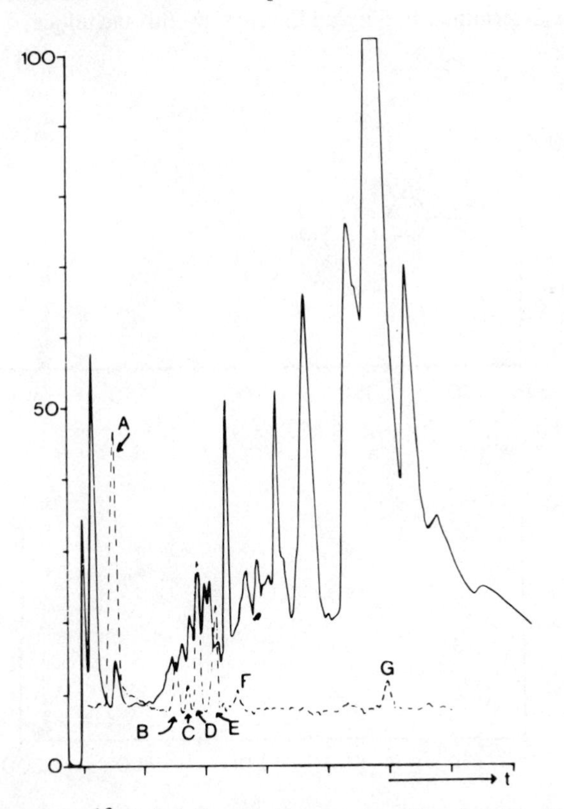

Figure 3.5 Detection of ^{13}C-labelled metabolites in urine after administration of ^{13}C-aspirin to man. Catalytic combustion to CO_2 and GC–MS measurement of $^{13}CO_2$: $^{12}CO_2$ ratio. Reproduced from Sano *et al.* (1976)

Some of the problems encountered in assay development adequate for pharmacokinetic studies with clonidine in man have been noted in the previous section. The low therapeutic dose, *ca* 0.3–1 mg, also presents difficulties in the identification of urinary metabolites. In the original investigations with ^{14}C-labelled clonidine, *para*-hydroxyclonidine was identified as a metabolite in animals and in a limited investigation in man (Rehbinder, 1970). The classical technique of reverse radioisotopic dilution employing milligram quantities of the predicted metabolite as carrier was employed. In our studies (Draffan *et al.*, 1976) we sought to re-examine the significance of hydroxylation in man but without the use of radiotracer. However, while chromatography was efficient as the permethylated derivative, instability of the metabolite resulted in low and very variable recovery from urine at the levels likely to be encountered following a therapeutic dose of the drug (<1 μg/ml). Accordingly, the deuterium-labelled drug, clonidine-d_4, was administered and synthetic unlabelled *para*-hydroxyclonidine employed as carrier. The additional specificity of chemical ionisation (CI) was required in selected ion-monitoring determination of the metabolite in crude extracts of urine after enzymic hydrolysis and methylation. Using isobutane CI (Figure 3.6), recording $m/e = 288$ (MH) as the reference channel and $m/e = 294$ (d_4, MH + 2) for labelled metabolite detection, levels could be determined to 5 ng/ml in urine. By this technique, 7.4 per cent of

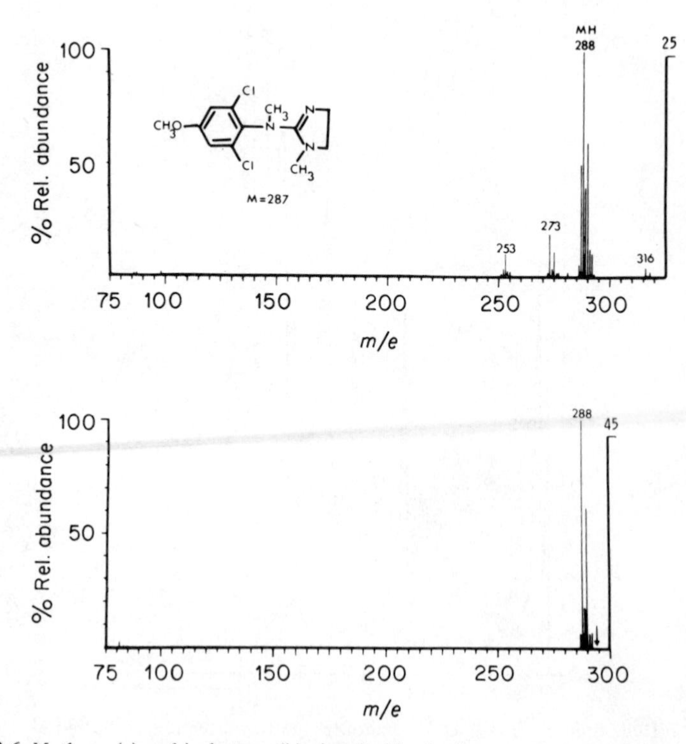

Figure 3.6 Methane (a) and isobutane (b) chemical ionisation mass spectra of the permethylated derivatives of *para*-hydroxyclonidine.

the 300 μg dose was recovered as *para*-hydroxyclonidine conjugates in 0–24 h urine, suggesting that this pathway of oxidative metabolism is unlikely to be a major factor in determining the elimination of clonidine in man.

For this type of study to be valid, labelled and unlabelled forms of the drug should be metabolically equivalent. In this case comparison of data for different individuals after d_4 and d_0 administration indicated that similar plasma concentrations were achieved and that the duration of the hypotensive effects was also comparable. The earlier peak level after d_4 administration (Figure 3.7) was accountable, since this material was dosed in solution, while clonidine-d_0 was taken as tablets. A further study in which the two compounds were dosed

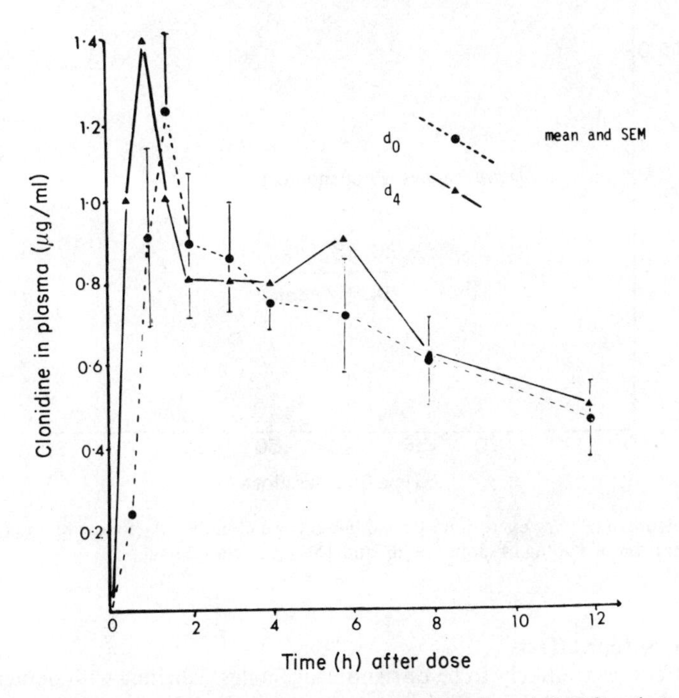

Figure 3.7 Comparison of the concentrations of clonidine-d_0 and clonidine-d_4 in human plasma after oral administration. One subject received 300 μg of clonidine-d_4 as a solution and 5 subjects 300 μg of clonidine-d_0 as tablets

orally as a mixture in solution also suggested similar pharmacokinetics (Figure 3.8). Plots of unchanged drug excreted in urine in unit time (ng/min) against time enabled estimation of the similar elimination half-lives of 11.5 and 11.3 h for the two forms of the drug. The fractions of the total dose recovered unchanged in urine were also comparable, being 28.4 per cent as the deuterated form and 26.3 per cent as the unlabelled form. This type of check, while not intended as a rigorous investigation of the possible existence of a deuterium isotope effect, is sufficient to justify the use of reverse dilution techniques in metabolite determination.

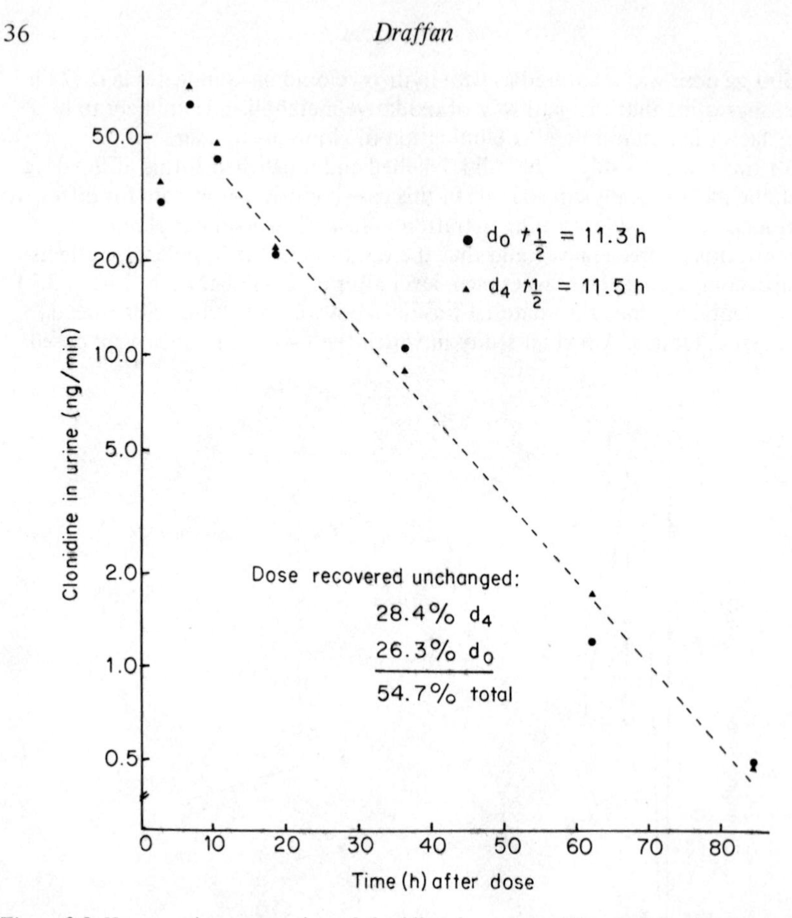

Figure 3.8 Human urinary excretion of clonidine-d_0 and clonidine-d_4 following the oral co-administration of 150 μg of clonidine-d_0 and 150 μg of clonidine-d_4

Deuterium isotope effects

Isotope effects are unlikely to be of significance unless labelling with deuterium (or tritium) is involved. In general, the reaction rate for cleavage of an X–D bond will be slower than that for an X–H bond. This will, of course, result in an observable effect only where X–H(D) fission is rate-determining. There is an extensive literature on deuterium isotope effects in biological systems, and for a review of this topic, and its relevance in drug metabolism, reference may be made to Blake, Crespi and Katz (1975). Kinetic effects are normally expressed in terms of a ratio of rate constants, k_H/k_D, and for primary isotope-effects (X–H or D cleavage), these are generally in the range 2–5. Higher values have been observed, and, on theoretical grounds, a k_H/k_D ratio approaching 10 is considered to be the maximum. Secondary isotope effects, in which deuterium is located near the reaction site, may also be observed, but here k_H/k_D ratios are much lower.

Deuterium isotope effects may be regarded either as a troublesome parameter where, as in the section below, comparative bioavailability studies are the objective, or as an invaluable aid in defining rate-limiting steps and pathways in metabolism. The role of deuterated drugs in this latter context is treated in some detail in the accompanying chapters by Horning *et al*. and by Jarman *et al.*

In a corollary to such investigations introduction of deuterium to increase metabolic stability and to enhance the bioavailability of a therapeutic agent represents a challenging area in drug design. An impressive example of this approach may be found in the development of the antibacterial agent 2-deutero-3-fluoro-D-alanine (Figure 3.9). Here the introduction of deuterium at the 2 position diminishes the metabolism of the amino acid, and since labelled and unlabelled forms have equivalent antibacterial activity, the isotope effect results in a product with an enhanced pharmacological effect on the basis of an equivalent dose (Kropp, Kahan and Woodruff, 1975).

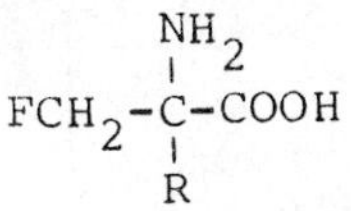

NH_2
$FCH_2-C-COOH$
R

Figure 3.9 Structures of 3-fluoro-D-alanine (R = H) and its deuterated analogue (R = ^2H)

Preliminary studies with lignocaine (Draffan, Gilbert and McGregor, 1977) provide a further example of an attempt to enhance bioavailability by deuterium labelling. Lignocaine is used intravenously for the treatment of ventricular arrhythmias, and in this mode of administration the plasma concentration of unchanged drug substantially exceeds those of two of its N-de-ethylated metabolites, 'EGX' (Figure 3.10; R = C_2H_5, R′ = H) and the corresponding bis-de-ethylated derivative 'GX' (Strong and Atkinson, 1972; Prescott, Adjepon-Yamoah and Talbot, 1976). The ratio of lignocaine to EGX, typically in the range 3–8, may reach 10. However, upon oral administration it is estimated that an average of only 35 per cent of the dose reaches the systemic circulation unchanged (Boyes and Keenaghan, 1971; Bending *et al*., 1976). This low oral bioavailability, consistent with a high hepatic first-pass elimination, would be expected to result in relatively higher levels of N-dealkylated metabolites in plasma in comparison with the intravenous route. Thus, similar concentrations of EGX and lignocaine have been observed by Bending *et al.* (1976) after oral dosing in man; however, in other studies (Garland, Trager and Nelson, 1974) this effect has been less significant.

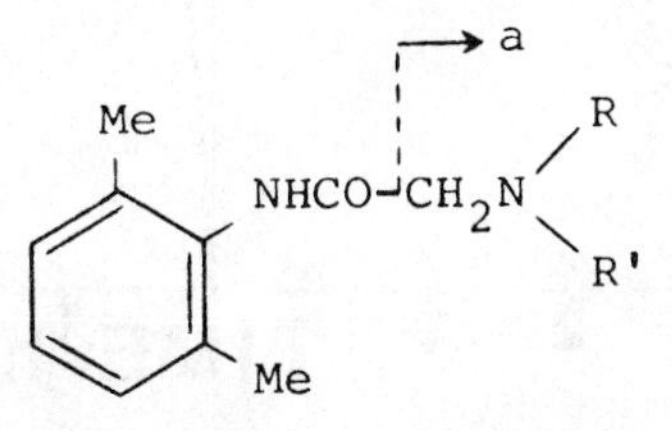

Figure 3.10 Structure of lignocaine (R = R′ = C_2H_5), lignocaine-d_{10} (R = R′ = $C_2{}^2H_5$), internal standard for quantitative measurement (R = C_2H_5, R′ = CH_3) and the mono-de-ethyl metabolite of lignocaine, EGX (R = C_2H_5, R′ = H)

It has been suggested that formation of metabolites may contribute to the side effects experienced by some patients taking oral lignocaine (Boyes *et al.*, 1971) and the drug is not now generally used orally in the treatment of arrhythmias. Consequently, there were two related objectives in deuterium labelling of the *N*-diethyl group (lignocaine-d_{10}) (Figure 3.10; R = R′ = $C_2{}^2H_5$): enhancement of the oral bioavailability and reduction in the concentration of possibly undesirable metabolites. A further incentive was provided by the observations of Nelson, Pohl and Trager (1975) that in rat liver microsomal preparations deuterium substitution in the α-methylene positions (lignocaine-d_4) resulted in a k_H/k_D of 1.49 and in the terminal methyl groups (lignocaine-d_6) a k_H/k_D of 1.52. These observations have been explained in terms of both primary and secondary isotope effects upon the N-dealkylation reaction.

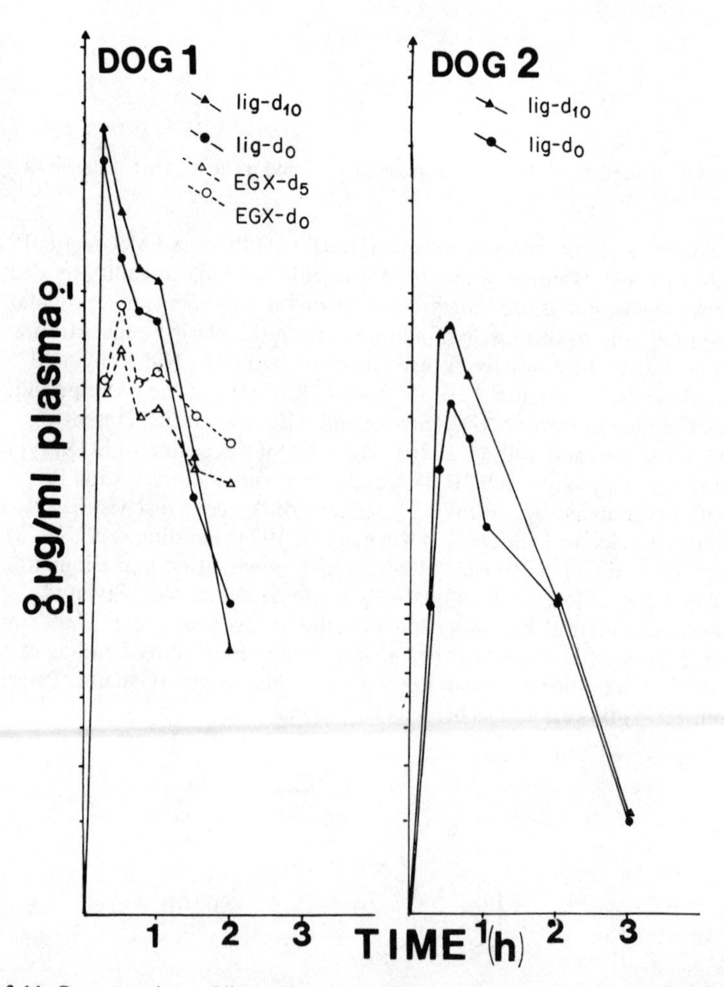

Figure 3.11 Concentrations of lignocaine, lignocaine-d_{10} and their monode-ethylated metabolites EGX and EGX-d_5 in dog plasma after oral co-administration of lignocaine and lignocaine-d_{10}

In our studies *in vivo* the labelled (d_{10}) and unlabelled (d_0) forms of the drug were determined by selected ion-monitoring GC–MS using a third substance (Figure 3.10; R = C_2H_5, R′ = CH_3) as the internal standard. Fragment 'a' was monitored at *m/e* 86 (d_0), 96 (d_{10}) and 72 for the internal standard. The two mono-de-ethylated metabolites, EGX-d_0 and EGX-d_5 were also determined by GC–MS after acetylation. Observations in two dogs, following co-administration of 1.5 mg/kg lignocaine and lignocaine-d_{10}, are illustrated in Figure 3.11. In both cases the plasma concentrations of the deuterated drug exceeded those of the unlabelled form, ratios of lignocaine-d_{10} to lignocaine-d_0 being 1.2–1.4 up to 1 h in dog 1 and at peak in dog 2, 1.9. Serial determination of the metabolites EGX-d_0 and EGX-d_5 confirmed the supposition of a reduction in the first-pass effect. Thus in dog 1 the relatively higher levels of lignocaine-d_{10} were associated with relatively lower levels of EGX-d_5, in comparison with the corresponding relationship of the unlabelled drug to its metabolite (Table 3.1). Again, this effect was more marked in dog 2 (not plotted) where at 0.75 hr, the time of peak drug concentration, the ratio of lignocaine-d_{10} to EGX-d_5 was 2.7 while that for lignocaine-d_0 to EGX-d_0 was 0.9.

Table 3.1 Plasma concentration ratios of lignocaine-d_0, lignocaine-d_{10} and their metabolites EGX-d_0 and EGX-d_5 in dog 1 after oral administration of a mixture of lignocaine-d_0 and lignocaine-d_{10} at 1.5 mg/kg

Time (hr)	lig-d_0/EGX-d_0	lig-d_{10}/EGX-d_5	lig-d_{10}/lig-d_0
0.25	5.31	7.15	1.22
0.50	1.36	2.58	1.37
0.75	1.70	2.80	1.32
1.00	1.50	2.60	1.36
1.50	0.55	1.10	1.36

These preliminary data have suggested a decrease *in vivo* of first-pass N-dealkylation and have prompted a fuller investigation of the pharmacokinetics of lignocaine-d_{10} in both dog and man. One point of interest thus far in studies involving both oral and intravenous dosing has been the close similarity of the plasma elimination half-life of the labelled and unlabelled compounds. One may provisionally interpret this on the basis that, while peak plasma concentrations are determined by the extent of the first-pass effect, the parameter determining eventual elimination is liver blood flow.

In the wider context, and particularly in view of the highly encouraging results obtained with deuterated 3-fluoro-D-alanine (noted above), further practical applications for deuterium in enhancing metabolic stability may be expected.

PHARMACOKINETIC STUDIES FOLLOWING CO-ADMINISTRATION OF LABELLED AND UNLABELLED FORMS OF A DRUG

In the preceding section co-administration of two forms of lignocaine was employed with the objective of demonstrating non-equivalence in the presence of a deuterium isotope effect upon metabolism. There are also important areas of application where the two forms of a drug are pharmacokinetically equivalent. In this situation it is clearly not advisable to consider deuterium labelling unless care is taken to place the label in a position remote from a site of metabolism. Isotopes of the heavier elements would otherwise be preferred in studies designed for comparison of the route of administration or of the formulation.

There are obvious advantages in conducting a comparative bioavailability study in which doses are given simultaneously. The subject acts in every sense as his own control, the normal procedure of completing the crossover one or two weeks later being no longer necessary. Since simultaneous measurements of drug concentration may also be made, the analytical effort is halved.

The use of this approach is illustrated in a determination of the extent of absorption of *N*-acetylprocainamide (NAPA), a candidate orally active antiarrhythmic agent (Strong *et al.*, 1975). This metabolite of procainamide has a longer half-life than that of the parent drug, justifying its development as a possible alternative to procainamide.

NAPA-^{13}C was administered intravenously while the unlabelled compound was taken orally and concentrations of both were then measured in plasma using a deuterated analogue, NAPA-d_5, as the internal standard. Figure 3.12 shows the plasma concentrations obtained in one such study in man, and a detailed pharmacokinetic analysis of these and of related data has been reported. A key observation was that oral absorption of NAPA was approximately 85 per cent, adequate for therapeutic purposes and similar to that for procainamide itself.

No applications have been reported in which two formulations, one of which is composed of labelled drug, are compared following simultaneous oral administration. This appears particularly attractive in the current climate of

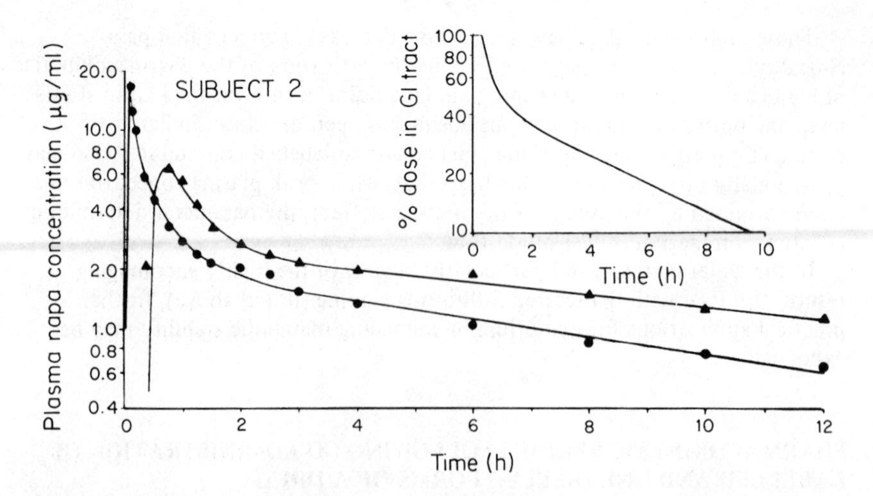

Figure 3.12 Human plasma concentrations of *N*-acetylprocainamide (NAPA) (▲) and its ^{13}C analogue (●) after co-administration of NAPA capsules orally, and NAPA-^{13}C intravenously. Determination of the absolute bioavailability of NAPA was obtained by using NAPA-d_5 as the internal standard. Reproduced from Strong *et al.* (1975)

growing insistence by most regulatory authorities upon demonstration in man of the bioavailability of novel formulations. However, the technical problems and costs involved in tableting a labelled batch of drug to an acceptable specification may preclude the use of this technique. A restriction to comparison of labelled drug in solution (or formulated as an elixir) with an unlabelled tableted formulations may be all that can be achieved in practice.

A further application, in which the simultaneous estimation of labelled and unlabelled drugs is required, is provided by the technique of 'pulse-labelling' during a multiple dosing regime. Thus, Figure 3.13 shows the plasma concentrations of propoxyphene-d_0 and propoxyphene-d_2 in a dog in which the labelled drug was given as a single oral dose on day 20 of chronic administration of propoxyphene-d_0 (Sullivan, Wood and McMahon, 1976). Pharmacokinetic interpretation depends upon prior demonstration of the equivalence of the two forms, and this was confirmed when, after a single oral dose of the mixture, the relative plasma concentrations remained constant. In the pulse study the most interesting observation was the much shorter plasma half-life of the labelled compound than that of propoxyphene-d_0. This suggested that the pulse dose had not equilibrated with 'deep' pools which, in effect, sustained the plasma concentration of propoxyphene-d_0. It was also observed that the concentration of propoxyphene-d_0 was immediately elevated following propoxyphene-d_2 administration, presumably because of displacement from tissue binding sites.

While perhaps of more academic interest than the prospects for bioavailability comparison using stable isotopes, the foregoing work with propoxyphene provides an excellent illustration of a novel approach to the investigation of the pharmacokinetics of a drug at steady state.

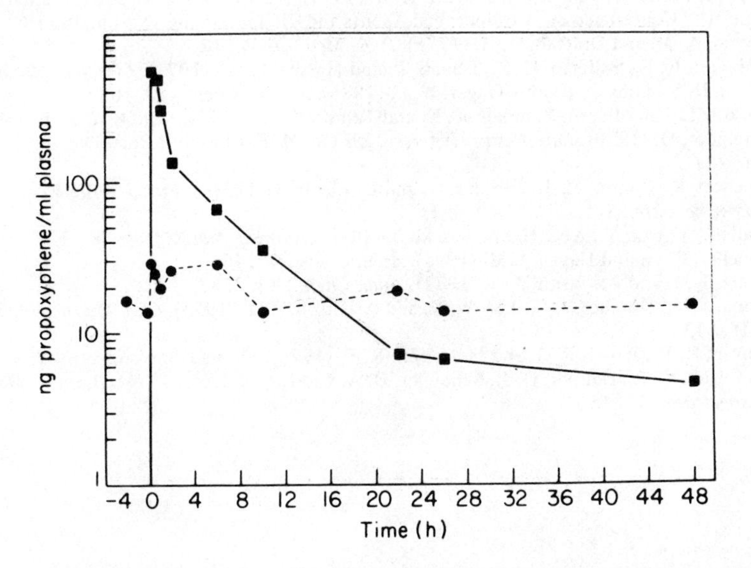

Figure 3.13 Dog plasma concentrations of propoxyphene-d_0 (•) and propoxyphene-d_2 (▪) following oral administration of a single pulse dose of propoxyphene-d_2 on day 20 of a multi-dose study with propoxyphene-d_0. Reproduced from Sullivan *et al.* (1976)

REFERENCES

Beckinsale, R. D., Freeman, N. J., Jackson, M. C., Powell, R. E. and Young, W. A. P. *Ion Phys.*, **12**, 299

Bending, M. R., Bennett, P. N., Rowland, M. and Steiner, J. (1976). Proceedings of the British Pharmacological Society Meeting, July, 1976, Abstract No. C25

Blake, M. I., Crespi, H. L. and Katz, J. J. (1975). *J. Pharm. Sci.*, **64**, 367

Boyes, R. N. and Keenaghan, J. B. (1971). *Lidocaine in the Treatment of Ventricular Arrhythmias* (ed. D. B. Scott and D. G. Julian), Livingston, Edinburgh, p. 140

Boyes, R. N., Scott, D. B., Jebson, P. J., Goodman, M. J. and Julian, D. G. (1971). *Clin. Pharmacol. Ther.*, **12**, 105

Brash, A. R., Baillie, T. A., Clare, R. A. and Draffan, G. H. (1976). *Biochem. Med.*, **16**, 77

Davies, D. S., Wing, L. M. H., Reid, J. L., Neill, E., Tippett, P. and Dollery, C. T. (1977). *Clin. Pharmacol. Ther.*, **21**, 593

Draffan, G. H., Bellward, G. D., Clare, R. A., Dargie, H. J., Dollery, C. T., Murray, S., Reid, J. L., Wing, L. M. H. and Davies, D. S. (1975). *Proceedings of the Second International Conference on Stable Isotopes* (ed. E. R. Klein and P. D. Klein), US Department of Commerce, Springfield, Virginia. Document code CONF-75 1027, p. 149

Draffan, G. H., Clare, R. A., Goodwin, B. L., Ruthven, C. R. J. and Sandler, M. (1974). *Adv. Mass Spectrom.*, **6**, 245

Draffan, G. H., Clare, R. A., Murray, S., Bellward, G. D., Davies, D. S. and Dollery, C. T. (1976). *Advances in Mass Spectrometry in Biochemistry and Medicine*, Vol. 2 (ed. A. Frigerio), Spectrum Publications, New York, p. 389

Draffan, G. H., Gilbert, J. D. and McGregor, J. S. (1977). Unpublished observations

Falkner, F. C., Sweetman, B. J. and Watson, J. T. (1975). *Appl. Spectrosc. Rev.*, **10**, 51

Garland, W. A., Trager, W. F. and Nelson, S. D. (1974). *Biomed. Mass Spectrom.*, **1**, 124

Hammar, C.-G., Holmstedt, B. and Ryhage, R. (1968). *Anal. Biochem.*, **25**, 337

Inaba, T., Tang, B. K. and Kalow, W. (1976). *Proceedings of the Second International Conference on Stable Isotopes* (ed. E. R. Klein and P. D. Klein), National Technical Information Service, US Department of Commerce, Springfield, Virginia. Document code CONF-75 1027, p. 163

Knapp, D. R., Gaffney, T. E. and McMahon, R. E. (1972). *Biochem. Pharmacol.*, **21**, 425

Kropp, H., Kahan, F. M. and Woodruff, H. B. (1975). Abstracts of 15th International Science Conference on Antimicrobial Agents and Chemotherapy, Washington DC

Lawson, A. M. and Draffan, G. H. (1975). *Prog. Med. Chem.*, **12**, 1

McMahon, R. E., Sullivan, H. R., Due, S. L. and Marshall, F. J. (1973). *Life Sci.*, **12**, 463

Nelson, D. S., Pohl, L. R. and Trager, W. F. (1975). *J. Med. Chem.*, **18**, 1062

Prescott, L. F., Adjepon-Yamoah, K. K. and Talbot, R. G. (1976). *Brit. Med. J.*, **1**, 939

Rehbinder, D. (1970). *Catapres in Hypertension* (ed. M. E. Conolly), Butterworths, London, p. 227

Roncucci, R., Simon, M.-J., Jacques, G. and Lambelin, G. (1976). *Eur. J. Drug Metab. Pharmacokin.*, **1**, 9

Sano, M., Yotsui, Y., Abe, H. and Sasaki, S. (1976). *Biomed. Mass Spectrom.*, **3**, 1

Schoeller, D. A. and Hayes, J. M. (1975). *Anal. Chem.*, **47**, 408

Strong, J. M. and Atkinson, A. J. (1972). *Anal. Chem.*, **44**, 2287

Strong, J. M., Dutcher, J. S., Lee, W.-K. and Atkinson, A. J. (1975). *Clin. Pharmacol. Ther.*, **18**, 613

Sullivan, H. R., Wood, P. G. and McMahon, R. E. (1976). *Biomed. Mass Spectrom.*, **3**, 212

von Unruh, G. E., Hauber, D. J., Schoeller, D. A. and Hayes, J. M. (1974). *Biomed. Mass Spectrom.*, **1**, 345

Section 2
Applications in pharmacology and toxicology

4

Applications of deuterium labelling in pharmacokinetic and concentration-effect studies of clonidine

D. S. Davies, E. Neill and J. L. Reid (Department of Clinical Pharmacology, Royal Postgraduate Medical School, Ducane Road, London W12 0HS, UK)

INTRODUCTION

Clonidine 2-(2,6-dichlorophenylamino)-2-imidazoline, Catapres (Figure 4.1), is a potent hypotensive agent widely used in the treatment of patients with high blood pressure (Conolly, 1975). Data from experiments in animals suggest that the hypotensive action of clonidine is predominantly due to stimulation of alpha-adrenoceptors in the central nervous system (Kobinger, 1975), resulting in a reduction of activity in sympathetic nerves (Schmitt and Schmitt, 1969). A similar mechanism probably operates in man, since clonidine does not lower blood pressure in patients with chronic cervical spinal cord transection (Reid *et al*., 1977). The use of clonidine in the clinic is limited because it consistently causes dry mouth and sedation which also probably arise from the drug's effects within the central nervous system (Rand, Rush and Wilson, 1969).

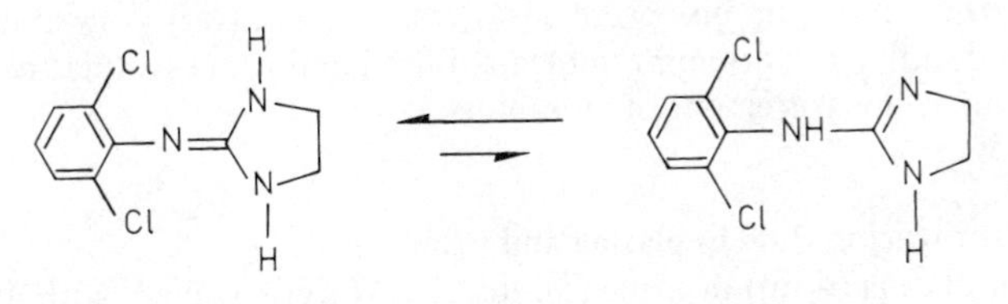

Figure 4.1 The structure of clonidine

Daily dose requirements for the control of blood pressure with clonidine show a twentyfold inter-individual variation (Conolly, 1975). Preliminary studies (Rehbinder, 1970) with ^{14}C-labelled clonidine suggested that plasma levels in man following a dose of 300 μg would not exceed 1–2 ng/ml, too low to be measured by the gas chromatographic assay using electron capture detection described by Cho and Curry (1969). In order to determine the factors controlling dose requirements for clonidine and to gain further insight into its mechanism of action, we developed a highly specific and sensitive assay based on stable isotope dilution and utilising combined gas chromatography–mass spectrometery (Draffan *et al.*, 1977). This has enabled us to study the action of clonidine in normotensive (Davies *et al.*, 1977) and hypertensive (Wing *et al.*, 1977b) subjects and to understand why some patients are resistant to high doses of the drug (Wing *et al.*, 1977a).

METHODS

Subjects

The subjects were normotensive volunteers from the Department of Clinical Pharmacology or hypertensive patients attending Hammersmith Hospital. All subjects gave their written informed consent to the procedures, which had been approved by the Ethics Committee of the Royal Postgraduate Medical School and Hammersmith Hospital.

Procedure

Normotensive subjects

The subjects were fasted overnight prior to the study and had not taken any other medication in the previous 7 days. Each subject was studied on two separate occasions at least 7 days apart. On one occasion 300 μg of clonidine hydrochloride was infused intravenously and on the other occasion the subject was given a 300 μg oral dose as the commercially available formulation. Blood pressure, heart rate, salivary flow and sedation were monitored for 24 h after each dose as previously described (Dollery *et al.*, 1976). Blood samples were obtained over 24 h and urine was collected for 3 days for the measurement of clonidine.

Hypertensive subjects

The effects of single (300 μg) and multiple (150 μg three times a day for 7 days) oral doses of clonidine on blood pressure, heart rate, salivary flow and sedation were studied in five hypertensive subjects. Blood and urine samples were obtained for the measurement of clonidine.

Measurement of clonidine in plasma and urine

Aliquots of plasma (4 ml) or urine (diluted 1 : 4) were 'spiked' with 400 ng of [4,4,5,5-$^2H_4$]-2-(2,6-dichlorophenylamino)-2-imidazoline (clonidine-d_4), which acts as an internal standard and carrier in the assay. Clonidine was extracted from alkalinised plasma or urine with ethyl acetate, and following back-extraction into acid the aqueous phase was adjusted to pH 11 and extracted with diethyl ether. The ether extract was evaporated to dryness under nitrogen and redissolved in 20 μl of trimethylanilinium hydroxide for flash methylation in the gas chromatograph, which was fitted with a 5 ft glass column of 3 per cent OV-17 on 100/120 Gas Chrom Q. The injection port temperature was 260° and the oven temperature 195°, the helium flow rate being 30 ml/min. A Finnigan 3200 quadrupole mass spectrometer, operated in the electron impact mode, was set to monitor *m/e* 261 for clonidine-d_4 and *m/e* 257 for clonidine as previously described (Davies *et al.*, 1977). Response ratios for clonidine and clonidine-d_4 were determined as peak heights using an interactive Finnigan Model 6000 data system. Calibration was linear for clonidine-d_0 over the range 0–25 ng/ml. The limit of detection was 0.15 ng/ml ± 10 per cent (SD). A typical trace is shown in Figure 4.2.

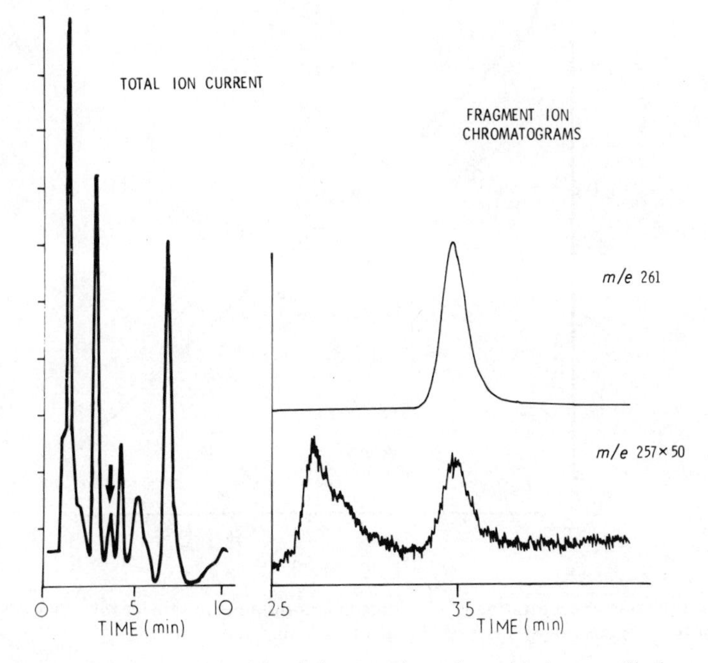

Figure 4.2 Quantification of clonidine in human plasma by stable isotope dilution and selected ion monitoring

RESULTS AND DISCUSSION

Pharmacokinetic and concentration-effect studies in normotensive subjects

Plasma concentrations of clonidine were similar after oral or intravenous dosing following the phases of absorption and distribution (Figure 4.3). These results indicate that the drug is well absorbed and is not extensively extracted by the liver.

A biexponential equation described the concentration–time profile of clonidine in plasma after intravenous dosing. This suggests that the drug is distributed into two compartments in the body: a rapidly equilibrating compartment which probably includes the vascular system and high blood flow tissues and a slowly equilibrating second compartment. The kinetic parameters of the two-compartment model are shown in Table 4.1. The

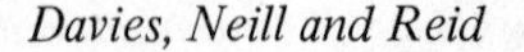

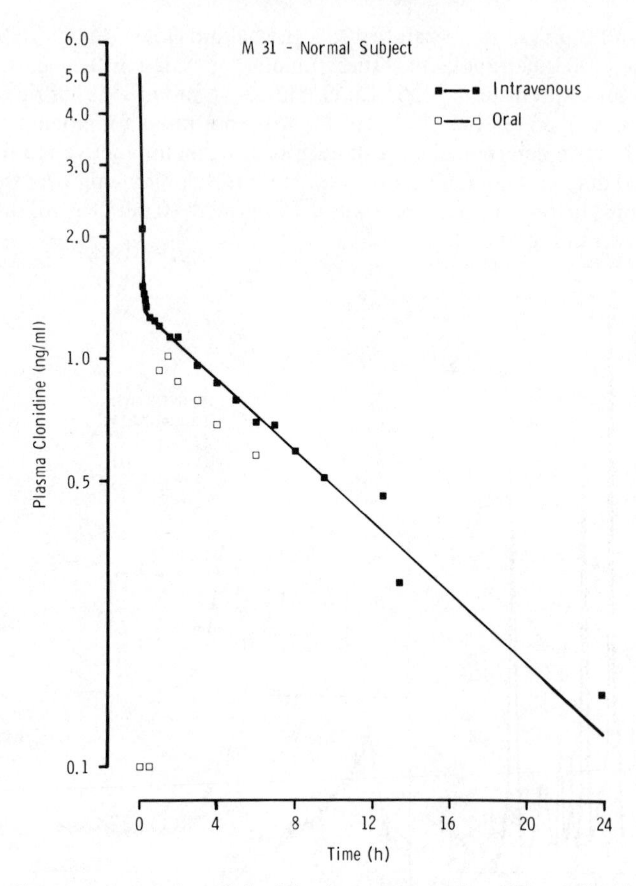

Figure 4.3 Plasma concentration of clonidine in a normotensive subject following oral (□—□) or intravenous (■—■) doses of 300 μg of clonidine hydrochloride

relatively large values for α and Vd_{ss} indicate rapid and extensive extravascular distribution.

The terminal half-lives after intravenous and oral dosing were not significantly different in the same subject (Table 4.2) and inter-subject variation was small (5.2–13.0 h). The parameters of the two-compartment model were used to calculate the amount of clonidine absorbed after oral dosing in each subject. These data showed the mean bioavailability of oral clonidine from the formulation used to be 75.2 ± 1.8 per cent (range 70.6–81.5 per cent), confirming that the drug is well absorbed and not subject to extensive 'first-pass' metabolism. Early studies with ^{14}C-labelled clonidine (Rehbinder, 1970) had shown that excretion in urine was a major route of elimination of the drug. In the present studies half of the clonidine reaching the systemic circulation was eliminated unchanged in urine after intravenous or oral dosing. Renal clearance of clonidine averaged 132 ml/min and showed considerable inter-subject variation (Table 4.3). It exceeded the calculated glomerular filtration rate in subjects 1 and 4, suggesting that tubular secretion of the drug occurs. Non-renal

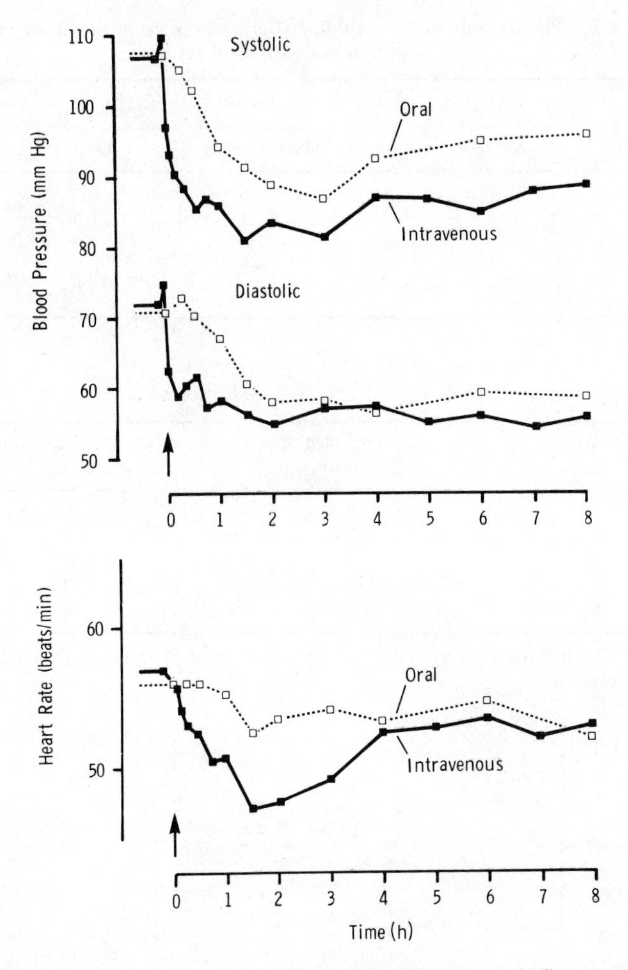

Figure 4.4 Mean systolic and diastolic blood pressure and heart rate in five normotensive subjects following an oral (□–□) or intravenous (■–■) dose of 300 μg of clonidine hydrochloride. Reproduced by permission of the Editor of *Clinical Pharmacology and Therapeutics*

Table 4.1 Two-compartment disposition constants for clonidine following intravenous administration in five normotensive subjects

Subject	α (min^{-1})	β (hr^{-1})	k_{12} (min^{-1})	k_{21} (min^{-1})	k_{10} (min^{-1})	Vd_{ss} (L)	Cl (ml/min)
1	0.187	0.102	0.151	0.024	0.013	191	325
2	0.079	0.060	0.049	0.028	0.003	136	136
3	0.065	0.090	0.045	0.015	0.006	101	171
4	0.314	0.102	0.267	0.033	0.016	184	312
5	0.024	0.066	0.012	0.011	0.002	136	150

Table 4.2 Plasma half-lives of clonidine after intravenous or oral dosing (300 μg) in five normotensive subjects

Subject	Plasma half-life (h)	
	Intravenous	Oral
1	6.9	6.7
2	11.1	11.4
3	7.6	5.2
4	6.9	6.7
5	10.1	13.0

Table 4.3 Renal clearance of clonidine

Subject	Average renal clearance (ml/min)	Non-renal clearance[a] (ml/min)
1	167	158
2	97	39
3	85	86
4	234	86
5	81	68

[a] The difference of total body clearance (Table 4.1) and average renal clearance.

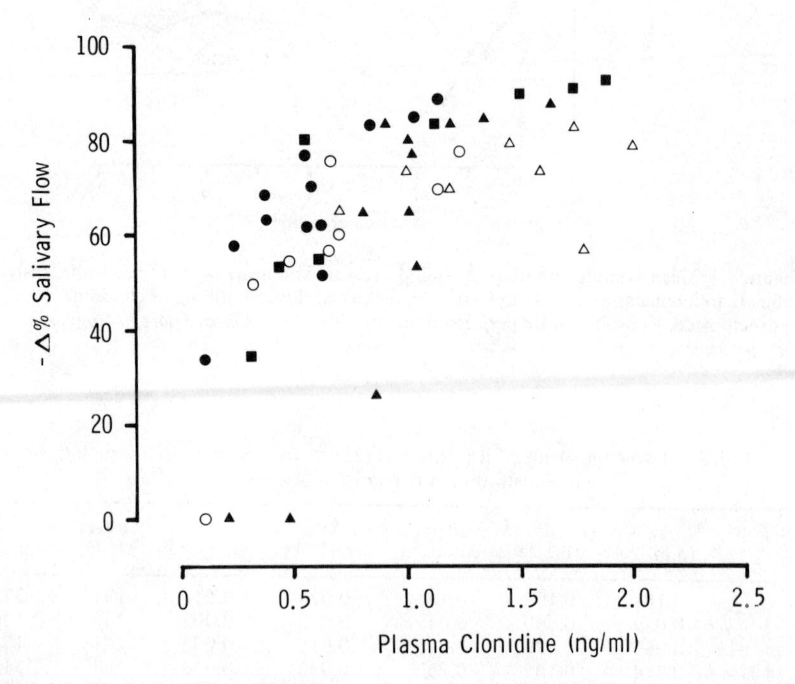

Figure 4.5 Change in salivary flow (percentage of control) against plasma concentration of clonidine. The different symbols represent results obtained in five subjects

clearance, which probably represents metabolism, ranged from 39 to 158 ml/min (Table 4.3). This considerable non-renal elimination probably explains why patients with renal failure have not been reported to be unusually sensitive to clonidine.

In these normotensive subjects the intravenous and oral doses produced similar falls in blood pressure, although the effect was more rapid by the former route (Figure 4.4). Similar observations were made in respect of sedation and reduction in salivary flow (Davies *et al.*, 1977). There was good correlation with plasma clonidine concentrations for sedation and salivary flow. For example, salivary flow decreased with increasing plasma concentration to 2.0 ng/ml and remained constant thereafter (Figure 4.5). Hypotensive action was related to plasma concentration below 2 ng/ml but showed considerable inter-subject variation. At concentrations above 2 ng/ml the hypotensive effect lessened and was almost abolished at 4.0 ng/ml in some subjects (Figure 4.6). This is probably

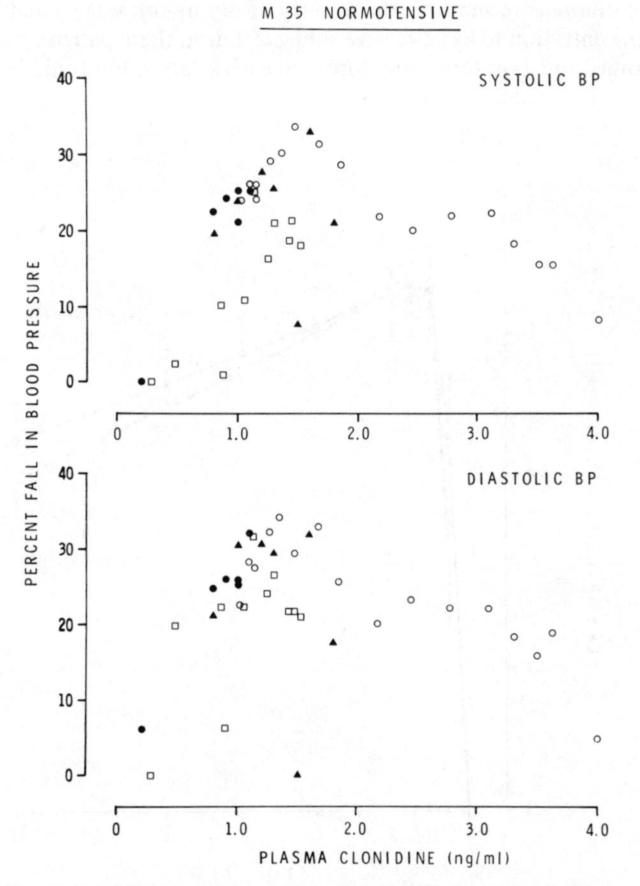

Figure 4.6 Percentage fall in systolic and diastolic blood pressure against plasma concentration of clonidine in subject 5. The different symbols represent results obtained in several studies in this subject. Reproduced by permission of the Editor of *Clinical Pharmacology and Therapeutics*

due to the increasing effect of stimulation of peripheral post-synaptic α-adrenoceptors leading to vasoconstriction (Kobinger and Walland, 1967). It was therefore important to study concentration-effect relationships of clonidine during chronic dosing in hypertensive subjects.

Pharmacokinetic and concentration-effect studies in hypertensive subjects
Plasma concentration–time profiles following single 300 μg oral doses of clonidine were similar in a small group of hypertensive and normotensive subjects (Figure 4.7). Plasma half-lives following single or multiple doses in the hypertensive subject did not differ significantly and there was no evidence for extensive accumulation of clonidine during chronic dosing (Wing *et al.*, 1977b). Single 300 μg doses of clonidine produced similar changes in sedation and salivary flow in normotensive and hypertensive subjects, but the absolute and percentage falls in blood pressure were greater in the latter group.

As in normotensive subjects, the severity of dry mouth was related to the plasma concentration in hypertensive subjects, but in these patients, unaccustomed to laboratory procedures, no such relationship could be demonstrated for changes in sedation score.

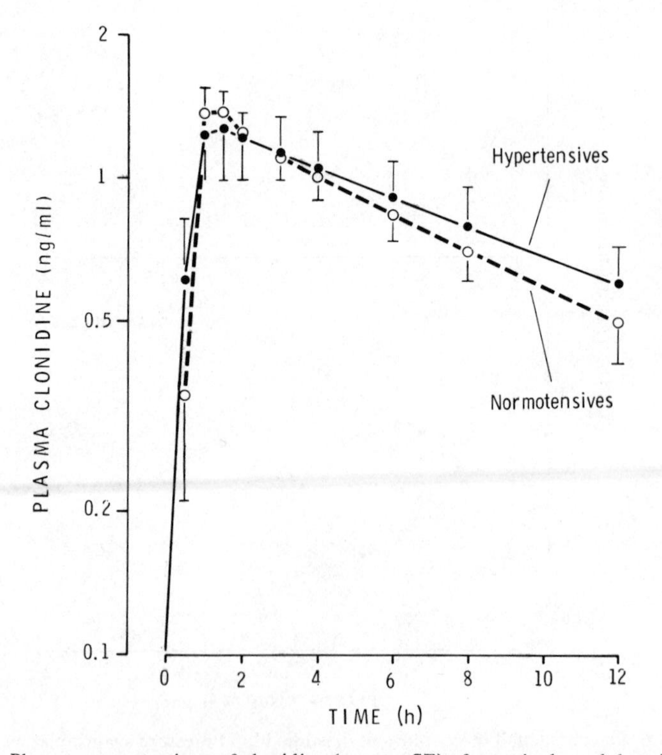

Figure 4.7 Plasma concentrations of clonidine (mean ± SE) after a single oral dose in five normotensive (○ · · · · ·○) and five hypertensive (●———●) subjects. Reproduced by permission from the *British Journal of Clinical Pharmacology*

The hypotensive effect became attenuated at plasma concentrations greater than 1 ng/ml (Figure 4.8), as it had done in normotensive subjects. The pressor effect of high doses of clonidine due to stimulation of peripheral post-synaptic α-adrenoceptors was to be expected and had been reported in man after intravenous administration (Mroczek, Davidov and Finnerty, 1973). The present results are surprising, since they suggest that effects of clonidine in the periphery, opposing the central hypotensive effect, become important at concentrations as low as 1 ng/ml. These results suggest that the optimal effect of clonidine on blood pressure may only be attainable over a very narrow range of plasma concentrations (Davies *et al.*, 1977; Wing *et al.*, 1977b).

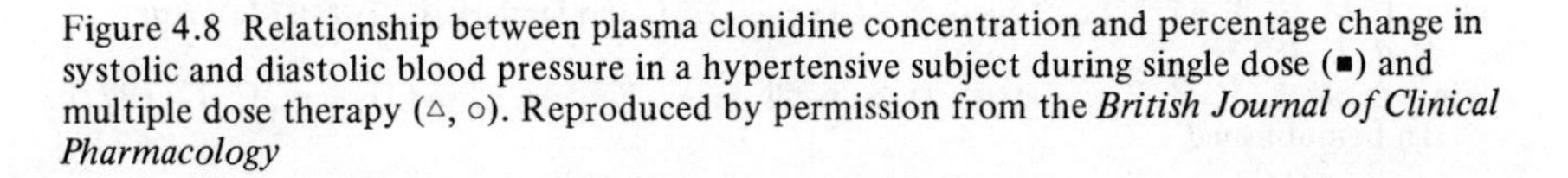

Figure 4.8 Relationship between plasma clonidine concentration and percentage change in systolic and diastolic blood pressure in a hypertensive subject during single dose (■) and multiple dose therapy (△, ○). Reproduced by permission from the *British Journal of Clinical Pharmacology*

This concept received further support in studies of two hypertensive patients 'apparently' resistant to clonidine (Wing *et al.*, 1977a). Blood pressure was not decreased in these patients following doses of 5.4 and 6.0 mg/day, when plasma concentrations were 26.2 and 14.4 ng/ml, respectively. However, several months after cessation of clonidine therapy, a single oral dose of 0.3 mg produced significant falls in blood pressure (30/22 mmHg and 88/41 mmHg) at clonidine concentrations of 1.4 and 0.9 ng/ml. It appears that the resistance to clonidine at the higher doses was due to excessive peripheral α-adrenoceptor stimulation.

CONCLUSIONS

A highly specific and sensitive assay for clonidine based on stable isotope dilution and using combined gas chromatography–mass spectrometry has enabled us to study for the first time the pharmacokinetics and concentration-effect relationships of the drug after the usual therapeutic doses. Our results suggest that the optimal hypotensive effect of clonidine may only be obtainable over a very narrow range of plasma concentration.

ACKNOWLEDGEMENTS

Dr J. L. Reid is supported by the Wellcome Trust as a Senior Fellow in Clinical Science; Elizabeth Neill is supported by a grant from Boehringer Ingelheim (UK) Limited, Bracknell, Berks., UK.

REFERENCES

Cho, A. K. and Curry, S. H. (1969). *Biochem. Pharmacol.*, **18**, 511

Conolly, M. E. (1975). *Central Action of Drugs in Blood Pressure Regulation* (ed. D. S. Davies and J. L. Reid), Pitman Medical, London, p. 268

Davies, D. S., Wing, L. M. H., Reid, J. L., Neill, E., Tippett, P. and Dollery, C. T. (1977). *Clin. Pharmacol. Ther.*, **21**, 593

Dollery, C. T., Davies, D. S., Draffan, G. H., Dargie, H. J., Dean, C. R., Reid, J. L., Clare, R. A. and Murray, S. (1976). *Clin. Pharmacol. Ther.*, **19**, 11

Draffan, G. H., Clare, R. A., Murray, S., Bellward, G. D., Davies, D. S. and Dollery, C. T. (1976). *Advances in Mass Spectrometry in Biochemistry and Medicine*, Vol. 2 (ed. A. Frigerio), Spectrum Publications, New York, p. 389

Kobinger, W. (1975). *Central Action of Drugs in Blood Pressure Regulation* (ed. D. S. Davies and J. L. Reid), Pitman Medical, London, p. 181

Kobinger, W. and Walland, A. (1967). *Europ. J. Pharmacol*, **2**, 155

Mroczek, W. J., Davidov, M. and Finnerty, F. A. (1973). *Clin. Pharmacol. Ther.*, **14**, 847

Rand, M. J., Rush, M. and Wilson, J. (1969). *Europ. J. Pharmacol.*, **5**, 168

Rehbinder, D. (1970). *Catapres in Hypertension* (ed. M. E. Conolly), Butterworths, London, p. 227

Reid, J. L., Wing, L. M. H., Mathias, C. J., Frankel, H. L. and Neill, E. (1977). *Clin. Pharmacol. Ther.*, **21**, 375

Schmitt, H. and Schmitt, Mme. H. (1969). *Europ. J. Pharmacol.*, **6**, 8

Wing, L. M. H., Reid, J. L., Davies, D. S., Dargie, H. J. and Dollery, C. T. (1977a). *Brit. Med. J.*, **1**, 136

Wing, L. M. H., Reid, J. L., Davies, D. S., Neill, E. A., Tippett, P. and Dollery, C. T. (1977b). To be published

5
Isotope effects on pathways of metabolism

M. G. Horning and K. Lertratanangkoon (Baylor College of Medicine, Houston, Texas, USA), K. D. Haegele (University of Texas Health Science Center, San Antonio, Texas, USA) and K. Brendel (University of Arizona, Tucson, Arizona, USA)

INTRODUCTION

In recent studies (Horning *et al.*, 1976) we have shown that the metabolism of a drug may be shifted by deuterium labelling if the drug is metabolised by multiple alternative pathways. Deuterated analogues of caffeine (1-trideuteromethyl-3,7-dimethylxanthine and 7-trideuteromethyl-1,3-dimethylxanthine) and antipyrine (2-trideuteromethyl-3-methyl- and 2-methyl-3-trideuteromethyl-1-phenyl-3-pyrazoline-5-one) were employed. The oxidation of the $-CD_3$ group was depressed and the metabolism was shifted to other pathways that did not involve cleavage of a carbon–deuterium bond. The change in metabolic pathways, like the primary isotope effect, occurs when deuterium is substituted at a principal site of metabolism and when oxidation is a primary mode of metabolism. However, this response, which we have observed in both *in vivo* and *in vitro* experiments, is different from the primary isotope effect, which is usually defined as the ratio of rate constants (k_H/k_D) or ratio of products formed [H] / [D]. For this reason we have used the term 'metabolic switching' to describe the effect of deuterium substitution on compounds that are metabolised by multiple alternative pathways rather than sequentially.

In our studies with antipyrine and caffeine, the changes in metabolism involved only the oxidation of an unlabelled methyl group (N-dealkylation) instead of a deuterated methyl group. Similar results, however, were observed by Cox *et al.* (1976) in studies with cyclophosphamides selectively labelled with deuterium, in which a β-elimination reaction was involved. If metabolic switching is a general response, it should also occur when the alternative pathways involve aromatic hydroxylation (epoxidation) or aliphatic hydroxylation. To test this point we selected methsuximide (*N*,2-dimethyl-2-phenylsuccinimide) for study because this drug is metabolised by N-demethylation, aromatic hydroxylation (epoxidation), aliphatic hydroxylation of the 2-methyl group and hydroxylation at carbon-3 of the heterocyclic ring; N-dealkylation and aromatic hydroxylation are the two major pathways (Horning *et al.*, 1973).

METHODS AND INSTRUMENTATION

Animal studies

Male Sprague–Dawley rats (180–250 g) received an intraperitoneal injection of either 10 mg of methsuximide, 10 mg of *N*-CD_3-methsuximide or 10 mg of a 1 : 1 mixture of methsuximide and *N*-CD_3-methsuximide dissolved in 0.1 ml of ethanol and 0.5 ml of propylene glycol. The rats were housed individually in metabolism cages with water but no food and 24 h urine samples were collected.

Methsuximide and α-methyl-α-phenylsuccinonitrile were obtained from Parke-Davis and Co., Ann Arbor, Michigan. *N*-Trideuteromethyl-2-methyl-2-phenylsuccinimide (*N*-CD_3-methsuximide) was synthesised by reacting deuterated methylamine with the acid obtained by hydrolysis of α-methyl-α-phenyl-succinonitrile. Analysis of the recrystallised *N*-CD_3-methsuximide by GC–MS indicated that the compound was homogeneous and that no unlabelled methsuximide was present in the labelled material.

Isolation of methsuximide and its metabolites

The drug and unconjugated metabolites were extracted from an aliquot (1/4) of a 24 h urine sample using ammonium carbonate–ethyl acetate as a salt–solvent pair (Horning *et al.*, 1974). The combined ethyl acetate extracts were taken to dryness under a nitrogen stream and redissolved in 1.0 ml of ethanol for derivatisation.

Since acidic conjugates of methsuximide are also excreted, the urine was hydrolysed enzymatically with Glusulase. An aliquot (1/4) of a 24 h urine sample was buffered with 0.5 g of sodium acetate and the pH adjusted to 4.5 with acetic acid. After adding 0.3 ml of the enzyme solution, hydrolysis was carried out for 18 h at 37°C. The aglycones liberated by enzymatic hydrolysis were extracted using ammonium carbonate–ethyl acetate as a salt–solvent pair. The combined ethyl acetate extracts were taken to dryness with a nitrogen stream and redissolved in 1.0 ml ethanol for derivatisation.

Derivative formation

The ethanolic solution containing the drug and metabolites was ethylated with 2 ml of an ethereal solution of diazoethane (Harvey *et al.*, 1974). After standing at room temperature for 30 min, the excess reagent and solvents were removed with a nitrogen stream. The residue was dissolved in 10 μl of pyridine and silylated with 20 μl of *N,O*-bis-trimethylsilylacetamide by heating at 60°C for 1 h. The *N*-desmethyl and phenolic metabolites are converted into *N*-ethyl and *O*-ethyl derivatives, respectively, by ethylation; silylation converts the non-aromatic hydroxyl groups into trimethylsilyl ethers.

Gas chromatography and gas chromatography–mass spectrometry

The gas chromatographic analyses of urinary methsuximide metabolites were performed on a glass W-column (12 ft x 4 mm) containing 5 per cent SE-30 on Gas Chrom Q. The analyses were temperature programmed (2°/min) from 150°. Methylene unit (MU) values were determined with *n*-alkanes as reference compounds. Mass spectra were recorded with an LKB 9000 instrument, fitted with a 9 ft x 2 mm glass coil column containing 1 per cent SE-30 packing, and using temperature-programmed analyses. The ion source temperature was 270°; the ionizing current was 60 μA; the electron energy was 70 eV.

Quantitative analyses were carried out by selected ion detection (SID) using a Finnigan 1015 mass spectrometer equipped with a chemical ionisation source and a Systems Industries data system. Glass coil columns (9 ft x 2 mm) packed with 1 per cent SE-30 on 80–100 mesh Gas Chrom Q were used. Methane was the carrier gas. The ions monitored are listed in Table 5.1.

Table 5.1

Succinimide derivatives	MH^+
N,2-Dimethyl-2-phenyl- (methsuximide)	204
N-CD_3-2-Methyl-2-phenyl-	207
N-Ethyl-2-methyl-2-phenyl-	218
N,2-Dimethyl-2-(4-hydroxyphenyl)-	248
N-CD_3-2-Methyl-2-(4-hydroxyphenyl)-	251
N-Ethyl-2-methyl-2-(4-hydroxyphenyl)-	262
N,2-Dimethyl-3-hydroxy-2-phenyl-	292
N-Methyl-2-hydroxymethyl-2-phenyl-	292
N-CD_3-2-Methyl-3-hydroxy-2-phenyl-	295
N-CD_3-2-Hydroxymethyl-2-phenyl-	295

In the quantitative studies equivalent samples of urine were extracted and derivatised. The final volume after derivatisation and the volume analysed by GC–MS was the same for the samples obtained from the methsuximide- and *N*-CD_3-methsuximide-treated rats.

RESULTS

The metabolites of methsuximide are excreted in urine as free (unconjugated) and as acidic metabolites that are hydrolysed by Glusulase. The major neutral (unconjugated) metabolites excreted in rat urine are *N*,2-dimethyl-2-(4-hydroxyphenyl(succinimide (III), two desmethyl metabolites (2-methyl-2-phenylsuccinimide (II) and 2-methyl-2-(4-hydroxyphenyl)succinimide (IV)) and unchanged methsuximide (I) (Scheme 5.1).

A comparison of the unconjugated metabolites excreted by a rat treated with methsuximide and with *N*-CD_3-methsuximide is shown in Figure 5.1. The rat treated with *N*-CD_3-methsuximide excreted very small amounts of the desmethyl metabolites II and IV; the amount of the phenol III excreted was greater than the amount excreted by the rat treated with unlabelled drug.

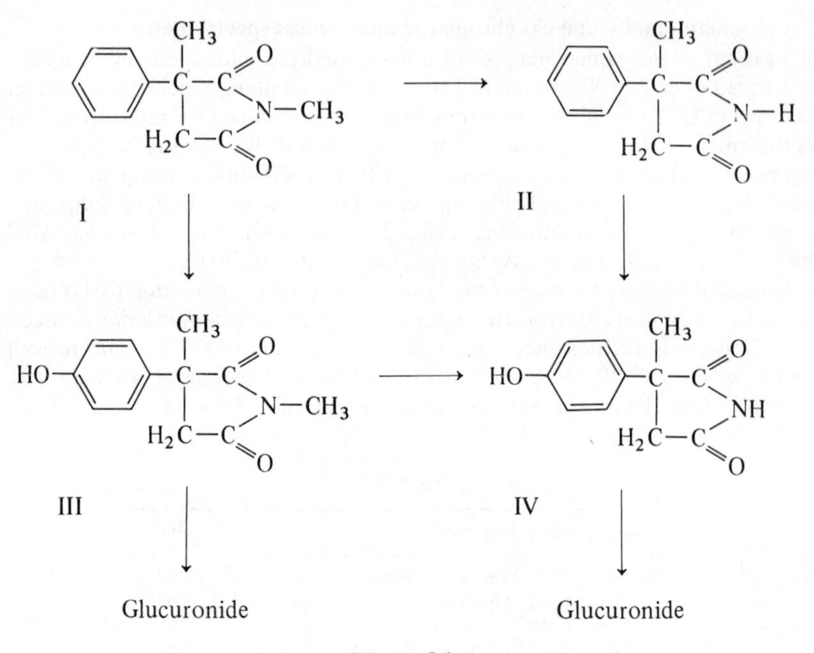

Scheme 5.1

Figure 5.2 shows a comparison of the metabolites present in hydrolysed urine following administration of labelled and unlabelled methsuximide. The urinary profiles were similar to those observed for the unconjugated metabolites; the major metabolites liberated by enzymatic hydrolysis were the phenols III and IV. Metabolites I–IV were quantified by SID in order to calculate the metabolite ratios. The results are summarised in Table 5.2.

The profile of urinary metabolites obtained following administration of a 1 : 1 mixture of methsuximide and *N*-CD_3-methsuximide is shown in Figure 5.3. In these experiments *N*-desmethylmethsuximide and 2-methyl-2-(4-hydroxyphenyl)succinimide are metabolites of methsuximide and also *N*-CD_3-methsuximide. The metabolites ratios have been calculated (Table 5.2), but it is not possible to distinguish the desmethyl metabolites formed from the labelled drug from desmethyl metabolites formed from unlabelled drug.

Several additional minor metabolites have been isolated from rat urine (Horning *et al.*, 1973) and identified as *N*,2-dimethyl-2-(3-hydroxyphenyl) succinimide (V), 2-methyl-2-(3-hydroxyphenyl)succinimide (VI), *N*-methyl-2-hydroxymethyl-2-phenylsuccinimide (VII), and *N*,2-dimethyl-3-hydroxy-2-phenylsuccinimide (VIII) (Scheme 5.2). The *m*- and *p*-structural assignments were made by analogy with the metabolism of phenobarbital (Harvey *et al.*, 1972; Maynert, 1972), phensuximide (Dudley, Bius and Grace, 1972) and phenytoin (Chang and Glazko, 1972). It is not possible to assign the position of the aromatic hydroxyl group on the basis of mass spectral analysis, and the possibility of an *o*-hydroxy rather than a *m*-hydroxy structure has not been eliminated.

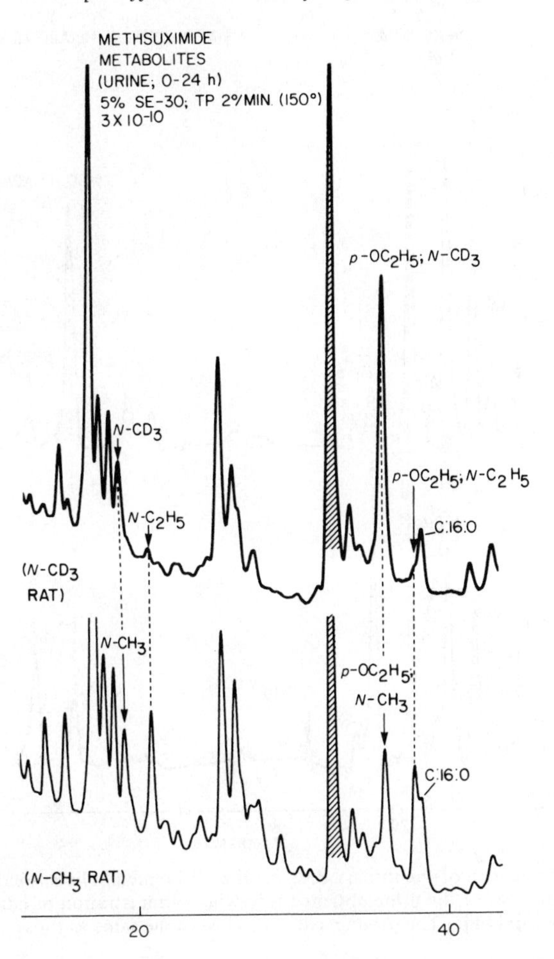

Figure 5.1 Comparison of the urinary drug profiles of neutral (unconjugated) metabolites obtained following administration of labelled methsuximide (N-CD_3) or unlabelled methsuximide (N-CH_3). The products identified (ethylated derivatives) were methsuximide (N-CH_3 and N-CD_3), desmethyl methsuximide (N-C_2H_5), p-hydroxymethsuximide (p-HO-C_2H_5, NCH_3 and p-HO-C_2H_5; N-CD_3) and desmethyl p-hydroxymethsuximide (p-HO-C_2H_5; N-C_2H_5) and the ethyl ester of palmitic acid. (C : 16 : 0)

DISCUSSION

Alterations in pharmacological activity resulting from changes in the rate of metabolism have been reported for several deuterated drugs (Blake, Crespi and Katz, 1975) but alterations in metabolic pathways have not been observed. However, the drugs employed in the earlier studies were metabolised sequentially; that is, by hydroxylation and subsequent conjugation. In our laboratory, changes in pathways of metabolism were observed for deuterated drugs metabolised by multiple alternative pathways. For example, the

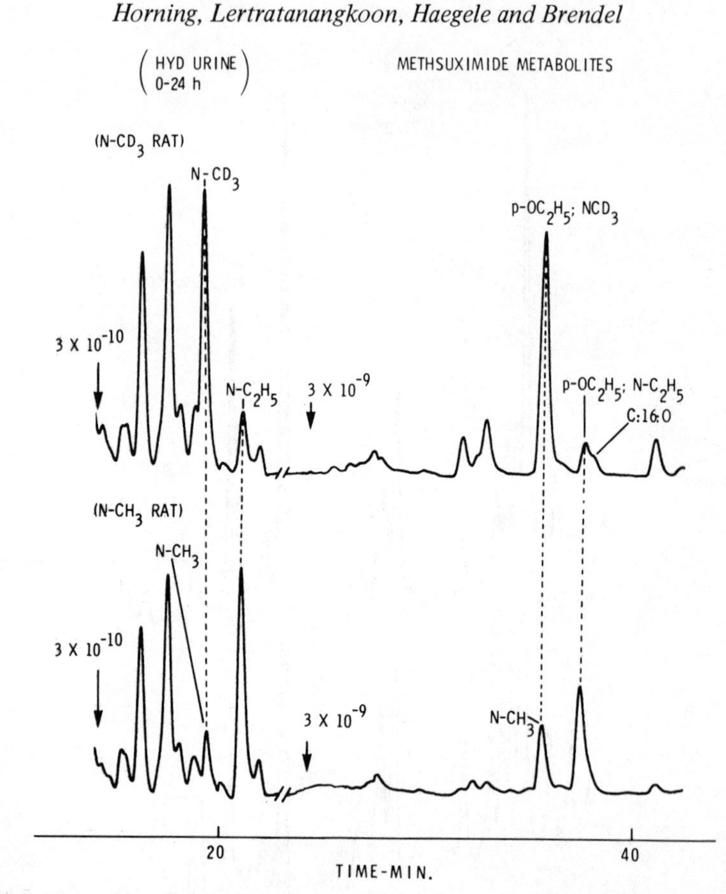

Figure 5.2 Comparison of the urinary drug profile of the metabolites isolated after enzymatic hydrolysis of the urine obtained following administration of labelled or unlabelled methsuximide. The products identified were the same as those described in Figure 5.1

metabolism of N-CD_3 analogues of caffeine and antipyrine was quite different from the metabolism of the unlabelled drug. Replacement of the hydrogens of the *N*-methyl group by deuterium not only decreased the rate of oxidation of the labelled methyl group, but also shifted the metabolism to other pathways.

Methsuximide, like antipyrine and caffeine, is metabolised by multiple

Table 5.2

Drug administered	Methsuximide (I) / Desmethyl (II)	4-HO-Methsuximide (III) / 4-HO-Desmethyl (IV)	4-HO-Methsuximide (III) / Desmethyl (II)
Methsuximide	0.45	0.73	2.0
N-CD_3-Methsuximide	6.8	8.0	11.0
Methsuximide + N-CD_3-Methsuximide (1:1)	2.0	1.9	4.7

Methsuximide metabolites

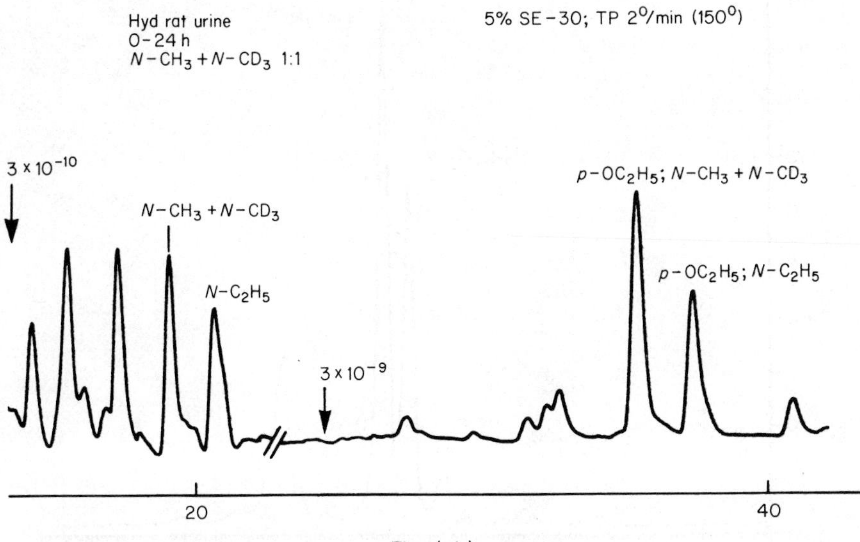

Figure 5.3 Comparison of the metabolites isolated after enzymatic hydrolysis of urine obtained following administration of a 1 : 1 mixture of methsuximide and *N*-CD_3-methsuximide to the rat. The products identified were the same as those described in Figure 5.1

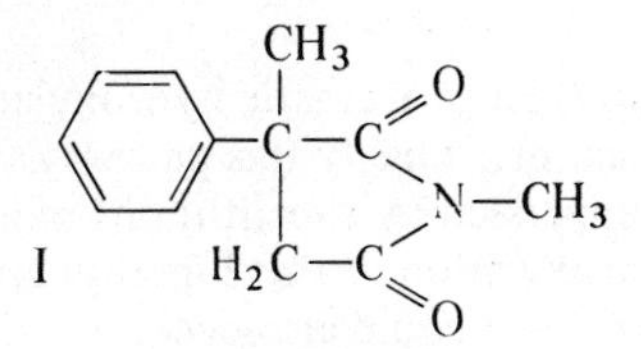

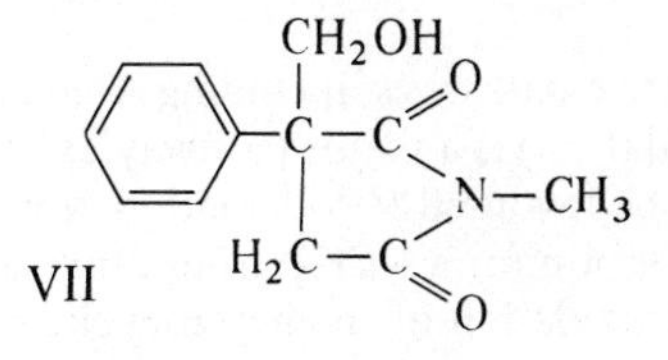

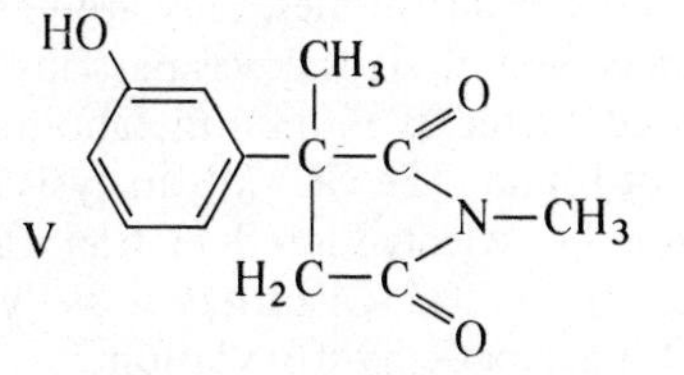

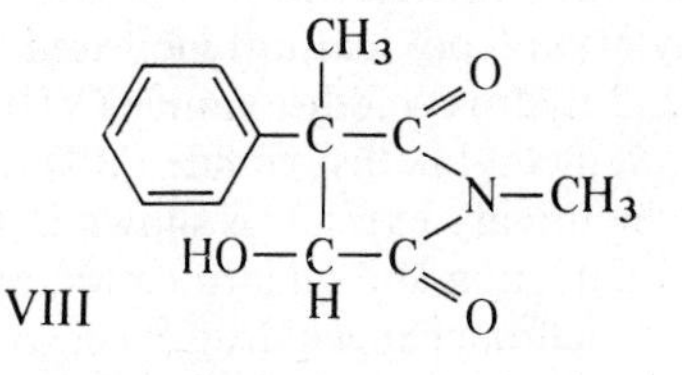

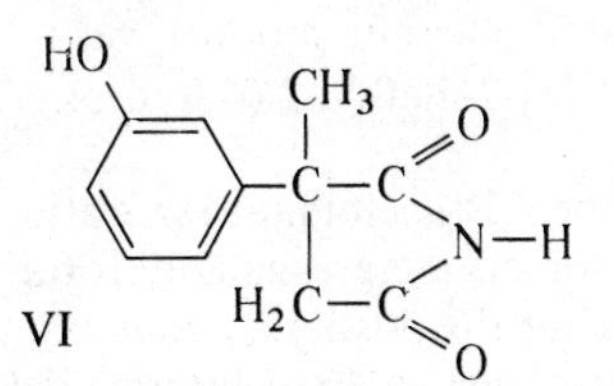

Scheme 5.2

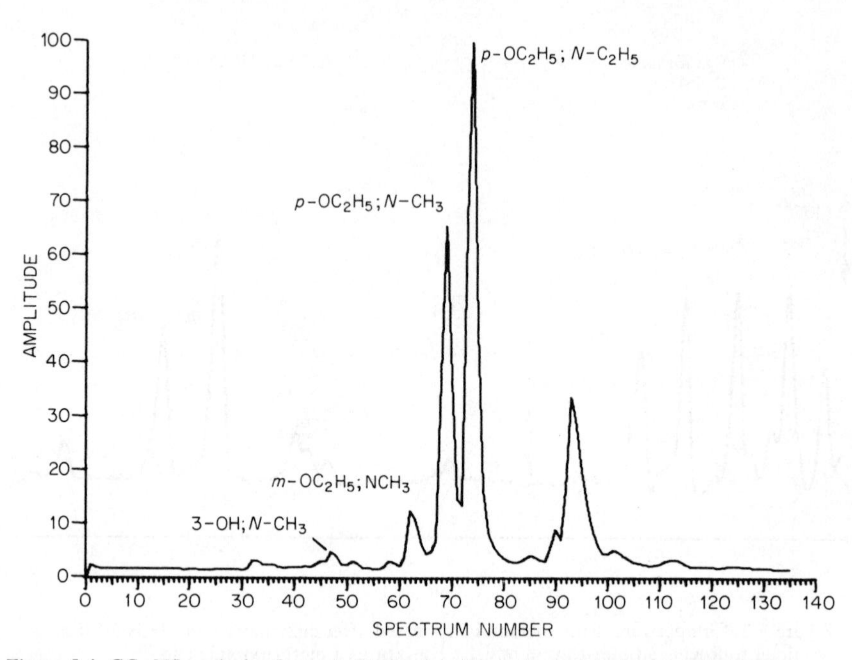

Figure 5.4 GC–MS analysis of metabolites isolated from hydrolysed urine. The analyses were carried out by selected ion detection with methane as the carrier gas. The ions monitored were 248, 251, 263, 292 and 295 (see Table 5.1)

alternative pathways, including *N*-demethylation. Since aromatic hydroxylation (epoxidation) is a major pathway of metabolism, presumably this pathway would be favoured if *N*-demethylation were suppressed by substitutions of a $-CD_3$ group for a $-CH_3$ group. Increased hydroxylation of the 2-methyl group or hydroxylation of the heterocyclic ring at carbon-3 could also occur.

Under the experimental conditions employed in our studies, only aromatic hydroxylation (epoxidation) increased when N-demethylation was partially blocked; 3-hydroxymethsuximide (VIII) was detected as a minor metabolite and 2-hydroxymethylmethsuximide (VII) was not found. The GC–MS analysis of one of the urinary extracts is shown in Figure 5.4; selected ion detection was used for detection and quantification of the metabolites. Quantitative analyses indicated that unchanged drug, a substrate for aromatic hydroxylation, accumulated and demethylation of the deuterated phenolic product was suppressed; as a result, more of the phenol (*N*-CD_3-2-methyl-2-(4-hydroxyphenyl)succinimide) was formed.

The *m*-phenol (*N*,2-dimethyl-2-(3-hydroxyphenyl)succinimide) was also excreted in urine (Figure 5.4). Both *m*- and *p*-isomers are presumably formed via an epoxide intermediate (Horning *et al.*, 1973) but the possibility that *m*-hydroxylation occurs by a mechanism not involving an arene-intermediate cannot be eliminated (Tomaszewski, Jerina and Daly, 1975).

It would be of interest to evaluate the pharmacological activity of deuterated methsuximide. Strong *et al.*, (1974) have suggested that the active anticonvulsant agent is *N*-desmethylmethsuximide because plasma concentrations of the *N*-desmethylmethsuximide are approximately 700 times (mg/l) the concentrations of methsuximide (μg/l). Furthermore, *N*-desmethylmethsuximide is almost as effective as methsuximide in protecting against pentylenetetrazol and electrically induced seizures (Nicholls and Orton, 1972; Chen, Portman and Ensor, 1951; Zimmerman, 1953). Because of the rapid N-demethylation of unlabelled methsuximide, it is difficult to distinguish between the pharmacological activity of the parent drug and the *N*-desmethyl metabolite. However, if *N*-desmethylmethsuximide is the active agent, *N*-CD_3-methsuximide should possess relatively little anticonvulsant activity because of the limited *N*-demethylation that occurs. In fact, the deuterated drug could be considered as a 'slow release' form of the active agent.

It seems unlikely that a deuterated anticonvulsant drug would be introduced into clinical use, since too little information is available concerning chronic ingestion of deuterated compounds. However, if selected deuteration improved the efficacy of a drug used in cancer therapy, clinical use might be considered, particularly if the risk/benefit ratio favoured the deuterated drug. This possibility, of course, has been considered by Dr Farmer and his colleagues (see Chapter 7, this volume).

CONCLUSIONS

In our studies deuterium was used as a tracer for mass spectrometric analyses and also as a means of altering the chemical and biological properties of a drug. Our results suggest that, under certain circumstances, a deuterated compound is not a good tracer for an unlabelled compound and that the corresponding tritium labelled compound would be even less satisfactory. If the breaking of a C–H, C–D or C–T bond is involved in a rate determining step and if the compound is metabolised by multiple alternative pathways, the substitution of deuterium or tritium for hydrogen may result in both qualitative and quantitative changes in the metabolism of the labelled drug. Conclusions based on measurements of the labelled compound only could be quite erroneous.

ACKNOWLEDGEMENT

This investigation was supported by National Institutes of Health Grant GM-16216.

REFERENCES

Blake, M. I., Crespi, H. L. and Katz, J. J. (1975). *J. Pharm. Sci.*, **64**, 367

Chang, T. and Glazko, A. J. (1972). *Antiepileptic Drugs* (ed. D. M. Woodbury, J. K. Penry and R. P. Schmidt), Raven Press, New York, p. 149

Chen, G., Portman, R. and Ensor, C. R. (1951). *J. Pharmacol. Exp. Ther.*, **103**, 54

Cox, P. J., Farmer, P. B., Foster, A. B., Gilby, E. D. and Jarman, M. (1976). *Cancer Treatment Rep.*, **60**, 483

Dudley, K. H., Bius, D. L. and Grace, M. E. (1972). *J. Pharmacol. Exp. Ther.*, **180**, 167
Harvey, D. J., Glazener, L., Stratton, C., Nowlin, J., Hill, R. M. and Horning, M. G. (1972). *Res. Commun. Chem. Pathol. Pharmacol.*, **3**, 557
Harvey, D. J., Nowlin, J., Hickert, P., Butler, C., Gansow, O. and Horning, M. G. (1974). *Biomed. Mass. Spectrom.*, **1**, 340
Horning, M. G., Butler, C., Harvey, D. J., Hill, R. M. and Zion, T. E. (1973). *Res. Commun. Chem. Pathol. Pharmacol.*, **6**, 565
Horning, M. G., Gregory, P., Nowlin, J., Stafford, M., Lertratanangkoon, K., Butler, C., Stillwell, W. G. and Hill, R. M. (1974). *Clin. Chem.*, **20**, 282
Horning, M. G., Haegele, K. D., Sommer, K. R., Nowlin, J., Stafford, M. and Thenot, J.-P. (1976). *Proceedings of the Second International Conference on Stable Isotopes in Chemistry, Biology and Medicine* (ed. R. Klein and P. Klein), US Atomic Energy Commission, Oak Ridge, Tennessee, p. 41
Maynert, E. W. (1972). *Antiepileptic Drugs* (ed. D. M. Woodbury, J. K. Penry and R. P. Schmidt), Raven Press, New York, p. 311
Nicholls, P. J. and Orton, T. C. (1972). *Brit. J. Pharmacol.*, **45**, 48
Strong, J. M., Abe, T., Gibbs, E. L. and Atkinson, A. J., Jr. (1974). *Neurology*, **24**, 250
Tomaszewski, J. E., Jerina, D. M. and Daly, J. W. (1975). *Biochemistry*, **14**, 2024
Zimmerman, F. T. (1953). *Am. J. Psychiatry*, **109**, 767

6
Use of stable-isotope-labelled compounds in pharmaceutical research

W. J. A. VandenHeuvel (Merck Sharp and Dohme Research Laboratories, Rahway, New Jersey 07065, USA)

Stable isotope labelling, especially in conjunction with mass spectrometry, plays an important role in the research leading to and the development of therapeutic agents. Internal standards for GC–MS assays, substrates for metabolism and biosynthesis studies, derivatisation reagents for metabolite structure determination, drugs with increased metabolic stability – labelling with stable isotopes has been found to afford each of the above. This presentation describes the use in pharmaceutical research of a number of deuterium-, ^{13}C- and ^{15}N-substituted compounds.

The characterisation of biotransformation products of drugs or other substrates can be facilitated by use of mass spectrometrically recognisable isotope clusters. Such signatures arise naturally with chlorine- and bromine-substituted compounds, but they can be created artificially by labelling with stable isotopes (Knapp, Gaffney and McMahon, 1972; Prox, Zimmer and Machleidt, 1973; Knapp *et al.*, 1976; Hsia *et al.*, 1976). As a radioactive label allows one to readily design an isolation procedure to yield a fraction known to contain a substrate-related species, the ideally labelled drug would possess a radioactive label (a tracer for isolation procedures) and a stable-isotope label (for immediate, certain, direct recognition of a spectrum as arising from a drug-related compound: Zimmer *et al.*, 1973; Braselton, Orr and Engel, 1973). We have found this to be very useful in a number of studies, as described below.

We (Mandel *et al.*, 1974) observed several years ago that, contrary to earlier reports (see Fuller, 1976, for a leading reference), methyltetrahydrofolic acid (MTHF) does not serve as a methyl source for the N-methylation of indoleamines. Incubation of ^{14}C-MTHF with *N*-methyltryptamine (NMT) in the presence of chick heart enzyme led to a radioactive product shown by TLC not to be *N,N*-dimethyltryptamine (DMT). In order to facilitate the structure proof of the reaction product, NMT labelled by exchange with deuterium in the benzene ring was employed as substrate in another incubation again using ^{14}C-MTHF (methyl label) as methyl source. The labelled NMT was used to provide a recognisable isotopic cluster in the mass spectrum of any substrate-derived product. A radioactive fraction was isolated by TLC. The mass spectra of the labelled substrate and a compound volatilised during the MS examination of the radioactive isolate are presented in Figure 6.1. The ion of *m/e* 44 is $[CH_2NHCH_3]^+$, arising from the side chain. Because the NMT substrate is an isotopic mixture, $^2H_1$-NMT and $^2H_2$-NMT being the major components, characteristic isotope clusters are observed at the M and M-44/M-43 regions. Comparable isotope clusters are observed in the two spectra in Figure 6.1, demonstrating the substrate-related nature of the compound in the isolate giving rise to the spectrum. The molecular ion cluster in the latter is 12, not 14, mass units greater than in the deuterium-labelled NMT, ruling out a simple methylation. A likely reaction is cyclisation with introduction of a methylene group between the side-chain nitrogen atom and the 2-position; this is supported by the absence of ions of *m/e* 44 or 58 (side-chain fragments from NMT and DMT, respectively). A reasonable structure for the product is 2,3,4,9-tetrahydro-2-methyl-1*H*-pyrido[3,4-*b*] indole. This compound was synthesised, and its mass spectrometric (Figure 6.2) and other properties were

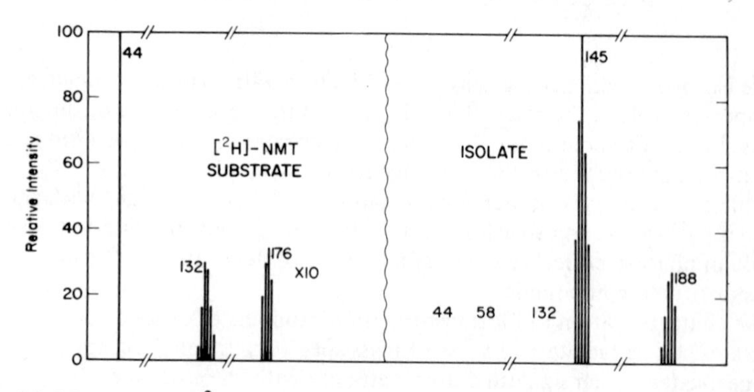

Figure 6.1 Mass spectra of ^{2}H-labelled NMT (left) and 2,3,4,9-tetrahydro-2-methyl-1*H*-pyrido[3,4-b] indole (right), an *in vitro* metabolite of the former

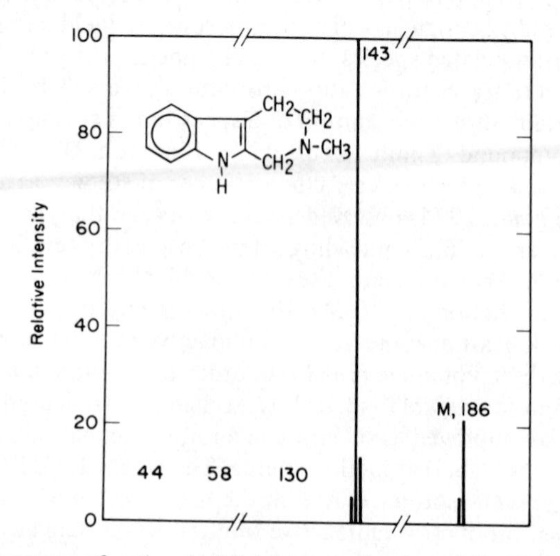

Figure 6.2 Mass spectrum of authentic 2,3,4,9-tetrahydro-2-methyl-1*H*-pyrido[3,4-b] indole

totally compatible with those of the isolate (Mandel *et al.*, 1974). Within a year other groups (see Fuller, 1976) also demonstrated that reaction with MTHF involves cyclisation rather than methylation.

DMT is a potent psychotomimetic agent that has been implicated by some workers in the aetiology of certain types of schizophrenia (Szara, 1956; Wyatt, Termini and Davis, 1971; Saavedra and Axelrod, 1972). Mandel *et al.* (1972) isolated from human lung an enzyme believed to convert *in vitro* NMT to DMT, *S*-adenosylmethionine-$^{14}CH_3$ serving as the methyl donor. As submicrogram amounts of DMT were produced by the *in vitro* reaction, and as unequivocal proof of structure was desired, DMT containing two deuterium atoms in the methylene group of the side chain immediately adjacent to the ring system (see below) was used as a carrier in a GC–MS approach to identify the reaction

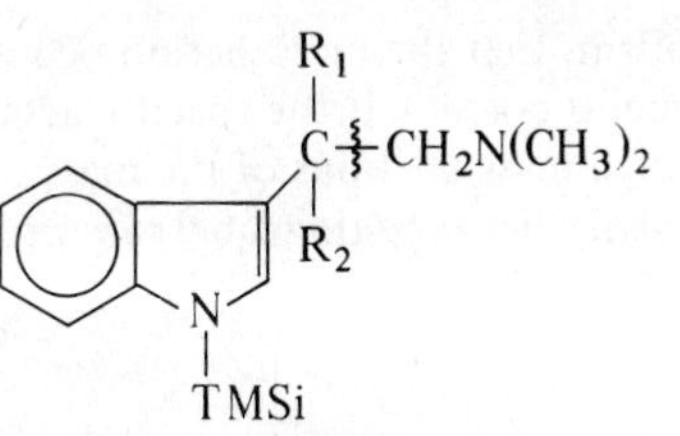

$R_1 = R_2 = H$, DMT
$R_1 = R_2 = {}^2H$, $[^2H_2]$-DMT

product. The mass spectra of DMT and $[^2H_2]$-DMT (as their TMSi derivatives) are such that the *m/e* values of the M and M-58 (loss of $CH_2N(CH_3)_2$) ions (260 *vs.* 262 and 202 *vs.* 204, respectively) differ by 2 mass units, and thus the labelled species can serve as a suitable carrier for the biosynthesised DMT. The base peak in each spectrum is *m/e* 58, the side-chain fragment. The mass spectrum taken when a mixture of trimethylsilylated DMT and $[^2H_2]$-DMT (ratio approximately 1 : 2) eluted from a GLC column is presented in Figure 6.3. GC–MS of the radioactive isolate (approximately 2 parts carrier to 1 part product) from the enzymic reaction gave a similar spectrum. In addition to the ions of *m/e* 262 and 204 from the carrier at the appropriate retention time,

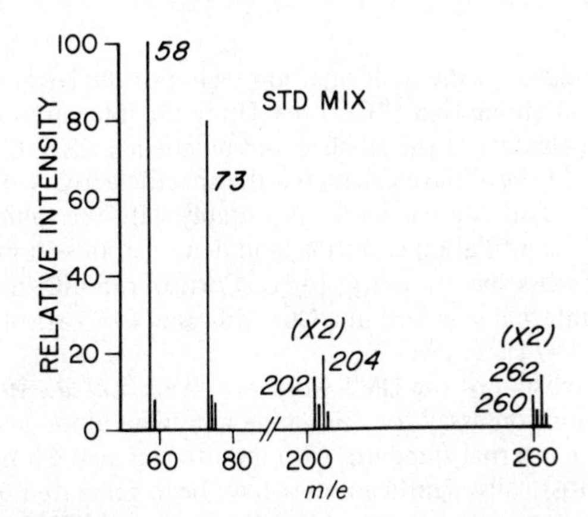

Figure 6.3 Mass spectrum obtained when a mixture of DMT and $[^2H_2]$-DMT (1 : 2) (TMSi derivatives) was subjected to GC–MS analysis. Reprinted from Walker *et al.* (1972), with permission. Copyright by Academic Press, Inc.

signals at *m/e* 260 and 202 were readily apparent, demonstrating the presence of DMT in the sample (Walker *et al.*, 1972). A signal of about 20 per cent relative intensity was observed at *m/e* 60, which is not seen in the spectra from DMT or [$^2H_2$]-DMT. As the specific activity of the *S*-adenosylmethionine-$^{14}CH_3$ was approximately 40 mCi/mmole, the side-chain fragment in the biosynthesised DMT should be significantly enriched in ^{14}C, and a readily apparent (by MS) proportion of these ions should be *m/e* 60,

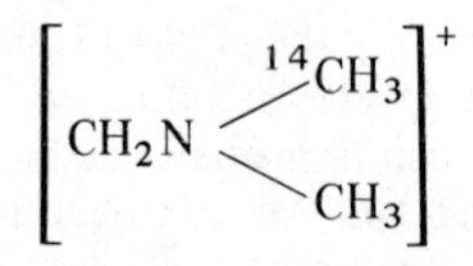

The presence of this ion confirms that the methylation occurs on the side chain.

MS recognition of ^{14}C label is possible if the specific activity is suitably great (Tennent *et al.*, 1976). The high mass portions of the mass spectra of CBZ and [^{14}C]-CBZ (high specific activity benzene ring label) are presented in Figure 6.4.

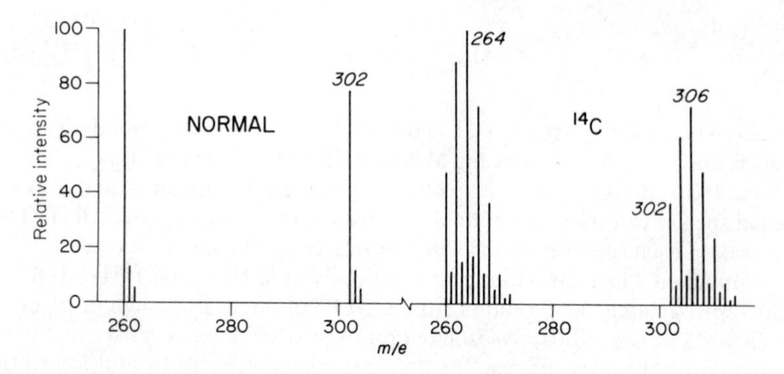

Figure 6.4 High-mass regions of mass spectra of CBZ (left) and [^{14}C]-CBZ (right). Reprinted from Ellsworth *et al.* (1976), with permission. Copyright by the American Chemical Society

The most intense signal in the molecular ion region of the latter is *m/e* 306, the M^+ of a species containing two ^{14}C atoms. Using the intensities of the signals in the molecular ion clusters of the labelled and unlabelled CBZ, Ellsworth, Mertel and VandenHeuvel (1976) have calculated the specific activity of the former to be 118 mCi/mmol. This compares very favourably with the value (116 mCi/mmol) obtained by liquid scintillation counting, and demonstrates an infrequently considered use of mass spectrometry. [^{14}C -Cortisol (55 mCi/mmol) has been employed as the internal standard in a GC–MS assay for cortisol in plasma (Björkhem *et al.*, 1974).

As a logical extension of the DMT work, we (Walker *et al.*, 1973) developed a GC–MS isotope dilution assay for this amine in human blood and plasma using [$^2H_2$]-DMT as the internal standard, with monitoring of the ions *m/e* 260 and 262. So far no statistically significant data have been generated to indicate that DMT levels are elevated in schizophrenics (Wyatt *et al.*, 1973; Lipinski *et al.*, 1974; Bidder *et al.*, 1974; Angrist *et al.*, 1976); indeed, the levels we have observed (<1 ng/ml plasma) do not confirm the high values reported earlier by others (e.g. Franzen and Gross, 1965; Narasimhachari *et al.*, 1971). An interesting study in which this methodology was used involved administration of DMT i.m. to normal human males (possessing experience with hallucinogenic drugs) and determination of their DMT blood levels for 2 h post dose (Kaplan *et al.*, 1974). The peak DMT level appeared about 10 min post dose and then dropped off rapidly, with the patients' subjective 'highs' and the DMT levels following a similar time course. The assay has also been employed to demonstrate *in vivo* production of DMT from NMT in the rabbit (Mandel *et al.*, 1976).

Preparation of the appropriate stable-isotope-labelled internal standard is frequently a problem in quantitative GC–MS work. In connection with our studies on the effect of certain drugs upon urinary excretion levels of 7α-hydroxy-5,11-diketotetranorprostane-1,16-dioic acid (I; the major human urinary metabolite of PGE_1 and PGE_2 (Hamberg and Samuelsson, 1969)), we undertook to prepare the ^2H (and ^3H)-labelled I for use as an internal standard. Our approach was to dose rabbits with [^2H,^3H]-15-keto-PGE_0 (see below; Rosegay, 1975), and isolate labelled I from the urine. The partial mass spectrum of the methyl ester-dimethoxime-TMSi ether derivative of the precursor (containing mainly three ^2H atoms/molecule) is presented in Figure 6.5. The

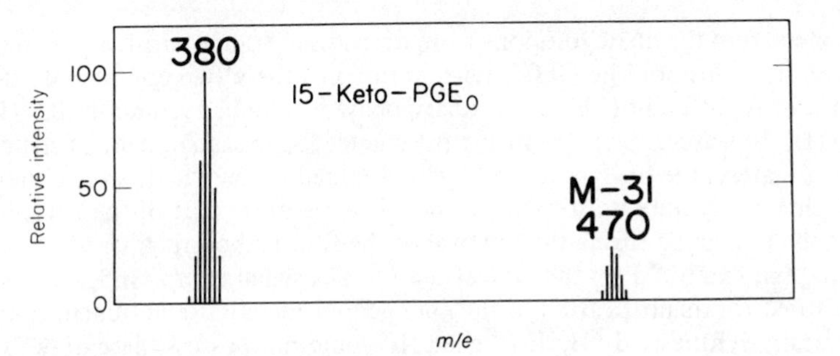

Figure 6.5 Partial mass spectrum of ^2H,^3H-labelled 15-keto PGE_0 as its methyl ester-dimethoxime-TMSi ether. Reprinted from VandenHeuvel *et al.* (1977), with permission. Copyright by Elsevier Publishing Co.

characteristic isotope clusters were expected to facilitate MS recognition of metabolites, and three were recognised by use of temperature-programmed GLC-repetitive scanning MS (VandenHeuvel *et al.*, 1977). These included 7α-hydroxy-5,11-diketotetranorprostanoic acid (a minor human urinary metabolite of PGE_2 (Hamberg and Wilson, 1973)), 5β,7α-dihydroxy-11-keto-tetranorprostanoic acid (II; the major urinary metabolite of a number of

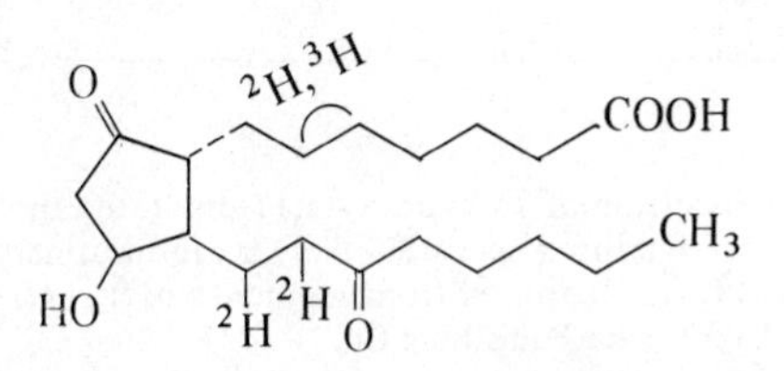

prostaglandins in the guinea-pig (Hamberg and Samuelsson, 1972)), and the desired I. The partial mass spectra of the methyl ester-methoxime-diTMSi ether derivative of the dihydroxy metabolite and the dimethyl ester-dimethoxime-TMSi ether of I are presented in Figures 6.6 and 6.7. As reduction in the length of the carbon chain during metabolism results in loss of one deuterium atom, the observed molecular ions are 2, not 3, mass units higher than those observed for the unlabelled compounds. The *m/e* values of the M-31 and M-(31 + 90) ions from derivatised II are the same as would be

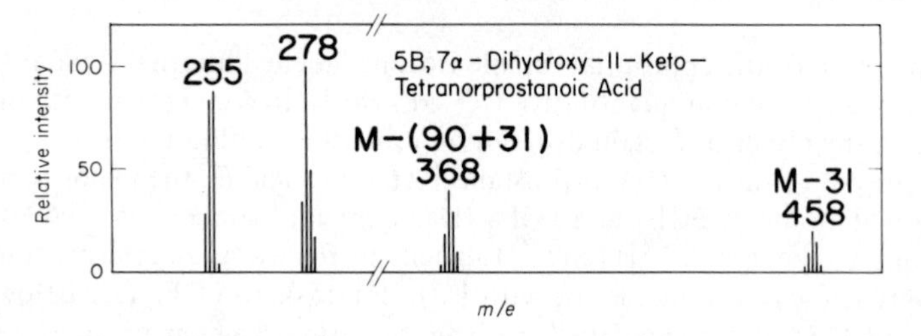

Figure 6.6 Partial mass spectrum of 5β,7α-dihydroxy-11-ketotetranorprostanoic acid (as its methyl ester-methoxime-diTMSi ether), a rabbit urinary metabolite of $^{2}H,^{3}H$-labelled 15-keto-PGE_0

expected from the analogous ions from derivatised tri-deutero-I, a possible source of confusion. The GLC retention times of these two compounds are significantly different (SE-30 stationary phase, C number values of 20.8 (II) and 24.0 (I)), however, greatly reducing the chance for misassignment of structure. Unfortunately, the biosynthesised I was obtained in low yield, was neither radiochemically pure nor of known specific activity (as part of the tritium is lost from the precursor during the removal of the four-carbon fragment from the side chain), and exhibited (in the derivatised form) a signal at *m/e* 365, the ion monitored for quantification of the endogenous metabolite in human urine. A chemically synthesised $^{2}H,^{3}H$-labelled III would not possess these drawbacks, and Rosegay and Taub (1976) have prepared the desired compound. The partial mass spectrum of this labelled prostaglandin metabolite, as its dimethyl

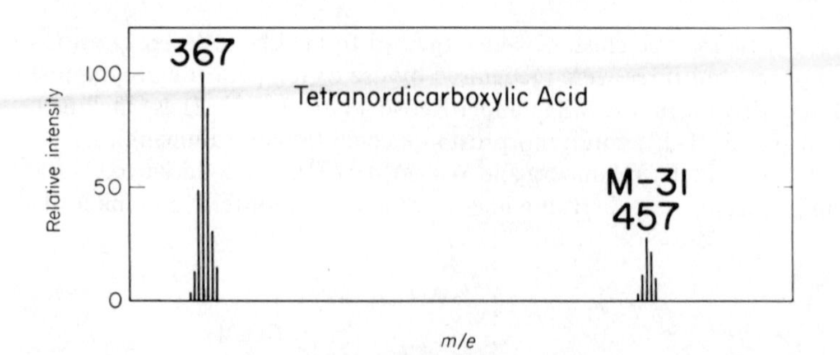

Figure 6.7 Partial mass spectrum of 7α-hydroxy-5,11-diketotetranorprostane-1,16-dioic acid (as its dimethyl ester-dimethoxime-TMSi ether), a rabbit urinary metabolite of $^{2}H,^{3}H$-labelled 15-keto-PGE_0. Reprinted from VandenHeuvel *et al.* (1977), with permission. Copyright by Elsevier Publishing Co.

ester-dimethoxime-TMSi ether derivative, is presented in Figure 6.8. The most intense signal in the clusters corresponds to a species containing seven ^{2}H atoms per molecule, and no signal is observed at *m/e* 365. We have demonstrated that the same isotope cluster is observed for the internal standard after it has been carried through the assay procedure, and thus the deuterium is truly stable.

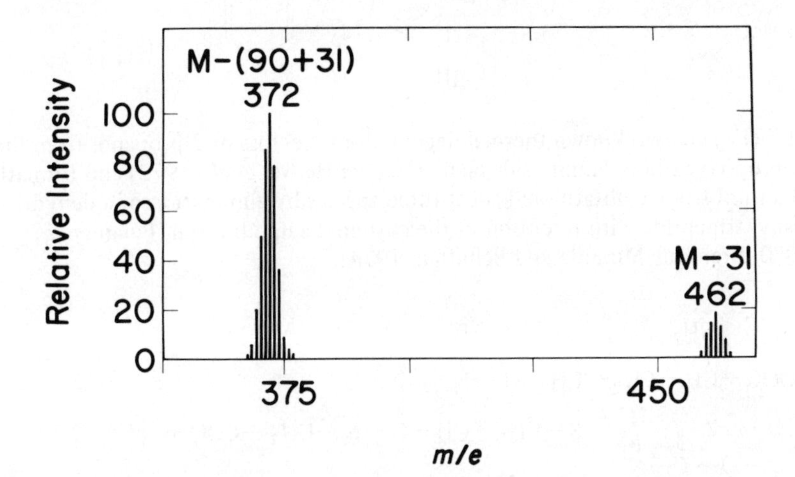

Figure 6.8 Partial mass spectrum of synthetic ^{2}H,^{3}H-labelled 7α-hydroxy-5,11-diketotetranorprostane-1,16-dioic acid as its dimethyl ester-dimethoxime-TMSi ether. Reprinted from VandenHeuvel *et al.* (1977), with permission. Copyright by Elsevier Publishing Co.

[^{2}H,^{3}H]-I has been employed in a study to determine the effect of diflunisal, a new analgesic, on urinary prostaglandin excretion (Steelman *et al.*, 1976). In normal males receiving 375 mg b.i.d. of this drug for 5 days the urinary levels of I were reduced by an average of about 70 per cent.

Incubation of the anthelmintic cambendazole, 2-(4′-thiazolyl)-5-isopropoxycarbonylaminobenzimidazole (CBZ; see below), with liver microsomes from rats

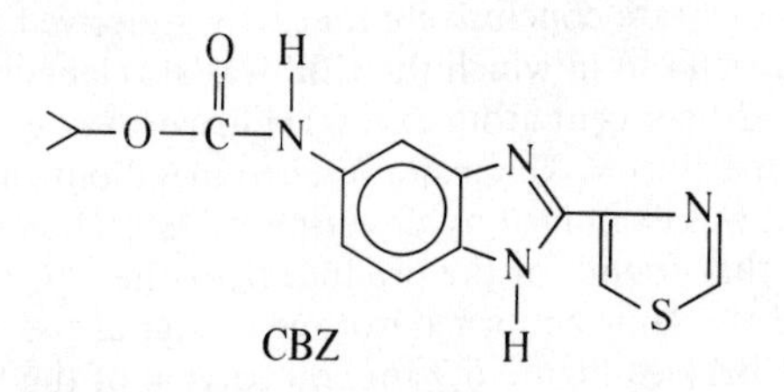

or hamsters leads to the formation of polar metabolites (Wolf *et al.*, 1974). Incorporation of glutathione (GSH) into the incubation mixture in an attempt to trap a reactive intermediate resulted in a 3.5-fold increase in the amount of polar metabolites formed. The major polar metabolite from ^{14}C (benzene ring)- and ^{3}H (isopropyl group)-labelled CBZ was isolated. Direct probe MS yielded data suggesting that a compound which vaporised at a probe temperature of about 250° possessed a molecular ion of *m/e* 274, and a possible structure for this species (III) is presented below. III could arise from a GSH-conjugate of

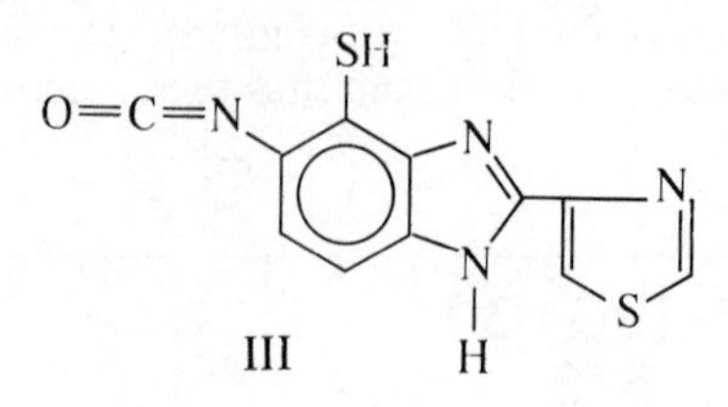

CBZ (IV) via two known thermal degradations, i.e. loss of 2-propanol from the isopropoxycarbonylamino side chain (VandenHeuvel *et al.*, 1972) and formation of a thiol from a glutathionyl-substituted species by elimination of a dehydro-alanyl tripeptide with retention of the cysteinyl sulphur atom (Polan *et al.*, 1970; Brent, de Miranda and Schulten, 1974).

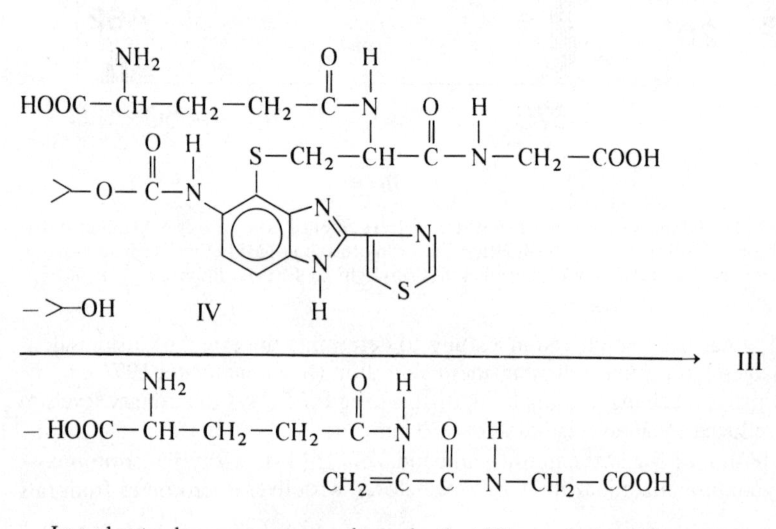

In order to demonstrate conclusively that III was derived from CBZ, an experiment was carried out in which the CBZ was also labelled with ^{13}C (thiazole ring, C-2, 20 per cent atom excess) (Ellsworth *et al.*, 1976) to yield a characteristic isotope cluster. The major *in vitro* metabolite arising from this triply labelled CBZ was examined by direct-probe MS. The resulting spectrum was identical with that found for the product from the $^{14}C,^{3}H$-labelled CBZ except it possessed the same abnormal isotope cluster as the ^{13}C (and $^{14}C,^{3}H$)-labelled CBZ (see Figure 6.9 for comparison of the molecular ion clusters) unequivocally characterising the volatilised compound as being drug-related. The somewhat greater intensity for the ion *m/e* 276 vis-à-vis *m/e* 304 can be accounted for by the presence of a second sulphur atom in the metabolite-related compound. When an aliquot of this metabolite was exposed to trimethylsilylation conditions and then subjected to direct-probe MS, the presence of a species with molecular and M-15 ions of *m/e* 346 and 331, respectively, was indicated, both ions displaying the enriched ^{13}C isotope. For

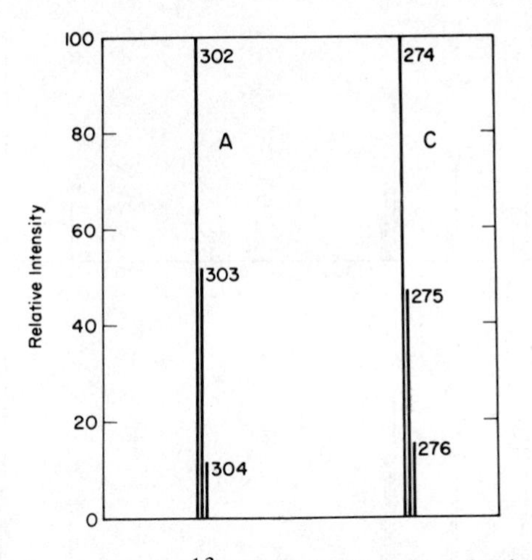

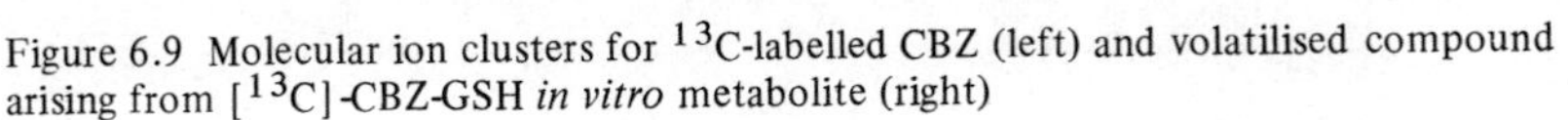
Figure 6.9 Molecular ion clusters for ^{13}C-labelled CBZ (left) and volatilised compound arising from [^{13}C]-CBZ-GSH *in vitro* metabolite (right)

comparison purposes, the mono-TMSi derivative of the isocyanate from cambendazole possesses signals at *m/e* 314 (M) and 299 (M-15), 32 mass units less.

Gas–liquid radiochromatography in combination with on-column methylation of the metabolite using trimethylanilinium hydroxide in methanol demonstrated the production of a volatile radioactive component. This derivatisation approach was then used with combined GC–MS. Under these conditions, CBZ forms a dimethyl product with a molecular ion of *m/e* 330 and an intense M-42 (M-C_3H_6) ion. The eluted compound from the isolate possessed an apparent molecular ion of *m/e* 376, with an intense fragment ion at *m/e* 287, M-89. Both of these ions exhibited enhanced ^{13}C isotope peaks (Figure 6.10), confirming that they arose from a drug-related compound. One can propose for the compound of molecular ion 376 a structure involving a trimethylated cambendazole molecule containing one additional sulphur atom; the fragment ion M-89 would arise from losses of 42 and 47 (SCH_3) a.m.u.

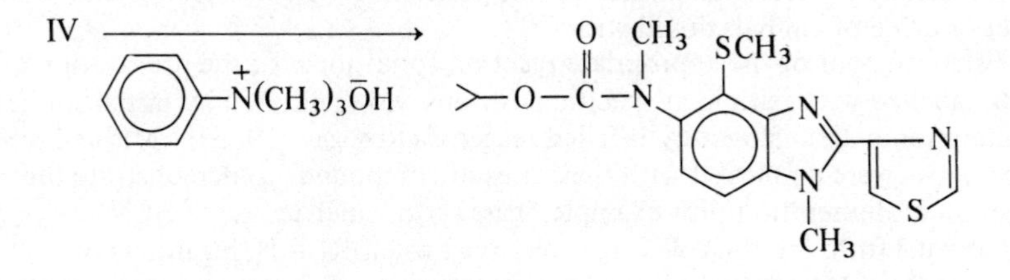

If IV were to undergo the two thermal degradations, the resulting thiol isocyanate (III) would possess a molecular weight of 274. The mono-TMSi derivative of III would possess a molecular weight of 346. Thermal scission of the CBZ–GSH S–C bond would occur after the trimethylsilylating reagent was pumped away in the mass spectrometer, and, hence, the thermally produced

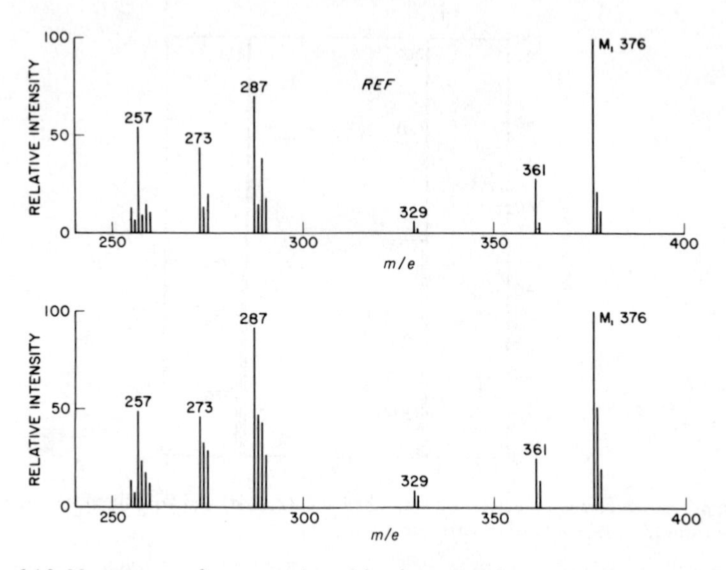

Figure 6.10 Mass spectra of compound resulting from on-column methylation and GC–MS of ^{13}C-labelled CBZ-GSH (bottom) and 4-methylmercapto-*N*,*N*′-dimethyl-CBZ (top)

thiol group would not be derivatised. In the case of the on-column GLC methylation, however, fragmentation of the parent molecule takes place at the top of the hot column in the presence of the methylating reagent, and, hence, the S–CH_3 group would be formed. 4-Methylmercapto-*N*,*N*′-dimethyl-cambendazole was synthesised and found to exhibit GC–MS behaviour (see Figure 6.10) indistinguishable (except for the ^{13}C content) from that of the compound resulting from on-column methylation of the isolate. These data suggest that the metabolite is a GSH conjugate of CBZ (with attachment at the 4 position) which degrades thermally with the sulphur atom of cysteine remaining attached to the CBZ molecule. Conjugation of GSH at the 4 position of CBZ was also suggested by experiments with drug labelled specifically with tritium at the 4, 6, 7, 2′ or 5′ positions. These compounds were subjected to incubation with liver microsomes (excess GSH); loss of tritium was greatest at C-4, indicating preferential attack at this position. No IV has been found in the liver or urine of animals dosed with CBZ.

Establishment of the appropriate reaction conditions for the preparation of CBZ labelled with tritium at specific positions was facilitated by initial studies on deuteration with analogously labelled reagents (Rosegay, 1973). NMR and MS techniques were employed with this series of compounds to demonstrate the position of deuteration. For example, mass spectrometric loss of HCN (27 a.m.u.) from the thiazole ring is observed when R_1 = H, regardless of whether R_2 is H or ^2H, but loss of ^{2}HCN (28 a.m.u.) is found when R_1 = ^2H.

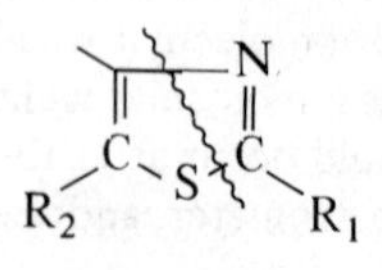

M-27 when R_1 = H
M-28 when R_2 = ^2H

Stable-isotope labelling of CBZ also played a key role in establishing the origin of the amide nitrogen atom in a urinary metabolite (see below) of this drug in the pig (VandenHeuvel *et al.*, 1974). The corresponding carboxylic acid

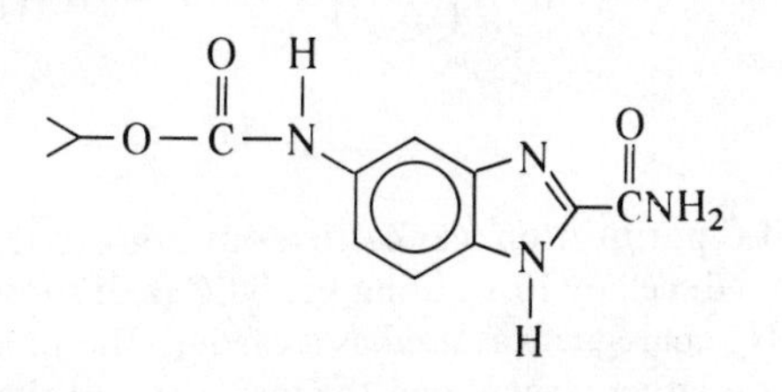

is also a metabolite (VandenHeuvel *et al.*, 1976), and it was thus of interest to ascertain whether the amide is formed directly from CBZ or from the acid via amidation. Use of CBZ labelled with ^{15}N in the thiazole ring should allow unequivocal demonstration of the source of the nitrogen atom. Retention of the label in the metabolite would require that the thiazole ring was the source of the amide nitrogen atom, whereas loss of label would require that it came from the body pool. [^{15}N]-thiazole-labelled (also ^{14}C-labelled in the benzene ring) CBZ (Figure 6.11) was administered to a rat, and a crude urine extract subjected to two-dimensional TLC. Elution of the appropriate radioactive zone gave a

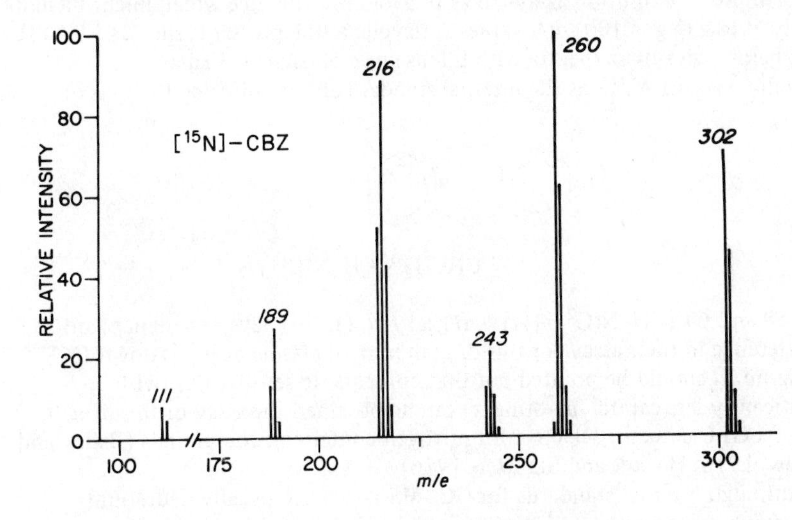

Figure 6.11 Mass spectrum of ^{15}N-labelled CBZ

fraction containing approximately 1 μg of the metabolite. This material was far too impure for determination of the isotope content in the metabolite by direct-probe MS, and it was decided that GC–MS using selective ion monitoring would be an appropriate route for obtaining the desired information. The metabolite possesses poor GLC properties, but on-column methylation using trimethylanilinium hydroxide in methanol converts it to a tetramethyl derivative (see below) with satisfactory GLC behaviour.

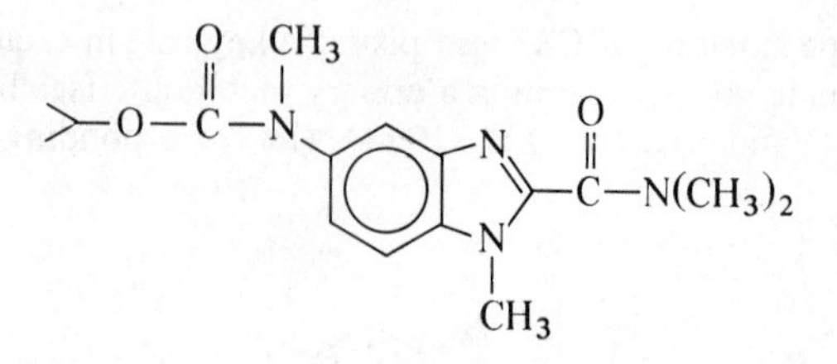

An additional TLC purification was carried out prior to the GC–MS experiment, and to reduce any loss during the TLC (and subsequent GLC) step, 5 μg of the 4,6,7-$^2H_3$ analogue was used as a carrier. The final TLC eluate was dissolved in the methylating reagent and the intensities of the ions of *m/e* 318 and 319 (M and M + 1 of the tetramethyl metabolite) monitored selectively during GC–MS. Presence of the carrier did not interfere with measurement of the ion intensity ratio in the metabolite, as the molecular ion of the methylated trideutero analogue is *m/e* 321. No difference was observed between the M/M + 1 intensity ratios of the ^{15}N-labelled CBZ and the metabolite, demonstrating that the thiazole N atom is retained in the latter. If desired, the [$^2H_3$]-amide could serve as a stable-isotope-labelled internal standard in a GC–MS isotope dilution assay (VandenHeuvel, Gruber and Walker, 1973).

Ions of high (e.g. >200) *m/e* values are usually chosen for monitoring in GC–MS isotope dilution assays so as to avoid interference which might be more likely at low (e.g. <100) *m/e* values. Tricyclic antidepressants such as ELAVIL (see below) give mass spectra with intense *m/e* 58 signals. Using *N,N*-di-C^2H_3-ELAVIL as the internal standard and monitoring the ions of

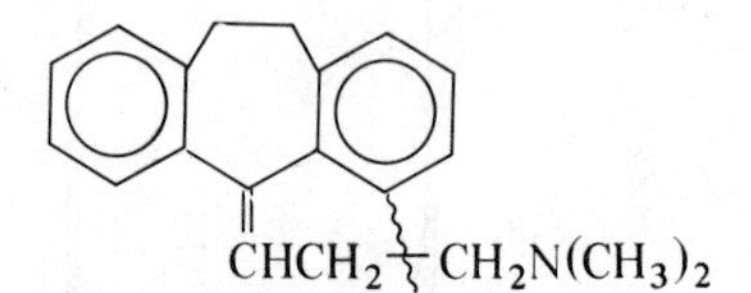

m/e 58 and 64 [$CH_2N(C^2H_3)_2$]$^+$, Biggs *et al.* (1976) have experienced no interference in their assay for this drug in human plasma at levels down to 10 ng/ml. It should be pointed out that comparable sensitivities (with significantly less capital investment) can be obtained for assay of this drug by use of a GLC detector selective for nitrogen-containing compounds (Bailey and Jatlow, 1976; Hucker and Stauffer, 1976).

Although internal standards for GC–MS assays are usually deuterium-containing compounds, we have employed ^{15}N-labelled uric acid as the internal standard in an assay for determining uric acid body pool size (Walker *et al.*, 1976). 2-[^{14}C]-Uric acid has also been used to measure body pool size (Sorensen, 1960), but when exposure to radiation is of concern the use of the stable isotope is mandatory. The spider monkey was chosen as the animal model for our study as it, like man, lacks the uricolytic enzyme uricase, and has a significant uric acid urine level. The method involves administration of a single dose of 1,3-[$^{15}N_2$]-uric acid to the animal. Urine is then collected at specified intervals and a partially purified specimen of uric acid isolated via ion exchange. This material is subjected to trimethylsilylation, and the intensity ratios for the molecular ions (*m/e* 456, 458) of the two TMSi derivatives are obtained by selective ion monitoring. The dilution of the labelled uric acid by biosynthesised uric acid thus determined, the magnitude of the uric acid body pool size is calculated.

[$^{15}N_2$]-uric acid was used in early studies (Benedict, Forsham and Stetten, 1949; Wyngaarden and Stetten, 1953) utilising a MS determination of the isotope composition of the N_2 gas obtained by chemical transformation of a purified isolate of uric acid. Use of [$^{15}N_2$]-uric acid with derivatisation and GC–MS avoids not only the radiation problem, but also the disadvantages of the N_2 gas approach for determination of ^{15}N content, i.e. lack of differentiation between N_2 from the uric acid and that from possible sample contaminants including atmospheric N_2. Although the N_2 gas approach is capable of yielding isotope ratios with great accuracy, MS on the actual compound of interest, while producing less accurate data, does not suffer from the above disadvantages. The GC–MS approach also allows the determination of isotope content on the molecular ion of the subject compound. Further, the great separating power of GLC combined with MS results in an excellent technique for obtaining isotope data on the microgram scale from only partially purified isolates.

The mass spectrum of the tetra-TMSi derivative of uric acid is characterised by a molecular ion of *m/e* 456 and intense fragment ions at *m/e* 441 (loss of methyl) and *m/e* 73 (trimethylsilyl cation). The satisfactory GLC properties (no column memory effects) and the intense molecular ion make this derivative an excellent choice for the assay. Figure 6.12 shows the molecular ion regions of the tetra-TMSi derivatives of unlabelled uric acid, [$^{15}N_2$]-uric acid, and a 1 : 1 mixture of the two isotopic species. Intensity values for ions *m/e* 456 and *m/e* 458 were obtained on a series of calibration mixtures and the ratios

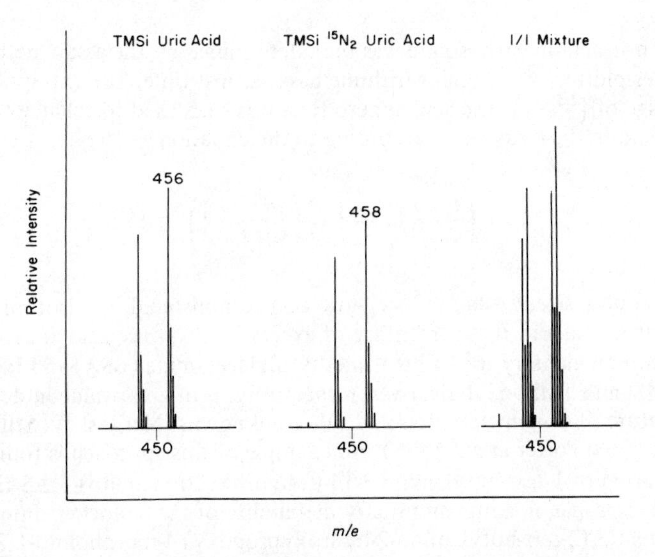

Figure 6.12 Partial mass spectra (M and M-15 region) of tetra-TMSi derivatives of uric acid, [$^{15}N_2$]-uric acid, and a 1 : 1 mixture of the two. Reprinted from Walker *et al.* (1976), with permission. Copyright by Academic Press, Inc.

I_{458}/I_{456} plotted against the percentage of $[^{15}N_2]$-uric acid; a linear relationship was obtained in the 0–25 per cent $[^{15}N_2]$-uric acid range.

An application of the method to the determination of the uric acid body pool size is shown in Figure 6.13. A spider monkey was dosed with 5 mg of $[^{15}N_2]$-uric acid (UA) and urine collected for 8 h in 1 h increments. The percentage of

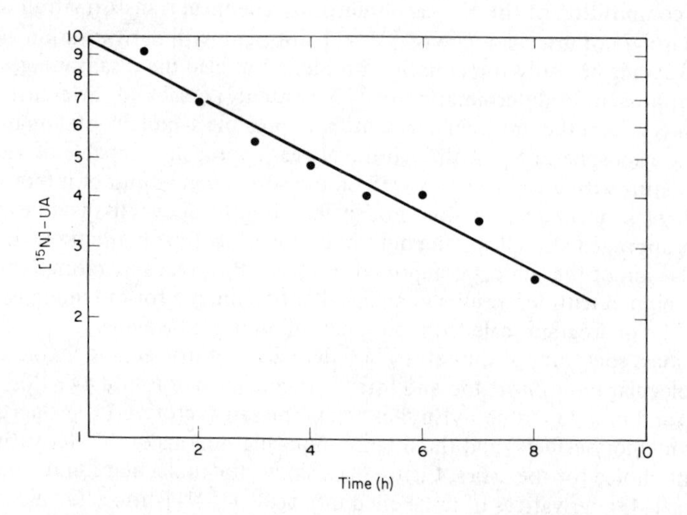

Figure 6.13 Plot of experimentally determined percentage of $[^{15}N_2]$-uric acid content of monkey urine *vs.* time following administration of $[^{15}N_2]$-uric acid to the animal. Reprinted from Walker *et al.* (1976), with permission. Copyright by Academic Press, Inc.

labelled uric acid in each isolate was then determined by the assay method and the values plotted on a semilogarithmic basis against time. The extrapolated percentage of $[^{15}N_2]$-uric acid at zero time was then used to calculate the amount of uric acid in the body pool, according to the equation

$$A = a\left(\frac{I_i}{I_0} - 1\right) = 5.0\left(\frac{99}{9.74} - 1\right) = 45.8 \text{ mg}$$

where A = pool size, a = mg $[^{15}N_2]$-uric acid administered, I_i = isotopic purity of $[^{15}N_2]$-uric acid and I_0 = percentage of excess $[^{15}N_2]$-uric acid at zero time.

The complementary use of bis-trimethylsilylacetamide (BSA) and BSA-d_{18} to form TMSi and TMSi-d_9 derivatives, respectively, is of great value in determining the structure of drug metabolites and other unknowns (McCloskey, Stillwell and Lawson, 1968; Porter *et al.*, 1975). An example of this approach is found in the identification of 1-*tert*-butylamino-3-[4-(2-hydroxyethylamino)-1,2,5-thiadiazol-3-yloxy]-2-propanol, a human urinary metabolite of the β-blocker timolol or Blockadren, 3-(2-*tert*-butylamino-2-hydroxypropoxy)-4-morpholino-1,2,5-thiadiazole (see below) (Tocco *et al.*, 1975). Use of radioactively labelled drugs greatly facilitates isolation of drug-related compounds in such studies, as gas-liquid radiochromatography indicates which of the numerous components

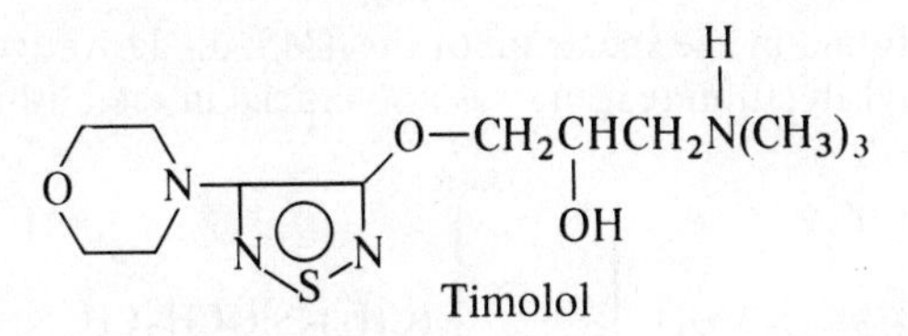

Timolol

eluting from the column is a metabolite and should be examined by combined GC–MS. The beauty of such a study is that a structure can be deduced by examining relatively crude isolates containing only microgram quantities of metabolite; indeed, in favourable cases one need not even examine the metabolite *per se*, but only its derivatives, to ascertain its structure. GC–MS analysis of the TMSi and TMSi-d_9 derivatives of this particular timolol metabolite gave the data shown in Table 6.1. The molecular weight of the

Table 6.1 Mass spectrometric data for TMSi and TMSi-d_9 derivatives of timolol metabolite (Tocco *et al.*, 1975)

m/e TMSi		*m/e* TMSi-d_9
434	M	452
419	M-Me	437/434
331	M-103 (112)[a]	340
318	M-116 (125)	327
305	M-129 (129)	323
233		242
218		224
202		211
103	$[TMSiOCH_2]^+$	112
86		86
57		57

[a] Values in parentheses are for TMSi-d_9 ions.

metabolite is 290, calculated as follows: 452–434 = 18; 18/9 = 2; 434 – (2 x 72) = 290. The ions of *m/e* 57, 86, 202 and M-116 (structures given below) are also found in the spectrum of timolol TMSi ether, requiring that the intact timolol side chain is present in the metabolite.

$[C(CH_3)_3]^+$ *m/e* 57

$[CH_2NHC(CH_3)_3]^+$ *m/e* 86

$[CH_2CH(OTMSi)CH_2NHC(CH_3)_3]^+$ *m/e* 202

M-CH_2CHOTMSi M-116

The ions of *m/e* 103 and M-103 suggest the presence of a hydroxymethyl group ($HOCH_2$) in the metabolite, and this is confirmed by the analogous ions

(m/e 112, M-112) found in the spectrum of the TMSi-d_9 derivative. Use of the perdeutero trimethylsilylation reagent was also crucial in establishing the

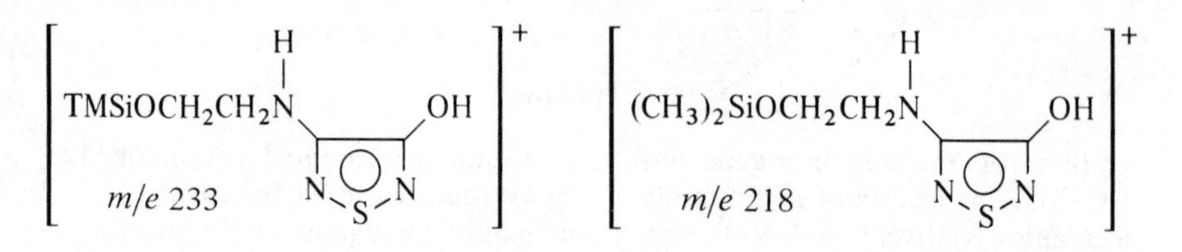

structure of the ions of m/e 233 and 218. The MS data demonstrate that the metabolite is an ethanolamine with the structure given below. The authentic

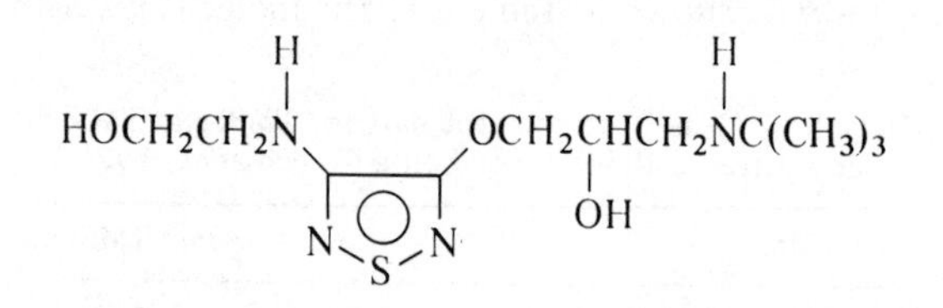

ethanolamine was synthesised (Ellsworth, 1974) and found to possess TLC and GC–MS (as TMSi derivatives) properties identical with those of the metabolite.

The kinetic isotope effect resulting from the substitution of deuterium for hydrogen has been employed in numerous studies on reaction mechanisms, including establishing pathways of biochemical transformations. As the carbon–deuterium bond is more stable than the carbon–hydrogen bond, substitution by deuterium will result in a significant decrease in the rate of reaction if the rate-determining step in a reaction involves breaking of a carbon–hydrogen bond (Blake, Crespi and Katz, 1975). This can be observed both physicochemically and pharmacologically. Thus the V_{max} values for the *in vitro* dealkylations of *p*-nitroanisole and *p*-nitrophenetole are 2–3 times greater than those for the analogous α-deuterated species, and carbon–hydrogen bond breaking in the alkyl group is the rate-determining step (Al-Gailany, Bridges and Netter, 1975). *In vitro* hydroxylation (mouse liver microsomes) of butethal (5-*n*-butyl-5-ethyl-barbituric acid) was found to be 1.6 times faster than that of the 3′-dideutero drug (Tanabe *et al.*, 1969). This substitution results in a doubling of the sleeping time in mice, and the biological half-life of the labelled species is 2.5 times greater than butethal itself. A number of examples of the effect of deuterium substitution upon biological properties of drugs are known (Blake *et al.*, 1975). In some cases no effect is desired. Hsia *et al.* (1976) have reported that in mice and rats methadone and methadone-C^2H_3 possess equivalent pharmacological properties. As the ratio of drug to its trideutero analogue in a mixture can be determined (by GC–MS), these authors suggest that detection of illicit methadone supplementation could be achieved by using the stable-isotope-labelled species as an *in vivo* marker. The suspected patient would be given a mixture of methadone and methadone-C^2H_3 of known composition, and any deviation from this composition, as reflected by a deviation in the isotope ratio found for urinary methadone, would indicate illegal supplementation.

As the rates of reaction of deuterium (and tritium)-substituted compounds can be strongly influenced by the kinetic isotope effect, one must be careful in drawing conclusions based on rates of displacement of a label. If breaking of a carbon–deuterium (or tritium) bond is the rate-determining step in a monitored reaction, it is not possible to extrapolate the data to the non-labelled substrate.

In no case has a deuterium-substituted compound been seriously considered for use as a therapeutic agent (Blake *et al.*, 1975). Kahan, Kropp and associates (Kahan and Kropp, 1975; Kropp, Kahan and Woodruff, 1975; Kahan *et al.*, 1975) have recently reported on the new antibacterial, 3-fluoro-D-alanine (FA) and its 2-deutero analogue. 2-Deutero-FA was selected for use in the combination antimicrobial MK-641/MK-642, the second component being the hemihydrate of sodium D-4[(2-oxo-3-penten-4-yl)-amino]-3-isoxazolidinone (a 'pro-drug' of cycloserine). The fluorine-containing compounds were synthesised via photofluorination by Kollonitsch and Barash (1976), using their newly invented method for substitutive fluorination of organic compounds. Substitution of deuterium for hydrogen at the 2 position of FA reduces *in vivo* metabolism and acute toxicity several-fold (Kropp *et al.*, 1975). It was suggested that the kinetic isotope effect occurs at the initial step of metabolism, the D-amino acid oxidase-mediated conversion of 2-deutero-FA to 3-fluoropyruvic acid. *In vitro* data supporting this proposal are presented in Table 6.2; the kinetic isotope effect is approximately 3. For comparison, Triggle and Moran (1966) observed a factor of 1.4 for the L-amino acid oxidase-mediated oxidation of tyrosine and α-[^{2}H]-tyrosine (95 per cent label).

Table 6.2 Rate of oxidation by D-amino acid oxidase (Kahan and Kropp, 1975)

Substrate	Rate of 3-fluoropyruvate formation	
	10^{-4} M	10^{-3} M
	n mol/min	
FA	1.0	6.7
2-Deutero-FA	0.33	2.17

Pharmacokinetic studies with the rhesus monkey are presented in Table 6.3, demonstrating the *in vivo* isotope effect most clearly in the reduced levels of serum fluoride ion. Fluoride was shown by the authors to be the predominant terminal metabolite resulting from secondary metabolism of 3-fluoropyruvic acid. As 2-deutero-FA exhibits the same antibacterial activity as the protio form of the fluoro amino acid (in both cases the result of inactivation of bacterial alanine racemase), the deuterium isotope effect leads to a significantly improved therapeutic index.

As is clear from the papers presented at this symposium, numerous stable-isotope-substituted drugs have been prepared and used to determine drug metabolism pathways and mechanisms, to examine isotope effects and to serve as internal standards for MS-based analytical procedures. Recognition of a stable-isotope-containing compound is not as easy as recognising a radioactively labelled compound. The background is negligible at the specific activities

Table 6.3 Single-subject crossover comparison of serum fluoride levels and urinary recoveries following administration[a] of FA and 2-deutero-FA (Kahan and Kropp, 1975)

Subject	Serum fluoride (p.p.m.)		Urinary recoveries	
	90–120 min	240 min	drug	fluoride (mol%)
No. 1 FA	2.56	2.2	54	7.46
2-deutero-FA	1.07	1.27	74	4.15
No. 2 FA	1.12	1.29	89	1.68
2-deutero-FA	0.53	0.61	100	0.50

[a] Rhesus monkeys; three doses of FA or 2-deutero-FA, 200 mg/kg i.m., alternated at 48 h intervals.

normally employed with the latter, thanks to the specificity of radiochemical techniques such as LSC, whereas MS methods used with stable isotopes are susceptible to interference from extraneous compounds present in the sample. Although a computer system can be programmed to recognise characteristic MS isotope ratios (Canada and Regnier, 1976), this is successful only if other components in the sample do not contribute to the intensities of the ions of interest. Direct-probe examination may well prove unsuccessful for indicating whether a stable-isotope-labelled drug metabolite is present in a crude isolate. This is in contrast to the ease of establishing whether such a sample contains a radiolabelled compound. Specificity of the MS approach may be enhanced by use of a gas chromatograph in combination with the mass spectrometer so as to reduce the background, but then a derivatisation step must often be introduced to allow of successful GLC. The resolving power of the GLC system can be maximised by use of capillary columns, and high-resolution MS would offer more specificity than low-resolution.

Use of stable isotopes in biological research has expanded markedly in the past 10 years. In situtations where use of radioisotopes is prohibited, stable-isotope labelling with MS-based recognition is invaluable. The development of methods to facilitate the detection and measurement of stable isotopes would increase the use of these tracer compounds.

ACKNOWLEDGEMENT

It is a pleasure to thank my many co-workers for their generous and invaluable contributions to this presentation.

REFERENCES

Al-Gailany, K. A. S., Bridges, J. W. and Netter, K. J. (1975). *Biochem. Pharmacol.*, **24**, 867
Angrist, B., Gershon, S., Sathananthan, G., Walker, R. W., Lopez-Ramoz, B., Mandel, L. R. and VandenHeuvel, W. J. A. (1976). *Psychopharmacology*, **47**, 29
Bailey, D. N. and Jatlow, P. I. (1976). *Clin. Chem.*, **22**, 777
Benedict, J. D., Forsham, P. H. and Stetten, D. (1949). *J. Biol. Chem.*, **181**, 183
Bidder, T. G., Mandel, L. R., Ahn, H.-S., VandenHeuvel, W. J. A. and Walker, R. W. (1974). *Lancet*, **i**, 165

Biggs, J. T., Holland, W. H., Chang, S., Hipps, P. P. and Sherman, W. R. (1976). *Pharm. Sci.*, **65**, 261
Björkhem, I., Blomstrand, R., Lantto, O., Lof, A. and Svensson, L. (1974). *Clin. Chim. Acta*, **56**, 241
Blake, M. I., Crespi, H. L. and Katz, J. J. (1975). *J. Pharm. Sci.*, **64**, 367
Braselton, W. E., Jr., Orr, J. C. and Engel, L. L. (1973). *Anal. Biochem.*, **53**, 64
Brent, D. A., de Miranda, P. and Schulten, H.-R. (1974). *J. Pharm. Sci.*, **63**, 1370
Canada, D. C. and Regnier, F. E. (1976). *J. Chromatog. Sci.*, **14**, 149
Ellsworth, R. L. (1974). Unpublished results
Ellsworth, R. L., Mertel, H. E. and VandenHeuvel, W. J. A. (1976). *J. Ag. Food Chem.*, **24**, 544
Franzen, F. and Gross, H. (1965). *Nature*, **206**, 1052
Fuller, R. W. (1976). *Life Sci.*, **19**, 625
Hamberg, M. and Samuelsson, B. (1969). *J. Am. Chem. Soc.*, **91**, 2177
Hamberg, M. and Samuelsson, B. (1972). *J. Biol. Chem.*, **247**, 3495
Hamberg, M. and Wilson, M. (1973). *Adv. Biosci.*, **9**, 39
Hsia, J. H., Tam, J. C. L., Giles, H. G., Leung, C. C., Marcus, H., Marshman, J. A. and LeBlanc, A. E. (1976). *Science, N.Y.*, **193**, 498
Hucker, H. B. and Stauffer, S. C. (1976). In preparation
Kahan, F. M. and Kropp, H. (1975). *Abstracts of 15th Inter-Science Conference on Antimicrobial Agents and Chemotherapy*, Washington, DC, Abstract No. 100
Kahan, F. M., Kropp, H., Onishi, H. R. and Jacobus, D. P. (1975). *ibid.*, Abstract No. 103
Kaplan, J., Mandel, L. R., Stillman, R., Walker, R. W., VandenHeuvel, W. J. A., Gillin, J. C. and Wyatt, R. J. (1974). *Psychopharmacology*, **38**, 239
Knapp, D. R., Gaffney, T. E. and McMahon, R. E. (1972). *Biochem. Pharmacol.*, **21**, 425
Knapp, D. R., Holcombe, N. H., Krueger, S. A. and Privitera, P. J. (1976). *Drug Metab. Disp.*, **4**, 164
Kollonitsch, J. and Barash, L. (1976). *J. Am. Chem. Soc.*, **98**, 5591
Kropp, H., Kahan, F. M. and Woodruff, H. B. (1975). *Abstracts of 15th Inter-Science Conference on Antimicrobial Agents and Chemotherapy*, Washington, DC, Abstract No. 101
Lipinski, J., Mandel, L. R., Ahn, H.-S., VandenHeuvel, W. J. A. and Walker, R. W. (1974). *Biol. Psychiat.*, **9**, 89
McCloskey, J. A., Stillwell, R. N. and Lawson, A. M. (1968). *Anal. Chem.*, **40**, 233
Mandel, L. R., Ahn, H.-S., VandenHeuvel, W. J. A. and Walker, R. W. (1972). *Biochem. Pharmacol.*, **21**, 1197
Mandel, L. R., Prasad, R., Lopez-Ramos, B. and Walker, R. W. (1976). *Res. Comm. Chem. Path. Pharmacol.* (in press)
Mandel, L. R., Rosegay, A., Walker, R. W., VandenHeuvel, W. J. A. and Rokach, J. (1974). *Science, N.Y.*, **186**, 741
Narasimhachari, N., Heller, B., Spaide, J., Haskovec, L., Fujimori, M., Tabushi, K., Himwich, H. E. (1971). *Biol. Psychiat.*, **3**, 9
Polan, M. L., McMurray, W. J., Lipsky, S. R. and Lande, S. (1970). *Biochem. Biophys. Res. Comm.*, **38**, 1127
Porter, C. C., Arison, B. H., Gruber, V. F., Titus, D. C. and VandenHeuvel, W. J. A. (1975). *Drug Metab. Disp.*, **3**, 189
Prox, A., Zimmer, A. and Machleidt, H. (1973). *Xenobiotica*, **3**, 103
Rosegay, A. (1973). Unpublished results
Rosegay, A. (1975). Unpublished results
Rosegay, A. and Taub, D. (1976). *Prostaglandins*, **12**, 785
Saavedra, J. M. and Axelrod, J. (1972). *Science, N.Y.*, **175**, 1365
Sorensen, L. B. (1960). *Scand. J. Clin. Invest.*, **12**, Suppl. 54
Steelman, S. L., Smit Sibinga, C. T., Schulz, P., VandenHeuvel, W. J. A. and Tempero, K. F. (1976). *XIII Intern. Conf. Int. Med., Helsinki*
Szara, S. (1956), *Experientia*, **12**, 441
Tanabe, M., Yasuda, D., LeValley, S. and Mitoma, C. (1969). *Life Sci.*, **8**, Part 1, 1123
Tennent, D. M., Kouba, R. F., Ray, W. H., VandenHeuvel, W. J. A. and Wolf, F. J. (1976). *Science, N.Y.*, **194**, 1059
Tocco, D. J., Duncan, A. E. W., deLuna, F. A., Hucker, H. B., Gruber, V. F. and VandenHeuvel, W. J. A. (1975). *Drug Metab. Disp.*, **3**, 361

Triggle, D. J. and Moran, J. F. (1966). *Nature,* **211**, 307
VandenHeuvel, W. J. A., Buhs, R. P., Carlin, J. R., Jacob, T. A., Koniuszy, F. R., Smith, J. L., Trenner, N. R., Walker, R. W., Wolf, D. E. and Wolf, F. J. (1972). *Anal. Chem.,* **44**, 14
VandenHeuvel, W. J. A., Carlin, J. R., Ellsworth, R. L., Wolf, F. J. and Walker, R. W. (1974). *Biomed. Mass Spectrom.,* **1**, 190
VandenHeuvel, W. J. A., Gruber, V. F. and Walker, R. W. (1973). *J. Chromatog.,* **87**, 341
VandenHeuvel, W. J. A., Gruber, V. F., Walker, R. W. and Wolf, F. J. (1977). *J. Chromatog., Biol. Appl.* (in press)
VandenHeuvel, W. J. A., Wolf, D. E., Trenner, N. R., Buhs, R. P., Ellsworth, R. L., Koniuszy, F. R., Carlin, J. R., Jacob, T. A., Walker, R. W. and Wolf, F. J. (1976). In preparation.
Walker, R. W., Ahn, H.-S., Albers-Schonberg, G., Mandel, L. R. and VandenHeuvel, W. J. A. (1973). *Biochem. Med.,* **8**, 105
Walker, R. W., Ahn, H.-S., Mandel, L. R. and VandenHeuvel, W. J. A. (1972). *Anal. Biochem.,* **47**, 227
Walker, R. W., VandenHeuvel, W. J. A., Wolf, F. J., Noll, R. N. and Duggan, D. E. (1976). *Anal. Biochem.* (in press)
Wolf, D. E., VandenHeuvel, W. J. A., Rosegay, A., Walker, R. W., Koniuszy, F. R., Jacob, T. A. and Wolf, F. J. (1974). *Fed. Proc.,* **33**, No. 3, March, Part 1
Wyatt, R. J., Mandel, L. R., Ahn, H.-S., Walker, R. W. and VandenHeuvel, W. J. A. (1973). *Psychopharmacology,* **31**, 265
Wyatt, R. J., Termini, B. and Davis, J. M. (1971). *Schizophrenia Bull.,* **4**, 10
Wyngaarden, J. B. and Stetten, D. (1953). *J. Biol. Chem.,* **203**, 9
Zimmer, A., Prox, A., Pelzer, H. and Hankwicz, R. (1973). *Biochem. Pharmacol.,* **22**, 2213

7
The use of deuterium labelling in mechanistic studies of metabolic activation

M. Jarman, P. B. Farmer, A. B. Foster and P. J. Cox (Chester Beatty Research Institute, Institute of Cancer Research, Royal Cancer Hospital, Fulham Road, London SW3 6JB, UK)

Stable isotopes have been used, in conjunction with mass spectrometry, in a variety of biomedical applications. Of these, the most widely exemplified has been the quantification of drugs and their metabolites in biological fluids (e.g. blood, urine), by mass spectrometry–stable-isotope dilution using isotopically labelled analogues as the reference standards. This technique has been extensively reviewed (Jenden and Cho, 1973; Horning *et al.*, 1973; Lawson and Draffan, 1975) and need not be detailed here except to note that, for reasons of cost and ease of synthesis, deuterium has been the most frequently employed isotope, although, in principle, any other stable isotope (e.g. ^{13}C, ^{15}N, ^{18}O, ^{34}S) can be used.

There are several other ways in which deuterium labelling has been used in studies of drug metabolism: (1) distinction between alternative pathways of metabolism, (2) deuterium isotope effects, (3) isotope cluster techniques, (4) comparative metabolism of enantiomers. In categories (3) and (4) stable isotopes other than deuterium could also be used.

The following review is intended to be illustrative, rather than comprehensive, with examples selected from the recent literature and from work in progress at the Chester Beatty Research Institute.

DISTINGUISHING ALTERNATIVE METABOLIC PATHWAYS

Frequently, the formation of a metabolite of a drug or other biologically active substance may be explained by more than one pathway. The most direct method of elucidating the operative pathway, namely isolation and characterisation of the intermediate metabolites, may be precluded if either (a) the intermediary metabolites are unstable or (b) they are rapidly metabolised, and therefore do not accumulate. For some drugs a study of the metabolism of specifically deuterated derivatives may solve the problem. The metabolism of the antitumour agent cyclophosphamide {2-[bis(2-chloroethyl)amino]-tetrahydro-2*H*-1,3,2-oxazaphosphorine-2-oxide} is illustrative.

Cyclophosphamide *per se* is not toxic to tumour cells but requires metabolic activation in the liver. *N,N*-Bis(2-chloroethyl)phosphorodiamidate (PAM; Figure 7.1) was shown to be the metabolite probably responsible for the tumour growth-inhibitory effect of the drug *in vivo* (Colvin, Padgett and Fenselau, 1973; Connors *et al.*, 1974a). Indirect evidence favoured a carbinolamine (4-hydroxycyclophosphamide; Figure 7.1, lower pathway) as an intermediary metabolite but initial formation of a lactol (upper pathway), also plausible on mechanistic grounds, merited consideration as an additional or alternative pathway to PAM. The carbinolamine was shown to be the preponderant, if not the exclusive, intermediate when monodeuterated acrolein (molecular weight, MW = 57) in preference to the dideuterated analogue (MW = 58) was found to be a by-product of the metabolism of cyclophosphamide which had been dideuterated at C-4 (Connors *et al.*, 1974b). Loss of deuterium should not have occurred if a 6-hydroxy derivative was involved. Subsequently, the relatively unstable 4-hydroxy-cyclophosphamide was synthesised (Takamizawa *et al.*, 1975) and compared with the corresponding metabolite, thereby validating the 4-hydroxylation pathway.

The same principle has been applied to the carcinogen dimethylnitrosamine (DMN), which requires metabolic activation in the liver yielding a product that methylates the nucleic acids therein. When DMN-d_6 is substituted for DMN (Lijinsky, Loo and Ross, 1968), the molecular weight, measured by mass spectrometry, of the principal methylation product (*N*-7-methylguanine)

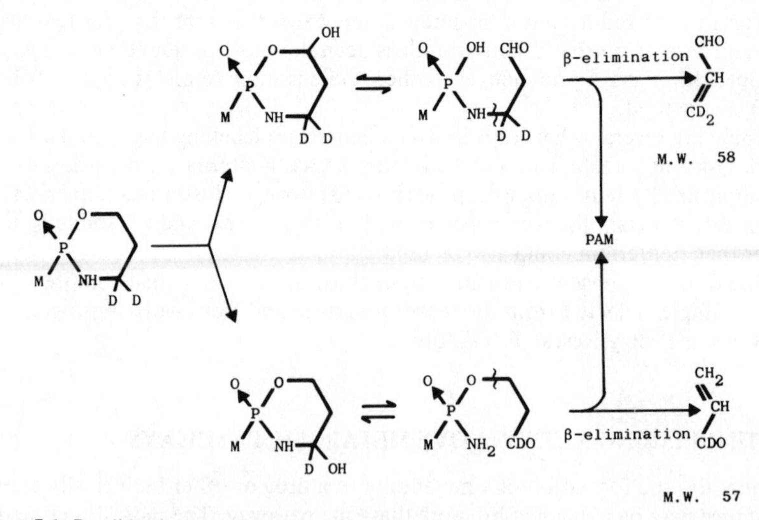

Figure 7.1 Possible alternative pathways for the metabolism of cyclophosphamide 4-d_2. M denotes the mustard ($-N(CH_2CH_2Cl)_2$) grouping and PAM phosphoramide mustard (*N,N*-bis(2-chloroethyl)-phosphorodiamidate)

isolated from the degraded nucleic acids was altered from 165 to 168, showing that the trideuteriomethyl group was transferred intact. This excluded an alternative mechanism involving alkylation via diazomethane (CH_2N_2 or CD_2N_2) which would have given 7-dideuteromethylguanine (MW = 167) from DMN-d_6.

A different principle has been used to distinguish alternative metabolic pathways associated with C-hydroxylation of benzene derivatives, namely the measurement of the magnitude of the NIH shift. A substituent (deuterium in the present examples) originally present at the site of hydroxylation may be partially transferred to an adjacent carbon atom (NIH shift) and the degree of deuterium retention is dependent on the nature of other substituents attached to the benzene ring (Daly, Jerina and Witkop, 1968). In general, where a substituent *para* to the site of hydroxylation possesses an ionisable hydrogen (e.g. –NH, –OH; Class 1 compounds), hydroxylation is accompanied by a small degree of retention (0–30 per cent) of the deuterium because re-aromatisation of the 2,5-cyclohexadienoid intermediate results in the loss of deuterium.

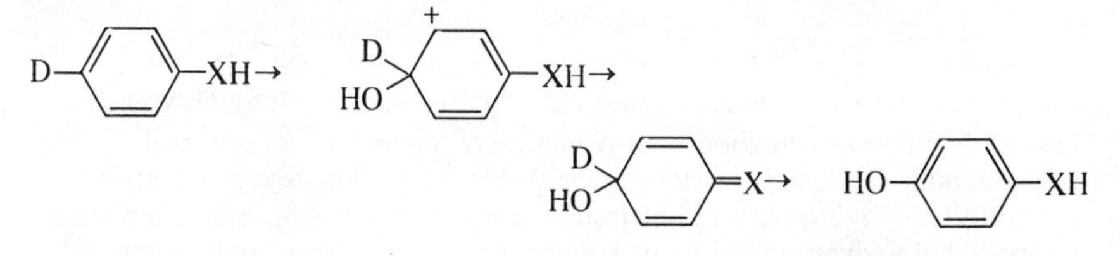

Substituents lacking an ionisable hydrogen (e.g. OCH_3, halogen, alkyl, carbonyl, dialkylamino; Class 2 compounds) are associated with much larger deuterium retentions (40–64 per cent) since the intermediate is 1,3-cyclohexadienoid, which rearranges with substantial migration of deuterium.

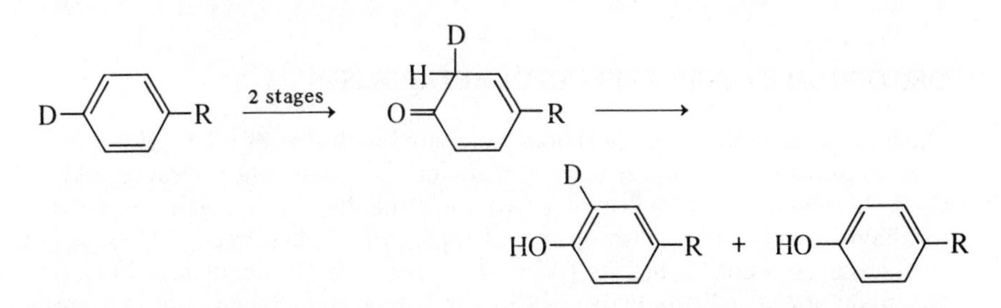

A major metabolite of the antitumour agent aniline mustard [*N,N*-bis(2-chloroethyl)aniline] in the rat is *N*-(2-chloroethyl)-4-hydroxyaniline, which might be formed by two alternative pathways (Farmer, Foster and Jarman, 1975) (Figure 7.2). Aniline mustard is a Class 2 compound, whereas *N*-(2-chloroethyl)aniline belongs to Class 1 and on microsomal *p*-hydroxylation of their 4-deuterated analogues, the extents of deuterium retention were 46 per cent and 11.5 per cent, respectively. Since deuterium retention in the overall *in vitro* conversion of *p*-deuterated aniline mustard into *N*-(2-chloroethyl)-4-hydroxyaniline was 40 per cent, this metabolite must have been formed via the intermediate *N,N*-di(2-chloroethyl)-4-hydroxyaniline, and not via *N*-(2-chloroethyl)aniline (Figure 7.2).

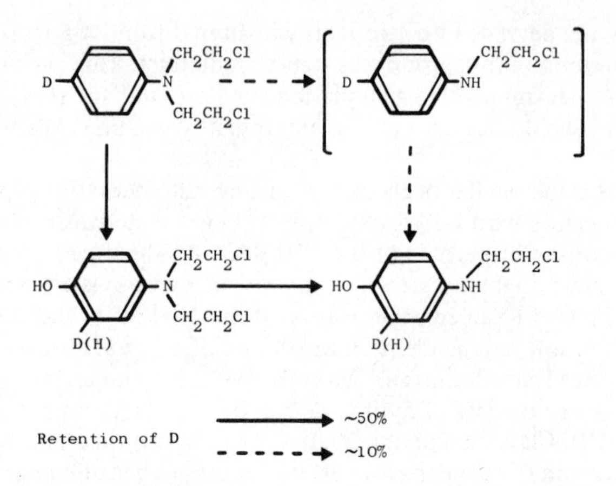

Figure 7.2 Alternative pathways for the metabolism of aniline mustard

Another example of the utilisation of the magnitude of the NIH shift to distinguish alternative metabolic pathways has been described by Daly *et al.* (1968). Microsomal metabolism of *N*-methyl-*N*-(phenyl-4-^2H) benzene-sulphonamide (a Class 2 compound) yielded both the *p*-hydroxy-derivative (*N*-methyl-*N*-4-hydroxyphenyl-benzenesulphonamide) and the *p*-hydroxylated N-desmethyl derivative (*N*-4-hydroxyphenyl-benzenesulphonamide) with deuterium retentions of 53 per cent and 2 per cent, respectively. Thus the end product must have been formed via the intermediate *N*-phenylbenzene-sulphonamide even though the latter was not detected as an intermediary metabolite.

DEUTERIUM ISOTOPE EFFECTS IN METABOLISM

The greater strength of the C–D bond compared with that of the C–H bond leads to isotope effects in the rates of metabolic processes where cleavage of a C–H/C–D bond is rate determining. Although the theoretical maximum value for the primary deuterium isotope effect k_H/k_D of 18 (Bigeleisen, 1949) has not been observed, values as high as 10 have been reported (Foster *et al.*, 1974). Metabolic *ortho*- and *para*-hydroxylations in the aromatic ring do not in general exhibit isotope effects, since they proceed by way of an epoxide intermediate, but there is growing evidence that *meta*-hydroxylation in particular can occur via direct cleavage of the C–H linkage. Thus chlorobenzene is metabolised to the *ortho*-, *meta*- and *para*-hydroxy derivatives. The synthetic 2,3- and 3,4-epoxides of chlorobenzene rearrange to *ortho*- and *para*-hydroxychlorobenzenes, respectively (Selander, Daly and Jerina, 1975). The *meta*-hydroxy derivative must therefore be formed metabolically by a pathway not involving an epoxide intermediate, e.g. direct insertion into the *meta* C–H bond of the active form of oxygen (oxene) generated by the cytochrome P450 system. Although the corroborating evidence for this mechanism, namely an isotope effect in the *meta*-hydroxylation of *meta*-deutero chlorobenzene, is lacking, significant isotope effects have been observed (Tomaszewski, Jerina and Daly, 1975) for metabolic *meta*-hydroxylation of methylphenylsulphone (k_H/k_D = 1.7), the corresponding sulphide, and nitrobenzene.

Deuterium isotope effects have been severally exemplified for metabolic transformations involving cleavage of aliphatic C–H linkages. A common pathway of drug metabolism is O-demethylation, and when this process is monitored on the basis of formaldehyde release and/or phenol formation, then parallel experiments must be performed for the methyl and trideuteromethyl derivatives. If mass spectrometry is the analytical technique, then a single experiment involving a mixture of two compounds is sufficient. An H/D isotope effect of ~2 has long been known (Mitoma *et al.*, 1967) for the O-demethylation of *p*-nitroanisole (Figure 7.3). An isotope effect of similar magnitude was observed (Foster *et al.*, 1974) for the microsomal O-demethylation of a mixture of *p*-di-(trideuteromethoxy)benzene and the corresponding dimethoxy derivative. However, for *p*-(trideuteromethoxy)anisole (Figure 7.3) the isotope effect was ~10, which means that the deuteromethyl group was virtually unaffected during metabolism. An apparently analogous situation involves the oxidation of 2,2′-bis(hydroxymethyl)biphenyl by Ag_2CO_3 on Celite (Kakis *et al.*, 1974). For a mixture of the bis(hydroxymethyl) and bis(hydroxy-dideuteromethyl) compounds an isotope effect of ~2 was observed but a value of 6–7 was obtained for the unsymmetrical compound 2-hydroxymethyl-2′-hydroxydideutero-methylbiphenyl (see Figure 7.3). Both the microsomal and Ag_2CO_3-mediated oxidations are heterogeneous, and it is tempting to draw an analogy between the binding of the substrate to the Ag_2CO_3 and the formation of a microsome–substrate complex, and to postulate that differences in binding efficiencies for the protium and deuterium analogues could account for the apparent anomalies. Indeed, Abdel-Monem (1975), in a discussion of the factors affecting the magnitude of isotope effects for

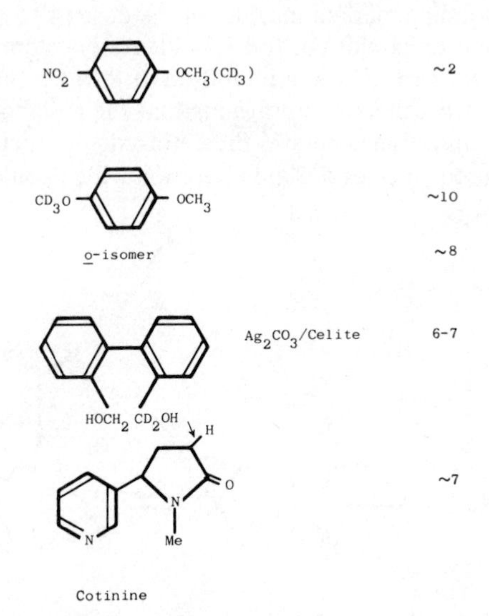

Figure 7.3 Isotope effects in microsomal and chemical oxidations

N-demethylation of *N,N*-dimethylamino derivatives, has pointed out that enzyme binding constants might be altered by deuteration. If the deuterated analogue is more strongly bound, this effect might counteract the isotope effect and lead to a low estimate of its value. The isotope effect for an $-N\begin{smallmatrix} CH_3 \\ CD_3 \end{smallmatrix}$ derivative may be larger than that for a mixture of $-N(CH_3)_2$ and $N(CD_3)_2$ derivatives.

Large isotope effects are not confined to the special situation in which the protons of one of a pair of symmetrically placed substituents are replaced by deuterium. Thus Dagne, Gruenke and Castagnoli (1974) have reported an isotope effect of 6–7 for the oxidative metabolism of the nicotine metabolite cotinine (Figure 7.3) to 3-hydroxycotinine consequent on 3,3-dideuteration.

If large isotope effects are associated with key metabolic transformations of a drug, then significant alterations in biological activity of the appropriately deuterated drug might be expected and a new means would be provided for probing the relationship of metabolism and biological activity. This approach is attractive because, unlike other substituents (e.g. halogen, methyl) which could be introduced to block metabolism at a particular site, the use of deuterium would not significantly change the partition coefficient or provide a new site for metabolism. We have used this approach in a systematic study (Connors *et al.*, 1974b; Cox *et al.*, 1976a) of the isotope effects operative for the various metabolic pathways for cyclophosphamide, and, in particular, for the 5,5-dideuterated analogues (Figure 7.4). A large isotope effect (~5.3) was observed (Cox *et al.*, 1976a) for the conversion of this analogue via the intermediate aldehyde derivative (see also Figure 7.1) into phosphoramide mustard and acrolein, the rate of formation of which was substantially reduced. There was a corresponding marked increase in the dose (47.5 mg/kg; cf. 3.6 mg/kg for cyclophosphamide) of the 5,5-dideuterocyclophosphamide required to effect 90 per cent growth inhibition of the experimental mouse tumour ADJ/PC6 *in vivo*, thereby confirming that the metabolic step which is retarded by 5,5-dideuteration generates the cytotoxic products.

An additional consequence of 5,5-dideuteration which could have

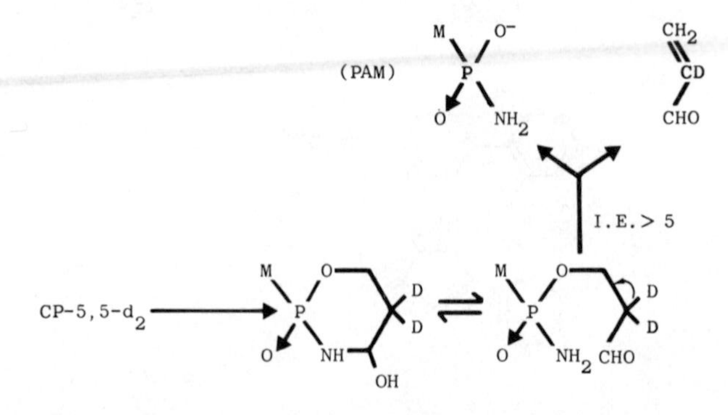

Figure 7.4 Conversion of cyclophosphamide-5-d_2 (CP-5,5-d_2) into phosphoramide mustard (PAM) and acrolein-d. IE denotes isotope effect (k_H/k_D); M = $-N(CH_2CH_2Cl)_2$

contributed to the observed reduction in drug potency is diversion of the metabolism of the intermediates 4-hydroxycyclophosphamide and aldophosphamide along detoxification pathways, leading to the relatively non-toxic metabolites (Struck *et al.*, 1971) 4-ketocyclophosphamide and carboxyphosphamide, respectively.

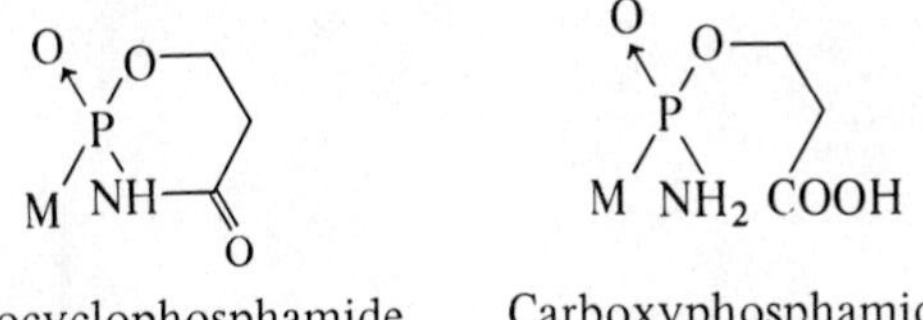

4-Ketocyclophosphamide Carboxyphosphamide

The change in metabolism pattern consequent on 5,5-dideuteration of cyclophosphamide is an example of metabolic switching. This term was introduced by Horning *et al.* as a result of study of the metabolism of deuterated derivatives of the anticonvulsive drug antipyrine. This work is described in detail elsewhere in this volume (pp. 55–64).

In a study of the anticancer drug CCNU (1-(2-chloroethyl)-3-cyclohexyl-1-nitrosourea) at present in progress in our laboratories, we are exploiting metabolic switching in relation to mode of action and tissue distribution. Metabolism of CCNU (Figure 7.5) affords various hydroxylated derivatives

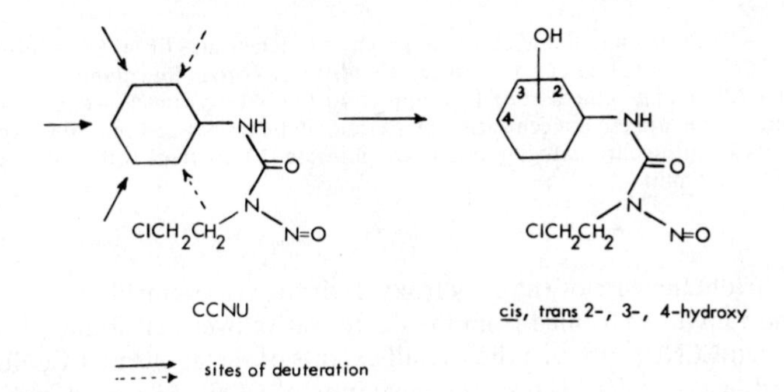

Figure 7.5 Metabolism of CCNU (1-(2-chloroethyl)-3-cyclohexyl-1-nitrosourea)

(May, Bosse and Read, 1974), and although the *in vivo* antitumour activities of some of these metabolites are comparable with that of the parent drug (Johnson, McCaleb and Montgomery, 1975), it is possible that tissue distribution, especially entry into the brain, might be adversely affected by metabolism. The contribution of these metabolites to the total activity of CCNU *in vivo* can be assessed by observing the changes (in either magnitude or type) of tumour inhibitory activity resulting from selective deuterium substitutions designed to produce metabolic switching. Thus selectively 2(6)-tetradeuterated (d_4) and 3,4(5)-hexadeuterated (d_6) analogues have been synthesised with the object of diverting metabolism either towards the formation of 3- and 4-hydroxy-CCNU or towards the 2-hydroxy derivative (Figure 7.5). Preliminary results on the metabolism by rat liver microsomes of CCNU alone (Figure 7.6a) and in admixture with the d_6 analogue (Figure 7.6b) revealed

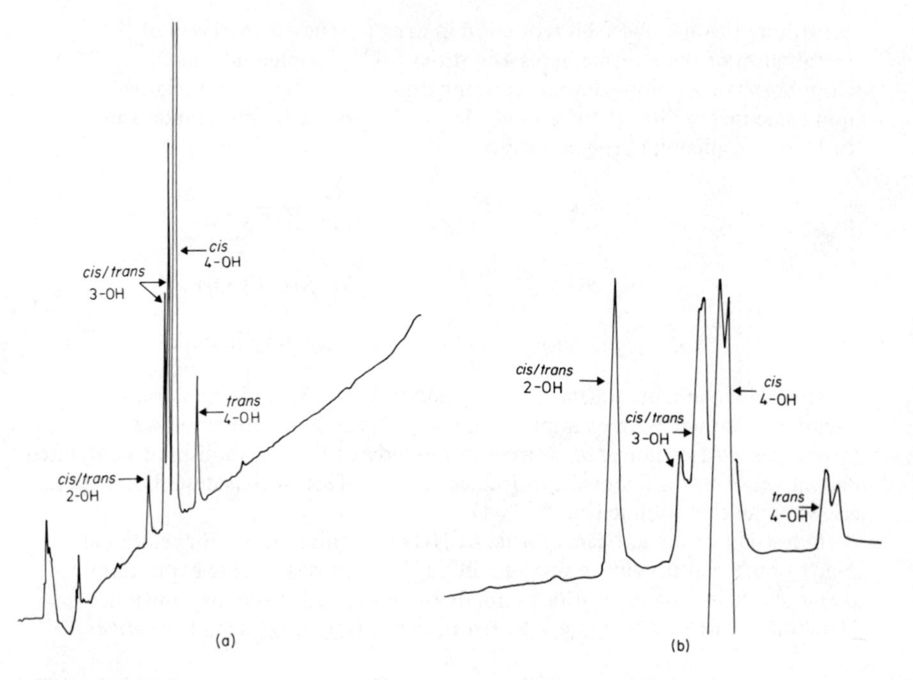

Figure 7.6 High-pressure liquid chromatography of metabolites of (a) CCNU, (b) CCNU + CCNU-3,4 (5)-d_6 (1 : 1 mixture). Conditions: Zorbax Sil column, 0.25 m x 7.9 mm i.d., eluted at *ca* 1 ml/min at 40° with dioxan:cyclohexane, initially 1 : 9 and increasing in dioxan concentration at 2%/min; detection at 254 nm. In (a) the trace represents an entire chromatographic run (50 min); in (b) the trace is that of an expanded region (11–22 min)

a *ca* fourfold increase of the 2-hydroxy derivative (presumably a *cis–trans* mixture) formed from the d_6 derivative over that formed from unlabelled CCNU, although the overall extents of metabolism of CCNU and CCNU-d_6 were similar. During the separation of CCNU (d_0 and d_6 species) and the corresponding hydroxy derivatives by HPLC, there was partial resolution of some of the hydroxy-CCNUs from their deuterated counterpart (Figure 7.6b). Resolution of perdeuterated and non-deuterated carboxylic acids has been reported previously (Tanaka and Thornton, 1976).

THE ISOTOPE CLUSTER TECHNIQUE

A mixture, preferably equimolar, of an unlabelled drug and a variant labelled with a stable isotope (and also any metabolites derived from this mixture and still retaining the stable isotope) will give a characteristic mass spectrum containing pairs or clusters of signals for the molecular ion or fragments in which the stable isotope is retained. These conspicuous clusters may help to locate the drug and/or its metabolites in the presence of other components in extracts of biological media (blood, urine). This technique was first described by Knapp,

Gaffney and McMahon (1972a) and Knapp *et al.* (1972b) and used to locate and identify a metabolite of nortriptyline in rat urine, following administration of the drug admixed with an N-CD_3 analogue. The usefulness of the isotope cluster method for the detection of nortriptyline metabolites in human urine and bile (Knapp *et al.*, 1972b) has also been evaluated. Although this method has limitations, it does obviate the need to administer radioactively labelled drugs as an aid to the detection of metabolites, which is clearly advantageous from an ethical viewpoint for studies of drug metabolism in man. In choosing deuterium as the stable isotope in this type of study, due consideration should be given to possible loss of the isotope through metabolism and to distortion of the metabolic profile through isotope effects.

DIFFERENTIAL METABOLISM OF ENANTIOMERS

When a drug molecule possesses one or more chiral centres, the pharmacological activity of the enantiomers may vary, and for some drugs it may be largely or exclusively associated with one enantiomer. Thus the activity of the psychotomimetic amine 1-(2,5-dimethoxy-4-methylphenyl)-2-aminopropane is associated with the *R*-enantiomer (McGraw, Calieri and Castagnoli, in press). One explanation for such differences may be differential metabolism and/or tissue distribution for the enantiomers or their metabolites. An elegant method of studying comparative metabolism of enantiomers is to administer a racemate in which one enantiomer is labelled with a stable isotope. The enantiomer ratios for recovered drug and metabolites can then be readily determined by mass spectrometry. Using this approach and with deuterium as the label, the differential metabolism of the enantiomers of the above-mentioned psychotomimetic amine (McGraw *et al.*, in press; Gal, Gruenke and Castagnoli, 1975) and of (–)- and (+)-propanolol (Ehrsson, 1976) has been studied. It is important to note that if deuterium is used as the stable isotope, then some knowledge of the metabolism of the compound under study is desirable, since the label should be located at a position in the molecule which is not involved in metabolism.

We have utilised deuterium labelling to study the differential metabolism of the enantiomers of cyclophosphamide, which is chiral by virtue of the asymmetric phosphorus atom. Metabolism of cyclophosphamide is stereoselective in man. Following conventional administration of the racemic (±) drug, the (–)-form accounts for up to 90 per cent of cyclophosphamide recovered from the urine of patients (Cox *et al.*, 1976b). This value was first determined by recovering the drug from *ca* 1 litre of urine and comparing its specific optical rotation with those (Kinas, Pankiewicz and Stec, 1975) for the synthetic enantiomers. Because of the low rotation (±2.3°, methanol) of the enantiomers of cyclophosphamide, determination of optical purity on the basis of $[\alpha]_D$ values requires relatively large amounts of material and the method is not suitable for small experimental animals (rats, mice) or for extensive studies in man. We have therefore utilised pseudoracemates of cyclophosphamide in which one enantiomer contained four deuteriums in the mustard group, i.e. $-N(CD_2CH_2Cl)_2$.

The relative proportions of the enantiomers of cyclophosphamide and two of its metabolites from the urine of mice following administration of the racemates (+)-d_0/(−)-d_4 and (+)-d_4/(−)-d_0 is shown in Figure 7.7. The metabolites were methylated prior to examination by mass spectrometry. Marked stereoselectivity was observed for the formation of 4-ketocyclophosphamide, the (+)-isomer preponderating.

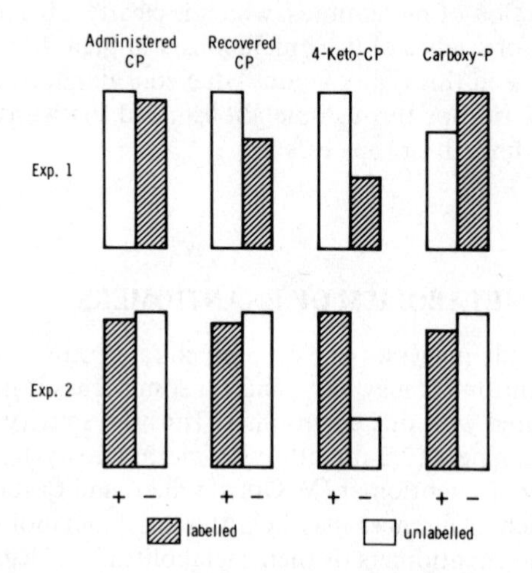

Figure 7.7 Metabolism of cyclophosphamide pseudoracemates in mice

CONCLUSION

It is now clear that there is great scope for stable isotope labelling in general and deuterium labelling in particular in studies of drug metabolism.

REFERENCES

Abdel-Monem, M. M. (1975). *J. Med. Chem.*, **18**, 427

Bigeleisen, J. (1949). *Science, N.Y.* **110**, 14

Colvin, M., Padgett, C. A. and Fenselau, C. (1973). *Cancer Res.*, **33**, 915

Connors, T. A., Cox, P. J., Farmer, P. B., Foster, A. B. and Jarman, M. (1974a). *Biochem. Pharmacol.*, **23**, 115

Connors, T. A., Cox, P. J., Farmer, P. B., Foster, A. B., Jarman, M. and MacLeod, J. K. (1974b). *Biomed. Mass Spectrom.*, **1**, 130

Cox, P. J., Farmer, P. B., Foster, A. B., Gilbey, E. D. and Jarman, M. (1976a). *Cancer Treatment Rep.*, **60**, 483

Cox, P. J., Farmer, P. B., Jarman, M., Jones, M., Stec, W. J. and Kinas, R. (1976b). *Biochem. Pharmacol.*, **25**, 993

Dagne, E., Gruenke, L. and Castagnoli, N. (1974). *J. Med. Chem.*, **17**, 1330

Daly, J., Jerina, D. and Witkop, B. (1968). *Arch. Biochem. Biophys.*, **128**, 517

Ehrsson, H. (1976). *J. Pharm. Pharmacol.*, **28**, 662

Farmer, P. B., Foster, A. B. and Jarman, M. (1975). *Biomed. Mass Spectrom.*, **2**, 107

Foster, A. B., Jarman, M., Stevens, J. D., Thomas, P. and Westwood, J. H. (1974). *Chem.-Biol. Interactions*, **9**, 327

Gal, J., Gruenke, L. D. and Castagnoli, N. (1975). *J. Med. Chem.,* **18**, 683
Horning, M. G., Nowlin, J., Lertratanangkoon, K., Stillwell, R. N., Stillwell, W. G. and Hill, R. M. (1973). *Clin. Chem.,* **19**, 845
Jenden, D. J. and Cho, A. K. (1973). *Ann. Rev. Pharmacol.,* **13**, 371
Johnston, T. P., McCaleb, G. S. and Montgomery, J. A. (1975). *J. Med. Chem.,* **18**, 634
Kakis, F. J., Fetizon, M., Douchkine, N., Golfier, M., Mourgues, P. and Prange, T. (1974). *J. Org. Chem.,* **39**, 523
Kinas, R., Pankiewicz, K. and Stec, W. J. (1975). *Bull. Acad. Pol. Sci. Ser. Sci. Chim.,* **23**, 981
Knapp, D. R., Gaffney, T. E. and McMahon, R. E. (1972a). *Biochem. Pharmacol.,* **21**, 425
Knapp, D. R., Gaffney, T. E., McMahon, R. E. and Kiplinger, G. (1972b). *J. Pharmacol. Exp. Ther.,* **180**, 784
Lawson, A. M. and Draffan, G. H. (1975). *Prog. Med. Chem.,* **12**, 1
Lijinsky, W., Loo, J. and Ross, A. E. (1968). *Nature,* **218**, 1174
McGraw, N. P., Caliery, P. S. and Castagnoli, N. (1976). *J. Med. Chem.* (in press)
May, H. E., Boose, R. and Reed, D. J. (1974). *Biochem. Biophys. Res. Commun.,* **57**, 426
Mitoma, C., Yasuda, D. M., Tagg, J. and Tanabe, M. (1967). *Biochim. Biophys. Acta,* **136**, 566
Selander, H. G., Daly, J. and Jerina, D. (1975). *Arch. Biochem. Biophys.,* **168**, 309
Struck, R. F., Kirk, M. C., Mellett, L. B., El Dareer, S. and Hill, D. L. (1971). *Molec. Pharmacol.,* 7, 519
Tanaka, N. and Thornton, E. R. (1976). *J. Am. Chem. Soc.,* **98**, 1617
Takamizawa, A., Matsumoto, S., Iwata, T., Tochino, Y., Katagiri, K., Yamaguchi, K. and Shiratori, O. (1975). *J. Med. Chem.,* **18**, 376
Tomaszewski, J. E., Jerina, D. M. and Daly, J. W. (1975). *Biochemistry,* **14**, 2024.

8
The quantitation by gas chromatography-chemical ionisation-mass spectrometry of cyclophosphamide, phosphoramide mustard and nornitrogen mustard in the plasma and urine of patients receiving cyclophosphamide therapy

I. Jardine, C. Fenselau, M. Appler, M.-N. Kan, R. B. Brundrett and M. Colvin (Departments of Pharmacology and Experimental Therapeutics and of Medicine, The Johns Hopkins University School of Medicine, Baltimore, Maryland 21205, USA)

Cyclophosphamide, the widely used antitumour drug, is not itself toxic, but must be converted by microsomal metabolism to active metabolites. Despite an early understanding of the requirement for metabolic activation, the drug was used clinically for more than 10 years before the structures of any of its hepatic metabolites were elucidated. When the metabolites were finally isolated and identified, their characterisation rested heavily on mass spectral analysis (Bakke, Feil and Zaylskie, 1971; Colvin, Padgett and Fenselau, 1973; Struck *et al.*, 1971). Subsequently mass spectrometry has been used in conjunction with stable-isotope labels to study the mechanism of alkylation of cytotoxic metabolites (Colvin *et al.*, 1976) and for experiments with metabolic switching (Jarman *et al.*, this volume pp. 85–95). Methodology has also been reported for assaying these cytotoxic metabolites using stable-isotope-labelled internal standards (Jardine *et al.*, 1975, 1976; Jarman *et al.*, 1975). In this report we will review the application of this methodology to assays of cyclophosphamide, phosphoramide mustard and nornitrogen mustard in blood and urine of patients, as well as in decomposition and protein-binding studies.

In Figure 8.1 is shown a summary of the important features of the human

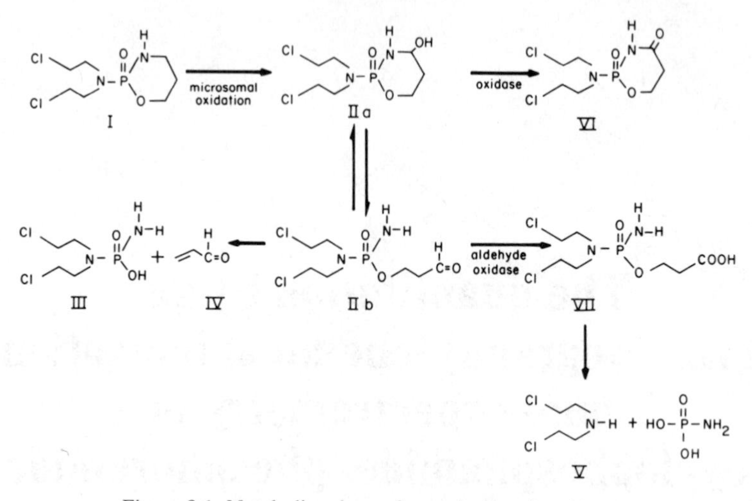

Figure 8.1 Metabolic scheme for cyclophosphamide

metabolism of cyclophosphamide (I). 4-Ketocyclophosphamide (VI) and carboxycyclophosphamide (VII) are major but inactive human metabolites. The metabolites in the active pathway include the tautomeric pair 4-hydroxy-phosphamide (IIa) and aldophosphamide (IIb) and also the active alkylating agent phosphoramide mustard (III). The parent compound itself is easily extracted into chloroform, along with small amounts of 4-ketocyclophosphamide and 4-hydroxycyclophosphamide. This has provided the basis for assays of cyclophosphamide itself for a number of years, either utilising radioactive isotope labels or a somewhat perfidious colour reaction. More recently, workers in England (Jarman *et al.*, 1975) have employed a stable-isotope-labelled internal standard to quantitate cyclophosphamide extracted in chloroform, introducing the sample into the mass spectrometer via the direct probe. The metabolites in the activation pathway, 4-hydroxycyclophosphamide and phosphoramide mustard, are much more difficult to assay. They occur at lower levels, are polar compounds, harder to purify and are thought to decompose easily. In short, these are good candidates for a selected ion monitoring assay using gas chromatography–mass spectrometry (GC–MS) and stable-isotope-labelled internal standards.

We designed the synthesis of our stable-isotope-labelled analogue with two tasks in mind: we wanted to use it as an internal standard for quantification and also to define the mechanism of the alkylation reactions of phosphoramide mustard and nornitrogen mustard as proceeding through a cyclic aziridinium ion or not. This work has been finished, and both metabolites were found to undergo alkylation through a symmetrical cyclic intermediate (Colvin *et al.*, 1976). Several principles dictated the synthetic route to be used. The label should not be lost by chemical exchange. The label should be specifically introduced and with high incorporation. In addition, we wanted to incorporate four or more atomic mass units so that the internal standard could be more easily distinguished from the compound being assayed, and, lastly, we chose to synthesise a labelled building block which could be incorporated into all of the metabolites of cyclophosphamide and into a number of other antitumour drugs as well. Figure 8.2 diagrams the synthetic approach used. Thus diethyliminodiacetate was reduced with lithium aluminium deuteride and the resulting tetradeuterated diol was converted with thionyl chloride to

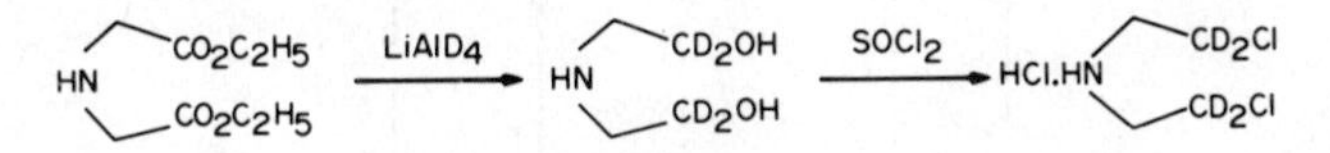

Figure 8.2 Synthesis of d_4-nornitrogen mustard

tetradeuterated nornitrogen mustard. Nornitrogen mustard is not only a metabolite of cyclophosphamide, it is also a common structural unit in all the other cyclophosphamide metabolites shown in Figure 8.1. Tetradeutero-nornitrogen mustard was elaborated by literature procedures to cyclophosphamide and to tetradeuterated phosphoramide mustard (Colvin *et al.*, 1976; Jardine *et al.*, 1976).

For quantitation of phosphoramide mustard we used the extraction procedure we had worked out earlier in seeking to confirm that the metabolite did indeed circulate in patient blood (Fenselau *et al.*, 1975). A 5–10 ml serum sample is percolated through a column of XAD-4 and recovered in methanol. The methanol is evaporated, and the residue is treated with diazomethane. This converts phosphoramide mustard to three derivatives. Although the formation of three derivatives has the disadvantage of lowering sensitivity somewhat, it has the advantage of providing a selection of three compounds to assay, with three different retention times. Thus we can choose that derivative to assay which can be detected with the least interference. The derivatised extract was introduced into our DuPont 491 mass spectrometer via a gas chromatograph interfaced through a jet separator. Chemical ionisation was used, and in fact we have found that this is necessary in order to obtain adequate selectivity in the serum extracts (Fenselau *et al.*, 1975). The use of chemical ionisation allows us to monitor protonated molecular ions at the highest mass possible. Fragmentation is reduced in contaminating components of the mixture as well as in the sample. For both these reasons, the possibility of interference by ions formed from endogenous material is decreased. A 6 ft glass chromatograph column was used, packed with 3 per cent OV-17 coated on Supelcoport. The temperature was held isothermally at 220°.

The multiplicity of peaks in the protonated molecular ion regions of phosphoramide mustard and d_4-phosphoramide mustard, shown in Figure 8.3, reflects the presence of two chlorine atoms in each molecule. That is, three molecular ions are detected for each compound. This has some interesting implications for selected ion monitoring. For example, we can monitor two molecular ions from the drug metabolite. In this case the intramolecular intensity ratios can be used to confirm the identity of the compound and the absence of interference. The intermolecular intensity ratios can be calculated in several combinations in order to obtain several values for the assay. Figure 8.4 shows standard curves obtained for nornitrogen mustard chromatographed as the trifluoroacetate derivative. Here two ratios have been calculated based on two different isotopic molecular species, and it can be seen that these curves coincide

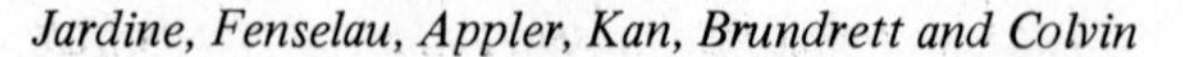

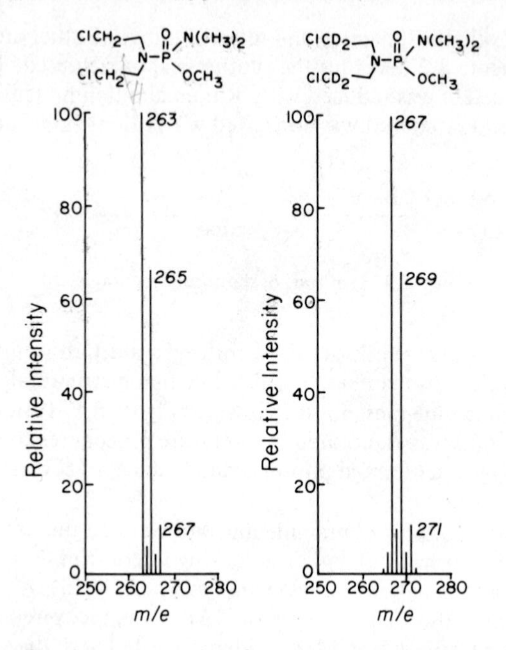

Figure 8.3 Protonated molecular ion regions from spectra of trimethyl derivatives of phosphoramide mustard and d_4-phosphoramide mustard

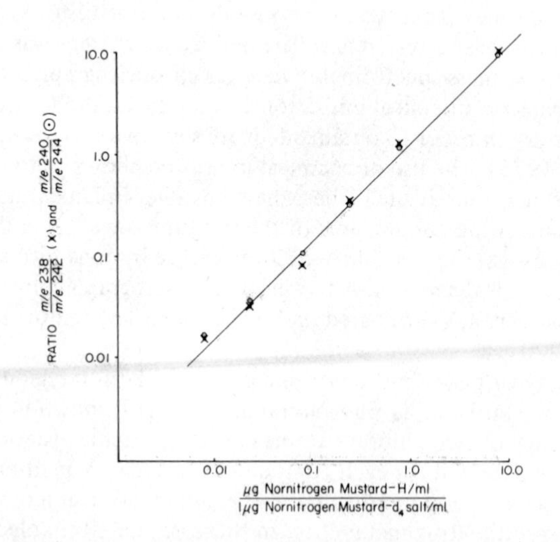

Figure 8.4 Calibration curves for nornitrogen mustard (trifluoroacetate)

quite well. The sensitivity range shown in this figure falls between 10 ng and 10 μg/ml of serum of nornitrogen mustard, to which 1 μg/ml of d_4-nornitrogen mustard has been added.

Figure 8.5 (a) illustrates the selected ion profiles of protonated molecular ions of a sample of phosphoramide mustard analysed under the GC–MS

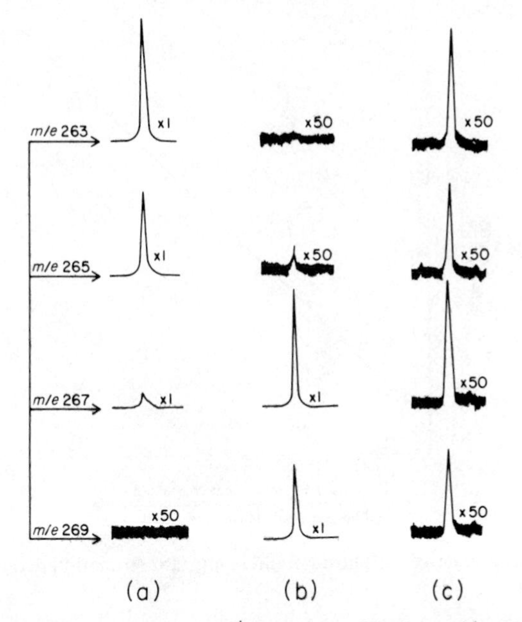

Figure 8.5 Selected ion profiles of $[M + H]^+$ ions in (a) phosphoramide mustard, (b) d_4-phosphoramide mustard and (c) a 1 : 1 mixture (trimethyl derivatives)

conditions described above. Signals are detected at *m/e* 263 and 265 in the expected 3 : 2 intensity ratio, while little or no signal is detectable at *m/e* 269, even when the amplification is set 50 times higher. The second set of profiles (b) were obtained by monitoring the same four masses when a sample of d_4-phosphoramide mustard was injected into the instrument. We see little or no signal at *m/e* 263, while the signals detected at *m/e* 267 and 269 occur in the expected intensity ratio of 3 : 2. The third set of profiles (c) in this figure are obtained from a 1 : 1 mixture of d_0- and d_4-phosphoramide mustard (trimethyl derivatives).

Figure 8.6 presents one of the standard curves measured for phosphoramide mustard. In this case samples ranging between 20 ng and 7 μg have been added to 1 ml of serum along with 1 μg of the deuterated internal standard, extracted, derivatised and assayed. The standard curve is calculated on the basis of the ratio of the ion profiles recorded at *m/e* 263 and *m/e* 267. The relationship between sample size and signal size suggests that this method does have accuracy suitable for the clinical assays intended. We found that the best accuracy is obtained if we work within a 1 : 10 and 10 : 1 ratio of internal standard to metabolite, and in the actual clinical studies we repeated assays if we found that the internal standard added did not fall within this range. Thus for each clinical measurement betweeen five and eight injections were usually required to obtain 3–5 valid ratios. Although the limit for simple detection was lower, the lower limit for quantitation in our work appeared to be around 30 ng added to 1 ml of plasma. Because we chose to monitor the trimethyl derivative of phosphoramide mustard, that having the gas chromatographic retention time which offered the least interference, we are really dealing with about 10 per cent of the sample or about 3 ng/ml.

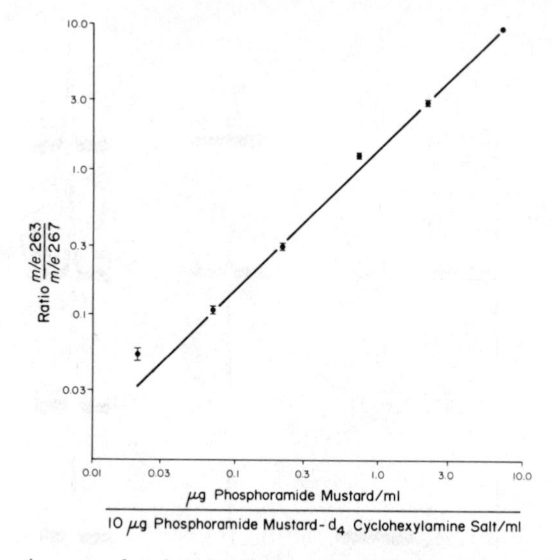

Figure 8.6 Calibration curve for phosphoramide mustard (trimethyl derivative); 1 ml plasma (20 ng–7 μg/ml)

It was necessary to determine the extent of decomposition in blood or urine processed under our extraction, derivatisation and assay conditions. The results of these experiments are shown in Table 8.1. A known amount of each of the metabolites listed was added to the physiological fluid along with a known amount of the labelled analogue we expected to be formed by decomposition. At the end of the work-up, the ratio between labelled and unlabelled decomposition product was measured. Thus we are measuring production directly and not disappearance of the precursor. The results in Table 8.1 show extensive decomposition of most of the metabolites to nornitrogen mustard. This suggests that a large component of the nornitrogen mustard values we measured in the patient samples was contributed by decomposition of other metabolites. Nonetheless the nornitrogen mustard levels that were measured have a certain consistency, as we will see, and may at least reflect the rise and

Table 8.1

Decomposition to nornitrogen mustard:	Plasma (%)	Urine (%)
Cyclophosphamide	11	7
4-Ketocyclophosphamide	19	4
Carboxyphosphamide	58	13
Phosphoramide mustard	11	18
Decomposition to phosphoramide mustard:	Plasma (%)	Urine (%)
Cyclophosphamide	0	0
4-Ketocyclophosphamide	0	0
Carboxyphosphamide	<1	<1

fall of blood levels of carboxycyclophosphamide. It also seems likely that nornitrogen mustard is formed *in vivo* by decomposition of other metabolites in both blood and urine. A second important point to be made about the measurements presented in Table 8.1 is that there is no decomposition of cyclophosphamide, 4-ketocyclophosphamide or carboxycyclophosphamide to phosphoramide mustard. This is consistent with the proposed scheme of metabolic production of phosphoramide mustard from aldophosphamide, and also validates the blood and urine levels subsequently measured of phosphoramide mustard. The third point to be made about Table 8.1 is that these studies confirm the need for an internal standard in assaying this family of metabolites. Extensive decomposition, especially of carboxyphosphamide, should make values measured by radioactivity or spectrophotometric assays low.

Protein binding of cyclophosphamide, phosphoramide mustard and nornitrogen mustard was also examined under conditions approximating those of our sample work-up procedure. Thus blood was collected from a patient 2 h after the start of an intravenous infusion of cyclophosphamide, 75 mg/kg. Levels of the three compounds were measured before and after filtration of plasma through an Amicon UM-10 membrane at 4°C. The values in Table 8.2 suggest that there is a fair amount of binding of all three of these compounds by protein

Table 8.2 Protein binding study

	Concentration (nmol/ml)		
	Plasma	Ultrafiltrate	Protein binding (%)
Cyclophosphamide	194	148	24
Phosphoramide mustard	33	20	39
Nornitrogen mustard	43	14	67

in the blood. This documents yet another problem encountered in attempting to assay total amounts of metabolites of cyclophosphamide, that can be overcome by the use of stable-isotope-labelled internal standards. It has been repeatedly established in the development of radioimmune assays that isotope-labelled internal standards equilibrate with drugs reversibly bound to protein in plasma within a few minutes of addition. Thus if we are using an isotope-labelled internal standard, we do not have to denature protein in our samples.

Table 8.3 summarises the amounts of the three compounds excreted through

Table 8.3 24 Hour urinary excretion of cyclophosphamide and metabolites

	Patient No. 2		Patient No. 3		Patient No. 4	
	mmol	(% of Dose)	mmol	(% of Dose)	mmol	(% of Dose)
Cyclophosphamide	2.8	(19.5)	5.4	(25.1)	2.0	(16.3)
Phosphoramide mustard	0.4	(2.8)	0.3	(1.2)	0.2	(1.8)
Nornitrogen mustard	1.4	(10.4)	2.4	(11.0)	1.8	(14.5)
Total	4.6	(32.7)	8.1	(37.3)	4.0	(32.6)

24 h by three patients who had received 60 or 75 mg/kg (4–5 g) of cyclophosphamide in 1 h intravenous infusions (Jardine *et al.*, 1977). These patients received furosimide concomitantly as a diuretic. The percentage of excretion is similar in the three patients and the total amount excreted is in line with earlier studies (Bagley, Botick and DeVita, 1973; Cohen, Jao and Jusko, 1971). Of the greatest interest here is the excretion of low but real amounts of phosphoramide mustard.

Figure 8.7 presents the blood levels in one of the patients of the three compounds as a function of time (Jardine *et al.*, 1977). 'Hour 1' is the end of the 1 h infusion of 60 mg/kg cyclophosphamide. This long infusion complicates the usual pharmacokinetic calculations, particularly that for the volume of

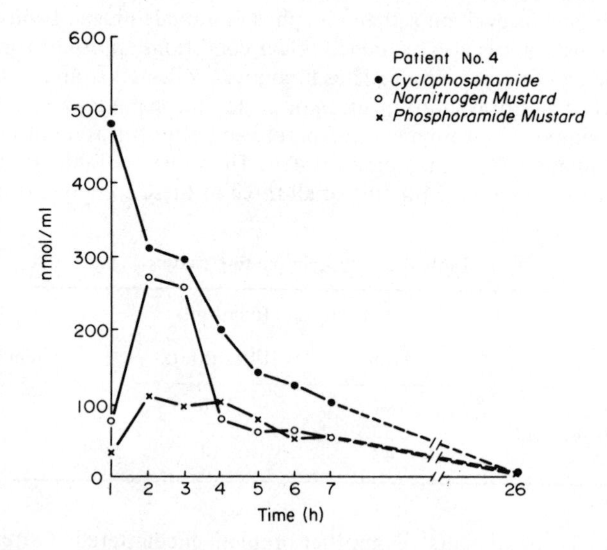

Figure 8.7 Plasma levels (patient No. 4) of phosphoramide mustard, cyclophosphamide and nornitrogen mustard as a function of time

distribution, but the time profile presented is still of interest. An hour 1 concentration of cyclophosphamide was measured at around 500 nmol/ml. The levels of cyclophosphamide fall to detectable but not quantifiable values at 24 h. Cyclophosphamide was not detectable in this patient's blood at 48 and 69 h. The nornitrogen mustard level peaked around 2 h at approximately 300 nmol/ml and fell rapidly. Phosphoramide mustard was measured at an initial value of about 50 nmol/ml. It reached a maximum level around 100 nmol/ml, which was sustained through 2–3 h, and it fell slowly to about 10 nmol/ml at 24 h.

Figure 8.8 presents the analogous measurements made for a second patient. Here the fine structure is a little different, but the levels and duration are generally similar. In particular, the initial concentration of phosphoramide mustard was found to be around 30 nmol/ml. Phosphoramide mustard reached a maximum concentration of approximately 50 nmol/ml, sustained from hour 3 through hour 5, and fell to about 10 nmol/ml at 24 h.

Figure 8.9 contains blood levels for the third patient. Here cyclophosphamide

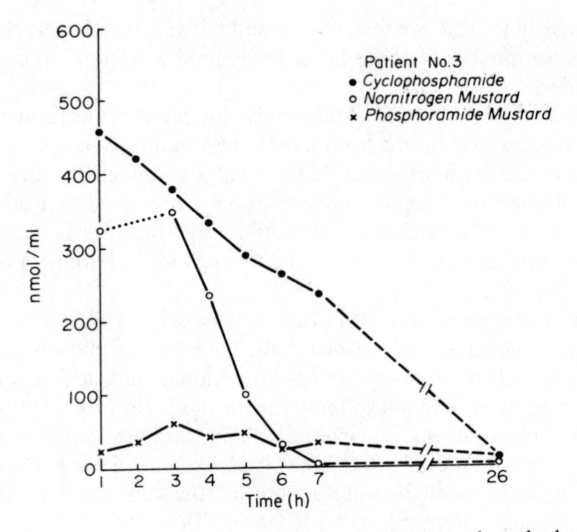

Figure 8.8 Plasma levels (patient No. 3) of phosphoramide mustard, cyclophosphamide and nornitrogen mustard as a function of time

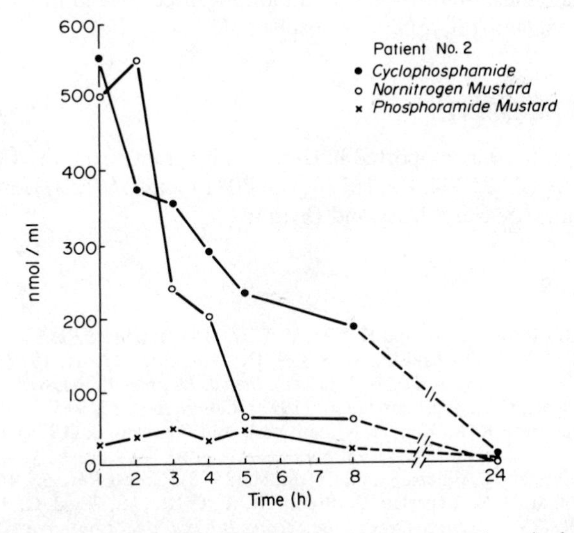

Figure 8.9 Plasma levels (patient No. 2) of phosphoramide mustard, cyclophosphamide and nornitrogen mustard as a function of time

was found to have an initial concentration of about 550 nmol/ml. This value is in general agreement with those of Bagley *et al.* (1973), who measured chloroform-extractable radioactivity, and with those of Jarman *et al.* (1975) when dosage differences in the latter work are taken into account. The plasma half-lives for cyclophosphamide in the three patients studied here are 7, 4.4 and 3.4 h. A similar range, 3–7 h, has been reported by others (Bagley *et al.*, 1973; Cohen *et al.*, 1971; Jarman *et al.*, 1975). The rise and fall of nornitrogen mustard primarily reflects, we feel, the rise and fall of carboxycyclophosphamide, because the decomposition of the latter compound was found to occur so readily in blood.

Returning to the focal point of this study, the initial concentration of phosphoramide mustard in this third patient was found to be approximately 25 nmol/ml. A maximum of about 50 nmol/ml is sustained for several hours, and the blood levels of phosphoramide mustard are reduced to around 10 nmol/ml at 24 h. These maximum levels are significant drug concentrations by most standards, and they are sustained through several hours in all three patients (Jardine *et al*., 1977).

It has reasonably been proposed (Takamizawa *et al*., 1975) that 4-hydroxycyclophosphamide, the metabolic precursor of phosphoramide mustard, enters cells more easily than phosphoramide mustard, and decomposes intracellularly to cytotoxic phosphoramide mustard. However, at high enough extracellular concentrations, phosphoramide mustard has also been found to kill a variety of mammalian cells in culture (Maddock *et al*., 1966). These concentrations lie between 30 and 80 nmol/ml, the same range we find for phosphoramide mustard in our patients' blood. Thus, the plasma levels measured here and found to be sustained through reasonably long time periods are cytostatically significant, and are consistent with the hypothesis that circulating phosphoramide mustard does play an important direct role in the pharmacological activities of cyclophosphamide.

ACKNOWLEDGEMENTS

The work reported was supported in large part by grants from the US Public Health Service, GM-21204, CA-16783, GM-70417 and CA-00103, and from the American Cancer Society, Maryland Division.

REFERENCES

Bagley, C. M. Jr., Botick, F. W. and DeVita, V. T. Jr. (1973). *Cancer Res.,* **33**, 226

Bakke, J. E., Feil, V. J. and Zaylskie, R. R. (1971). *Agric. Food Chem.,* **19**, 788

Cohen, J. L., Jao, J. Y. and Jusko, W. J. (1971). *Brit. J. Pharmacol.,* **43**, 677

Colvin, M., Padgett, C. A. and Fenselau, C. (1973). *Cancer Res.,* **33**, 915

Colvin, M., Brundrett, R. B., Kan, M. N., Jardine, I. and Fenselau, C. (1976). *Cancer Res.,* **36**, 1121

Fenselau, C., Kan, M.-N., Billets, S. and Colvin, M. (1975). *Cancer Res.,* **35**, 1453

Jardine, I., Kan, M.-N. N., Fenselau, C., Brundrett, R., Colvin, M., Wood, G., Lau, P.-Y. and Charlton, R. (1975). *Proceedings of the Second International Conference on Stable Isotopes* (ed. E. R. Klein and P. D. Klein), Oak Brook, Illinois, U.S.A., National Technical Information Service Document CONF-751027, p. 138

Jardine, I., Brundrett, R., Colvin, M. and Fenselau, C. (1976). *Cancer Treat. Rep.,* **60**, 403

Jardine, I., Fenselau, C., Appler, M., Kan, M.-N., Brundrett, R. B. and Colvin, M. (1977). *Cancer Res.* (in press)

Jarman, M., Gilby, E. D., Foster, A. B. and Bondy, D. K. (1975). *Clin. Chim. Acta,* **58**, 61

Maddock, C. L., Handler, A. H., Friedman, O. M., Foley, G. E. and Farber, S. (1966). *Cancer Chemother. Rep.,* **50**, 629

Struck, R. F., Kirk, M. C., Mellett, L. M., El Dareer, S. and Hill, D. L. (1971). *Molec. Pharmacol.,* **7**, 519

Takamizawa, W. A., Matsumoto, S., Iwata, T., Tochino, Y., Katagiri, K., Yamaguchi, K. and Shiratori, O. (1975). *J. Med. Chem.,* **18**, 376

9

Pharmacokinetics of pethidine and norpethidine in the rat: an application of selected ion monitoring using deuterated internal standards

C. Lindberg and C. Bogentoft (Department of Organic Pharmaceutical Chemistry, Biomedical Center, Box 574, S-751 23 Uppsala, Sweden),
B. Dahlström and L. Paalzow (Department of Pharmaceutical Pharmacology, Biomedical Center, Box 573, S-751 23 Uppsala, Sweden)

INTRODUCTION

During recent years there has been an increasing interest in the relationship between the pharmacokinetics and the effects of various drugs. A better understanding of such relationships is a fundamental basis for establishing a correct therapeutic dosage regimen and the adjustment of this to individual variations in, for example, drug metabolism and response (Wagner, 1975; Jellett, 1976).

Pethidine (meperidine) was introduced in 1939 as a potent analgesic and is still frequently used for the relief of severe pain. However, until recently (Klotz *et al.*, 1974; Mather *et al.*, 1975) pharmacokinetic studies of this drug were lacking and many questions are still to be answered. Recent investigations have revealed a discrepancy between the kinetics of morphine analgesia and the pharmacokinetics of the drug (Dahlström, Paalzow and Paalzow, 1975). The present study was undertaken in order to evaluate the pharmacokinetics of pethidine and one of its metabolites, norpethidine, and relate these to the time course of analgesia in the rat.

Several methods using gas chromatography with flame ionisation detection have been described for the determination of pethidine in biological fluids (Knowles, White and Ruelius, 1973; Chan, Kendall and Mitchard, 1974), but these methods lack sufficient sensitivity and specificity to allow small (<1 ml) blood samples to be used. Hartvig *et al.* (1977) have worked out a gas chromatographic method with electron capture detection capable of quantifying both pethidine and norpethidine in 0.1 ml of plasma. Since a less time-consuming procedure was required, we have designed a method based on gas chromatography–mass spectrometry with selected ion monitoring and using deuterium-labelled internal standards for the simultaneous determination of pethidine and norpethidine.

INTERNAL STANDARDS

Pethidine labelled with four deuterium atoms in the piperidine ring ($[^2H_4]$-pethidine) and norpethidine with five aromatic hydrogen atoms substituted with deuterium ($[^2H_5]$-norpethidine) are used as internal standards (Figure 9.1). Deuterium-labelled analogues of pethidine have been used in our laboratory for tracing pethidine metabolites (Lindberg *et al.*, 1975) and the syntheses of these compounds have been described previously (Lindberg, Bogentoft and Danielsson, 1974). $[^2H_5]$-Norpethidine was prepared by de-methylation of $[^2H_5]$-pethidine with cyanogen bromide.

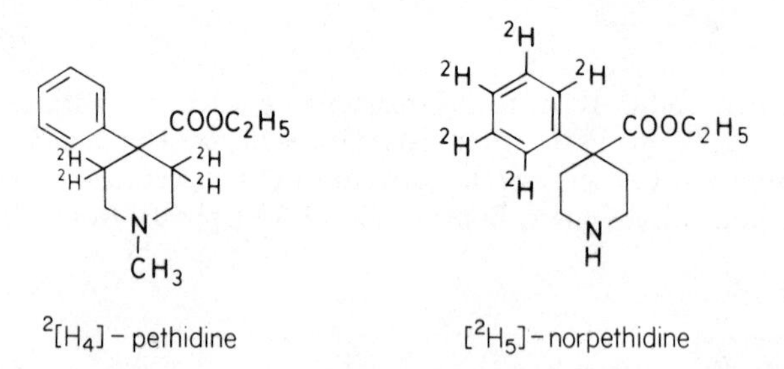

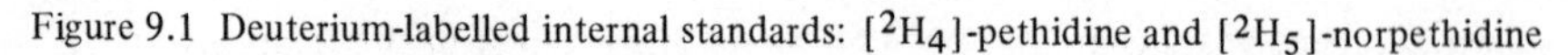
Figure 9.1 Deuterium-labelled internal standards: $[^2H_4]$-pethidine and $[^2H_5]$-norpethidine

During selected ion monitoring, the mass spectrometer is focused on *m/e* 247 and 251 (M^+, 6.8 per cent Σ_{40}, for pethidine and $[^2H_4]$-pethidine, respectively; see Figure 9.2) and on *m/e* 241 and 246 ($M^+ - 88$, 10.6 per cent Σ_{40}, for TFA-norpethidine and TFA-$[^2H_5]$-norpethidine, respectively; see Figure 9.3). The choice of these masses allows reasonable sensitivity and specificity to be obtained while switching of the instrument is kept within a narrow mass range.

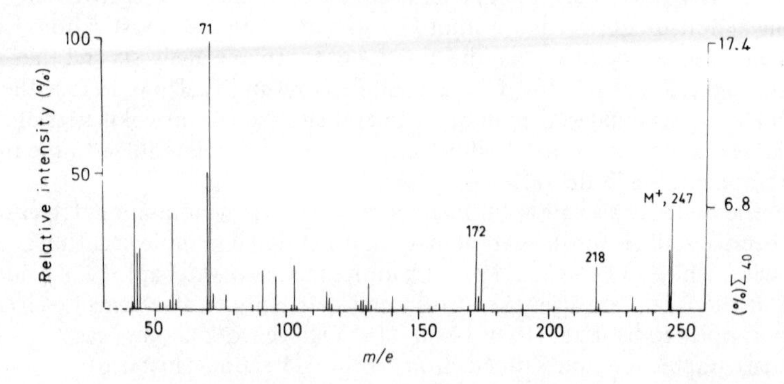

Figure 9.2 Mass spectrum of pethidine

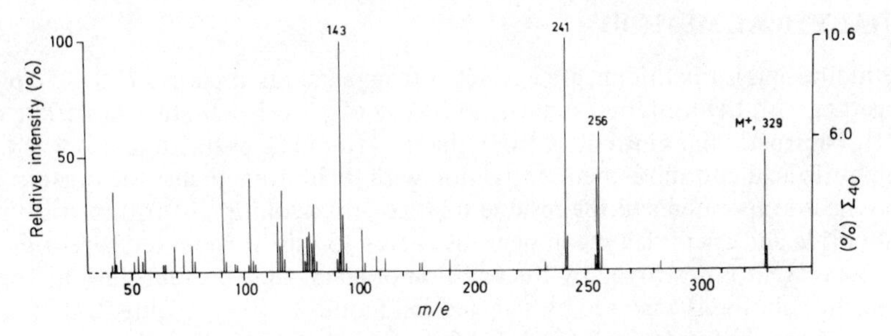

Figure 9.3 Mass spectrum of trifluoroacetylated norpethidine

Norpethidine labelled in the piperidine ring was considered unsuitable as internal standard since two of the deuterium atoms will be lost in the ion of interest. This ion ($M^+ - 88$) is formed by the expulsion of the ester side chain plus a methyl group containing the β-carbon from the piperidine ring. Corresponding fragmentations seem to be mandatory on amide derivatives of norpethidine but have not been observed in the case of pethidine and related compounds.

The isotopic purity of the internal standards was determined from mass spectra obtained from samples introduced into the instrument via the direct inlet probe. The peak height ratios of *m/e* 247/251 and of *m/e* 241/246 were less than 0.001. However, during selected ion monitoring, a significant blank was observed when the internal standards were analysed. The reason for this phenomenon is not presently understood. Selected ion-current profiles from a sample of plasma containing pethidine and norpethidine are shown in Figure 9.4.

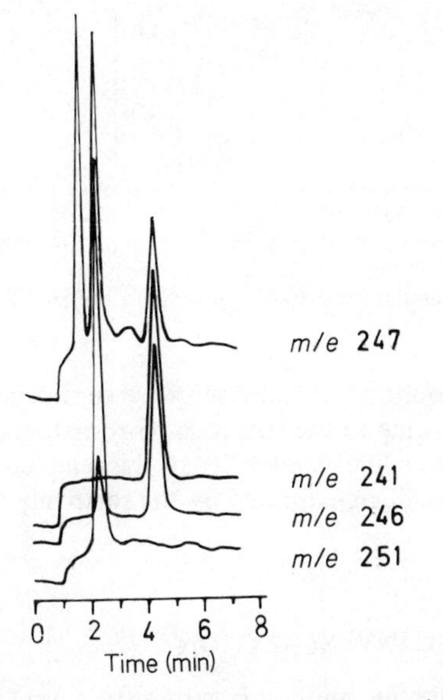

Figure 9.4 Selected ion-current profile from plasma sample containing 355 ng/ml of pethidine and 190 ng/ml of norpethidine. The signals from *m/e* 246 and 251 are attenuated 2.5 times relative to those from *m/e* 241 and 247

ANALYTICAL METHOD

Pethidine and norpethidine are extracted from alkalinised plasma (50–250 μl) together with their internal standards (100 ng of [$^2H_4$]-pethidine and 50 ng of [$^2H_5$]-norpethidine) into dichloromethane. After back-extraction into 0.1 N sulphuric acid and subsequent extraction with dichloromethane, the organic solvent is evaporated and the residue treated with a solution of trifluoroacetic anhydride and triethylamine in benzene. After 15 min at room temperature, excess reagent is destroyed by the addition of an alkaline buffer solution. The benzene solution is analysed by selected ion monitoring (see Figure 9.4) using a Pye 104 gas chromatograph (column: 3 per cent OV-17 on Gas Chrom Q, 200°) interfaced to an LKB 2091 instrument via a two-stage jet separator. The mass spectrometer is operated at an ionising energy of 50 eV and with a trap current of 50 or 100 μA.

Pethidine and norpethidine concentrations are determined from standard curves, constructed by plotting the peak height ratios of *m/e* 247/251 and *m/e* 241/246, respectively, against concentration. Linear standard curves were obtained in the concentration ranges 100–600 ng/ml for pethidine and 50–300 ng/ml for norpethidine (see Figure 9.5) by adding known amounts to 100 μl of plasma and treating the solutions as described above.

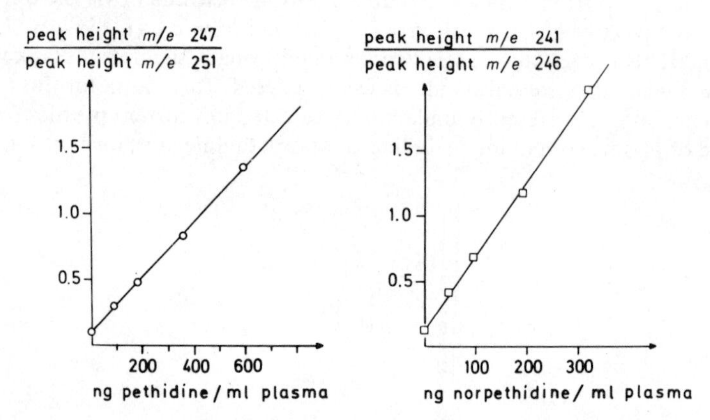

Figure 9.5 Standard curves for pethidine (left) and norpethidine (right)

The method was found to be more sensitive for norpethidine than for pethidine, probably owing to the better gas chromatographic properties of the TFA derivatives and to a higher intensity of the selected ion. The sensitivity of the method is, to some degree, limited by the relatively high blank values.

PHARMACOKINETIC INVESTIGATION

The mass fragmentographic method is currently in use for the determination of plasma concentrations of pethidine and its metabolite, norpethidine, in the rat, in order to evaluate the pharmacokinetics of these compounds. Male

Sprague–Dawley rats, implanted with a chronic catheter in the vena jugularis, are administered analgesic doses (10 mg/kg) of pethidine intravenously and serial blood samples (0.1–0.5 ml) are withdrawn from each rat over the following 6 h period. The blood samples are centrifuged and the plasma is stored at -18° until analysed.

The plasma concentrations of pethidine were found to decline rapidly in a three-exponential manner (Figure 9.6). By utilising a digital computer program (NONLIN), the time course of the pethidine concentration in plasma was found to be adequately described by a three-compartment mamillary model. The half-lives of the α, β and γ phases were 3.3, 21.5 and 109 min, respectively.

The norpethidine levels reached a maximum of approximately 175 ng/ml after 30 min. The decline of the plasma levels of this metabolite appeared to be mono-exponential with a half-life of 108 min.

The pharmacokinetic data will be used in connection with a study of the time course of analgesia in the rat following administration of pethidine and norpethidine.

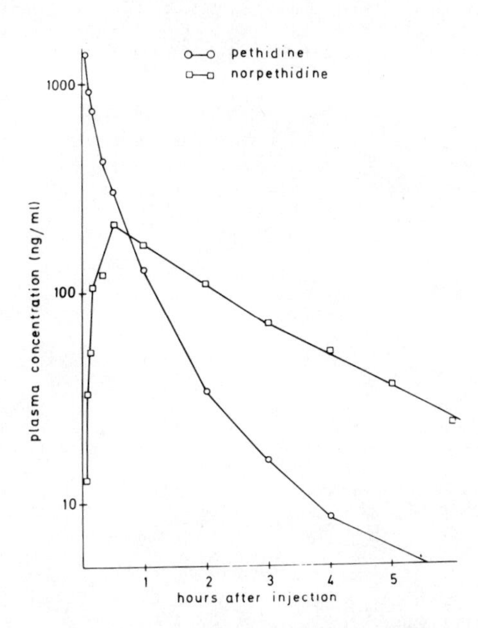

Figure 9.6 Plasma concentration–time curves for pethidine and norpethidine following an intravenous dose (10 mg/kg) of pethidine. Each point represents the mean value from four animals

ACKNOWLEDGEMENTS

We are greatly indebted to Dr Bo Karlén, Department of Drugs, National Board of Health and Welfare, for allowing us access to the mass spectrometer. Part of this work was supported by the Swedish Medical Research Council (Project Nos. O3X-3760 and 7OE-3743).

REFERENCES

Chan, K., Kendall, M. J., and Mitchard, M. (1974). *J. Chromatog.*, **89**, 169
Dahlström, B., Paalzow, G. and Paalzow, L. (1975). *Life Sci.*, **16**, 11
Hartvig, P., Karlsson, K.-E., Lindberg, C. and Boréus, L. O. (1977). *Europ. J. Clin. Pharmacol.*, **11**, 1
Jellett, L. B. (1976). *Drugs*, **11**, 412
Klotz, U., McHorse, T. S., Wilkinson, G. R. and Schenker, S. (1974). *Clin. Pharmacol. Ther.* **16**, 667
Knowles, J. A., White, G. R. and Ruelius, H. W. (1973). *Anal. Lett.*, **6**, 281
Lindberg, C., Bogentoft, C. and Danielsson, B. (1974). *Acta Pharm. Suec.*, **11**, 201
Lindberg, C., Bogentoft, C., Bondesson, U. and Danielsson, B. (1975). *J. Pharm. Pharmacol.*, **27**, 975
Mather, L. E., Tucker, G. T., Pflug, A. E., Lindop, M. J. and Wilkerson, C. (1975). *Clin. Pharmacol. Ther.* **17**, 21
Wagner, J. G. (1975). *Fundamentals of Clinical Pharmacokinetics*, Drug Intelligence Publ., Illinois, USA

10

Drug metabolism and pharmacokinetic studies in man utilising nitrogen-15- and deuterium-labelled drugs: the metabolic fate of cinoxacin and the metabolism and pharmacokinetics of propoxyphene

R. L. Wolen, B. D. Obermeyer, E. A. Ziege, H. R. Black and C. M. Gruber, Jr.
(Lilly Laboratory for Clinical Research, Wishard Memorial Hospital,
Indianapolis, Indiana 46202, USA)

INTRODUCTION

During the past several years, stable isotope utilisation has entered explosively into the clinical pharmacology laboratory for two major reasons: (1) the development of relatively simple, reliable instrumentation for gas chromatography–mass spectrometry (GC–MS) and (2) the availability of highly enriched stable-isotope-labelled starting materials. The techniques involved in mass spectrometry have been reviewed (Hammar *et al.*, 1969; Gordon and Frigerio, 1972; Finnigan *et al.*, 1974), as have the chemical aspects of stable isotope use (McCloskey, 1975). These reviews only scratch the surface of an enormous body of literature. Our application of these techniques to problems in clinical pharmacology forms the basis of this presentation as an encouragement to others.

Clinical pharmacology seeks information concerning absorption, distribution, metabolism and excretion of many drugs. The risk-to-benefit ratio in human studies mandates that every effort be made to minimise the number of subjects exposed and the duration of exposure. The use of stable isotopes offers a unique opportunity to increase the assurance of successful experimentation and reduce exposure. As a drug is developed, it is necessary to evaluate the effects of various formulations on kinetic parameters. Stable isotope labelling and mass spectrometric measurement permit small groups of subjects to provide statistically valid data; crossover studies may be replaced by a single treatment.

In our laboratory, GC–MS was accomplished using a quadrupole mass spectrometer with a chemical ionisation source. No separator was needed since the methane carrier gas also served as the ionisation reagent gas. Data reduction was accomplished with a minicomputer data system. All studies were carried out in informed, fasted, normal male volunteers who freely participated. All drugs were administered orally with 180 ml of water, and the subjects remained upright for at least 2 h post drug. Blood samples were drawn at designated intervals into heparinised, evacuated tubes; the plasma was recovered and stored at –20° until analysed. Urine samples were collected without preservative and aliquots stored as above.

CINOXACIN METABOLISM

Cinoxacin, 1-ethyl-1,4-dihydro-4-oxolo[1,3] dioxolo[4,5-*g*] cinnoline-3-carboxylic acid, is a potent antimicrobial agent (Wick *et al.*, 1973). A study designed to establish the major metabolic pathways in man using a stable isotope is outlined in Figure 10.1. The drug was synthesised to incorporate two atoms of ^{15}N utilising $K^{15}NO_3$ and $Na^{15}NO_2$ as the isotope source resulting in

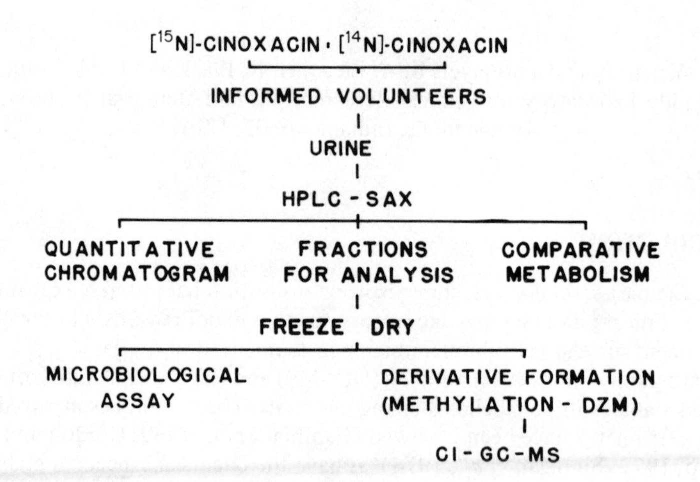

Figure 10.1 Procedure for determining metabolites of cinoxacin in urine. Comparative metabolism refers to animal urines treated in a similar way but not reported

the labelling shown in Figure 10.2. For use in experimental work, the labelled and unlabelled drugs were co-precipitated at approximately a 1 : 1 ratio by weight. The chemical ionisation spectrum (Figure 10.3) of the mixture indicates the presence of strong (M + 1) ions and small (M + 29) ions typical of methane chemical ionisation spectra; the distinct doublets at an *m/e* difference of 2 are also apparent.

High-pressure liquid chromatography (HPLC), as developed by the Oak Ridge National Laboratory (Scott, 1968), was utilised for urine fractionation. The

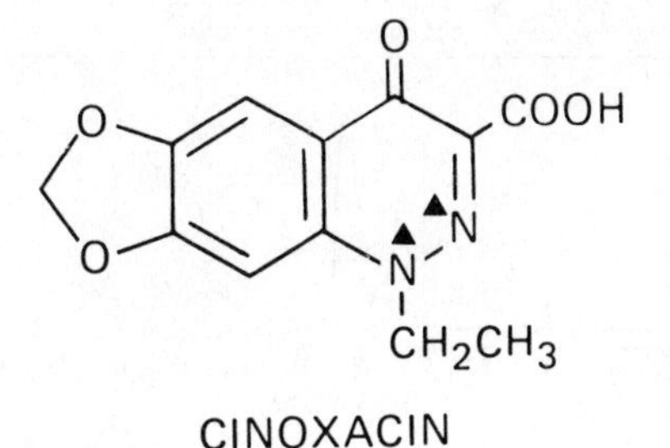

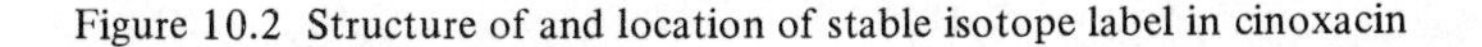

Figure 10.2 Structure of and location of stable isotope label in cinoxacin

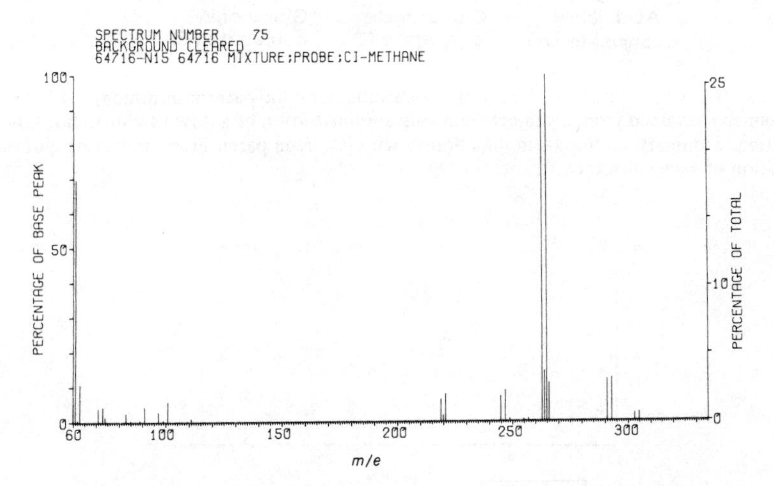

Figure 10.3 Chemical ionisation (methane) spectrum of a 1 : 1 by weight mixture of $^{15}N_2$-cinoxacin and unlabelled drug. Spectrum was obtained by solid probe insertion of sample and represents the methyl ester derivative produced by treatment with diazomethane

system uses Aminex A-17 strong anion exchange resin, an ammonium acetate buffer gradient at pH 4.4 varying from 0.015 to 6 M and a stepwise temperature gradient. Column effluent was monitored at 254, 280, and 360 nm. The 360 nm wavelength proved most useful, since the cinoxacin structure has a high specific absorbance and urine exhibits relatively little endogenous interference at this wavelength.

Fractions recovered from the HPLC were freed of buffer by lyophilisation, and examined by mass spectrometry either by direct probe analysis or by GC–MS following derivatisation with diazomethane or permethylation (Leclercq and Desiderio, 1971). Prior to derivatisation, the fractions were treated with glucuronidase or subjected to acid hydrolysis in order to free conjugates for identification.

The upper portion of Figure 10.4 presents a typical chromatographic record

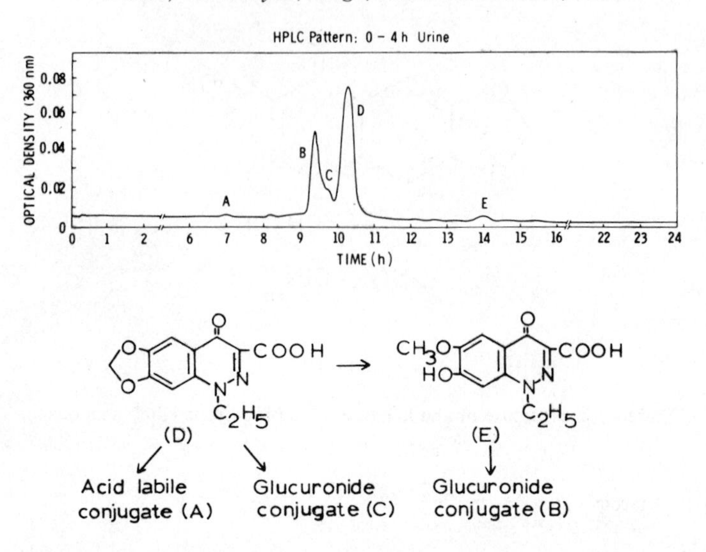

Figure 10.4 Upper portion: High-pressure chromatographic pattern in urine (0–4 h specimen) obtained from a subject following administration of a dose of cinoxacin. Lower portion: Summary of cinoxacin metabolism with letters in parentheses indicating elution position of each substance

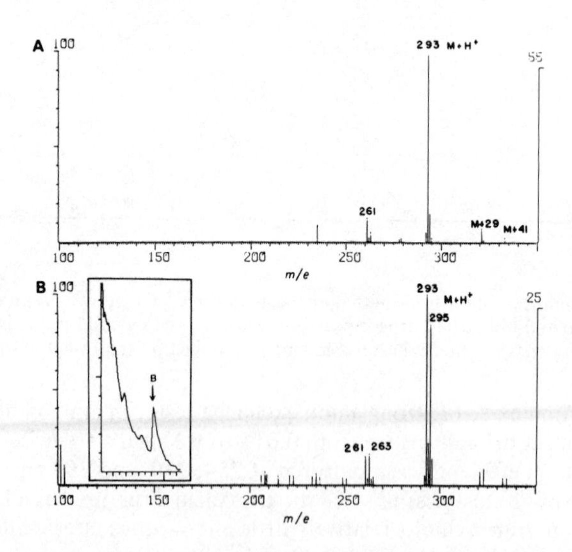

Figure 10.5 (A). Chemical ionisation spectrum of metabolite E (synthetic material). The structure is shown in Figure 10.4. The compound was permethylated to form the methyl ester-methoxy derivative. (B). Mass spectrum of material recovered from subject urine after oral administration of a mixture of $^{15}N_2$- and unlabelled cinoxacin. Material was permethylated prior to analysis. Note particularly the presence of the doublets representing the mixture. (Insert B). Total ion recording from GC–MS analysis of urinary metabolite E. The conditions were a 37 cm x 2 mm id column of 3 per cent Dexsil 400 on 100/120 mesh Supelcoport at the following temperatures: injector 250°; oven 220°; ion source 180°. The arrow indicates the location at which mass spectrum (B) was obtained

for the 0–4 h urine of one subject. Fractions from designated areas A, B, C, D and E in the chromatogram were examined. The fractions recovered were not single-component materials. Figure 10.5 presents typical results; fraction E of the chromatogram (see Figure 10.4) was permethylated and examined by GC–MS.

The mass spectra of the material at position B of the total ion recording (see insert Figure 10.5) contained the doublet features shown in mass spectrum 5B. The presence of the doublet established cinoxacin as the origin of this material; it was then compared with an authentic spectrum for the projected metabolite 1-ethyl-1,4-dihydro-7-hydroxy-6-methoxy-4-oxo-3-cinnolinecarboxylic acid (Figure 10.5A). The spectrum confirms the proposed structure, as do the elution positions in both liquid and gas chromatography. When the concentration of the metabolite was insufficient to permit complete spectra to be obtained, it was possible to confirm the presence of the metabolite by mass fragmentography as demonstrated in Figure 10.6. The ratio of *m/e* 295/293 is characteristic of the dosage form, as is that of *m/e* 263/261. The doublet ratios coupled to liquid and gas chromatographic elution positions provide confirmation of the presence of metabolite and take advantage of the increased sensitivity of the mass fragmentographic technique.

Figure 10.4 summarises the initial studies. There are two possible structures

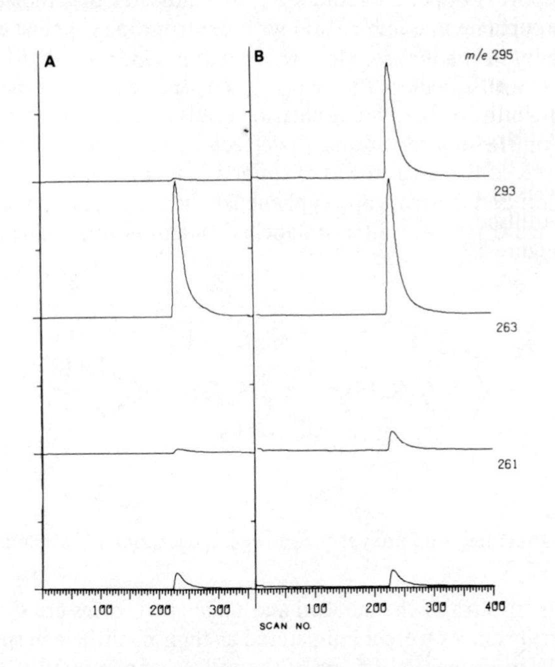

Figure 10.6 Mass fragmentographic recording of permethylated metabolite E of cinoxacin. Recording A is from an injection of 25 ng of derivatised synthetic material. Recording B is from injection of a portion of a sample recovered from human urine after cinoxacin administration. Sample was recovered from the HPLC elution at the location of the metabolite. The sample used was insufficient to provide full spectra for positive identification. Note particularly the presence of the *m/e* 293/295 doublet resulting from the presence of $^{15}N_2$ labelling

which could result from the splitting of the dioxolo ring and these would have identical molecular weights. Synthesis of each of the structures allowed of comparisons to establish the identity of the metabolite. Several minor metabolites which appear in the chromatogram have not been structurally identified. The quantitative conversion to either the conjugates or the identified metabolites varied among the subjects; however, in each case the intact drug was present in the largest quantity. The study illustrates the utility of the stable isotope and GC–MS in metabolite identification.

D,L-PROPOXYPHENE STUDY

Propoxyphene, 4-dimethylamino-3-methyl-1,2-diphenyl-2-butanol propionate, is used medicinally as both the α-form D-isomer (dextropropoxyphene) and the α-form L-isomer (laevopropoxyphene). Dextropropoxyphene is a widely used analgesic activity in the mouse. More recent studies indicate that laevo-propoxyphene both enhances the analgesic activity of the D-isomer in the first-pass effect (Perrier and Gibaldi, 1972), which results in lower circulating levels of the drug than would be expected from complete absorption of a given dose.

A recent report (Cooper and Anders, 1975) indicates that the administration of laevopropoxyphene in combination with dextropropoxyphene enhances analgesic activity in the mouse. More recent studies indicate that laevo-propoxyphene both enhances the analgesic activity of the D-isomer in the rat and causes a large increase in circulating levels of this form (Murphy *et al.*, 1976). Based on these observations, the effect of laevopropoxyphene on the circulating levels of the D-form was examined in man.

The study utilised dextropropoxyphene labelled with seven deuterium atoms as shown in Figure 10.7 as well as unlabelled D- and L-forms. The chemical

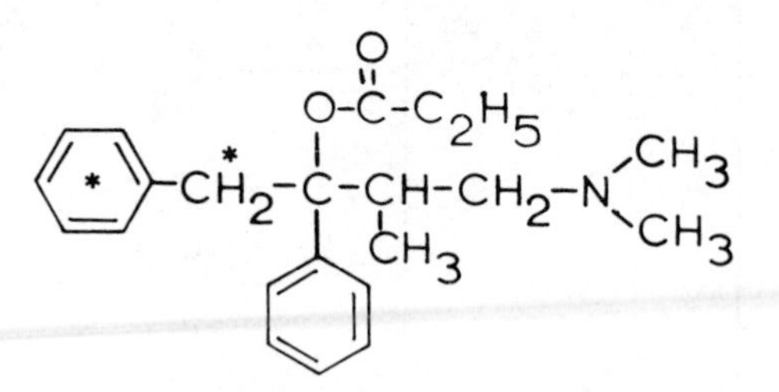

Figure 10.7 Structure of propoxyphene showing the location (*) of the deuterium label

ionisation mass spectra of the labelled and unlabelled forms are shown in Figure 10.8. All test substances were administered as their naphthalene sulphonic acid salts. Dosage forms consisted of one of the following as a finely divided suspension: (1) 100 mg d_7-dextropropoxyphene napsylate; (2) 100 mg d_7-dextropropoxyphene plus 100 mg of unlabelled D-form; or (3) 100 mg d_7-dextropropoxyphene plus 100 mg of unlabelled L-form.

Three volunteers took part in the study. Drug was administered on a crossover basis at intervals of 3 weeks to reduce any possible effects of residual

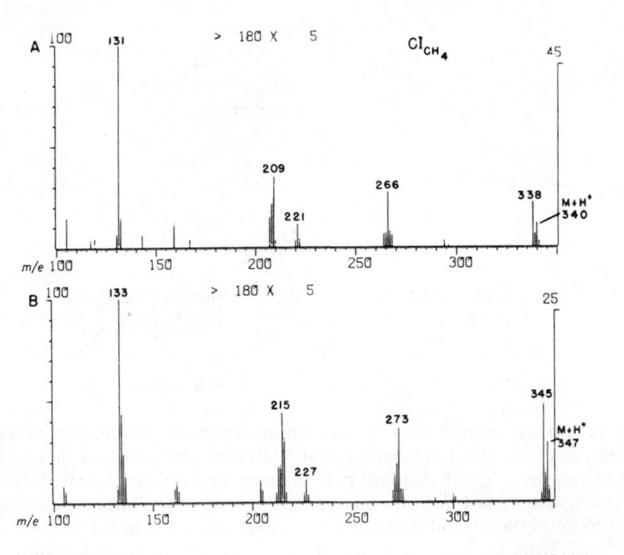

Figure 10.8 Chemical ionisation spectra of unlabelled (A) and d_7-labelled (B) propoxyphene. The spectra illustrate the loss of varying amounts of the label in different fragments

drug. Blood samples were drawn at 0, 0.5, 0.75, 1, 1.25, 1.5, 1.75, 2, 2.25, 2.5, 3, 4, 6, 8, 12, 16, 24, 32 and 40 h post medication. Samples were analysed for propoxyphene and its major metabolite, *N*-desmethylpropoxyphene, by mass fragmentography using a method described earlier (Wolen, Ziege and Gruber, 1975) which was modified by the inclusion of two internal standards. The two standards were α-D-4-pyrrolidino-1,2-diphenyl-3-methyl-2-acetoxybutane hydrochloride (pyrroliphene hydrochloride) and α-D-1,2-diphenyl-3-methyl-4-methylamino-2-propionoxybutane (norporpoxyphene) containing three deuterium atoms in the terminal methyl group of the propionyl ester moiety. The former standard has been used previously as an internal standard for gas chromatographic analysis of propoxyphene (Wolen and Gruber, 1968; Nash *et al.*, 1975). Fragments at *m/e* 266 and 273 were monitored for unlabelled and labelled propoxyphene, respectively, with the fragments at *m/e* 308 and 315 for the corresponding norpropoxyphene detection. The internal standards were measured at *m/e* 292 for pyrroliphene and 311 for the deuterated norpropoxyphene standard. Proposed structures of the fragments are presented in Figure 10.9.

The ratios of the areas under the *m/e* 266 and 273 curves for propoxyphene to the 292 value for pyrroliphene were calculated and compared with standard curves generated with known concentrations of unlabelled and labelled drug to determine the plasma concentrations. A similar procedure was utilised to determine the concentrations of norpropoxyphene for both the proteo and deutero forms. Utilising these procedures, it was possible to determine the concentration of each form of drug and metabolite in the presence of the other. The estimated errors appear to be no greater than those found when assaying for a single form of the drug.

The results of the analysis for the deuterium-labelled D-propoxyphene are

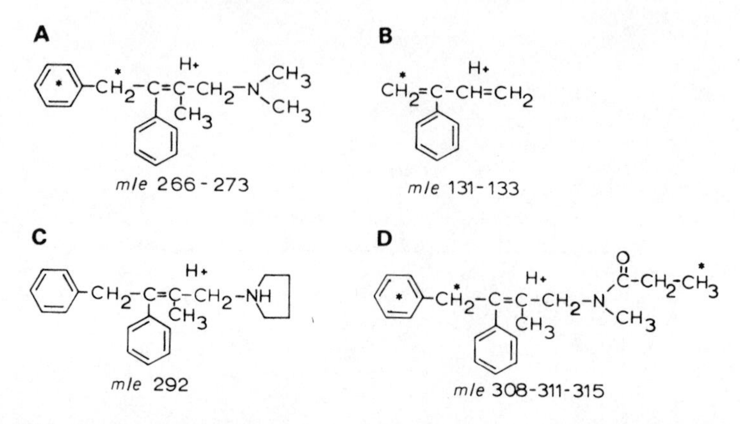

Figure 10.9 Fragments utilised for analysis of propoxyphene and norpropoxyphene by chemical ionisation GC–MS. Fragments (A) and (B) represent propoxyphene; (C) represents the fragment from the internal standard pyrroliphene; and (D) represents the fragments from norpropoxyphene used both for analysis (308/315) and as internal standard (311). The location of deuterium is indicated by an asterisk. Note that in the norpropoxyphene the benzyl labelling is from the administered drug, whereas the propionyl labelling is in the internal standard only

presented in Figures 10.10 and 10.11. The addition of either the D- or L-form of the drug to the initial dose results in a higher circulating level of the labelled form (Figure 10.10). This is in agreement with the results in rodents (Murphy *et al.*, 1976) and consistent with the theory surrounding the first-pass effect (Perrier and Gibaldi, 1972), which predicts that systemic availability of propoxyphene napsylate increases as the dose increases. The increase seen does not appear to be significantly greater with the L-form than with an equivalent amount of the D-form, which is again in agreement with the earlier animal data. The small number of subjects and the normal variation among subjects – found especially with drugs exhibiting large first-pass effects – establish a clear trend,

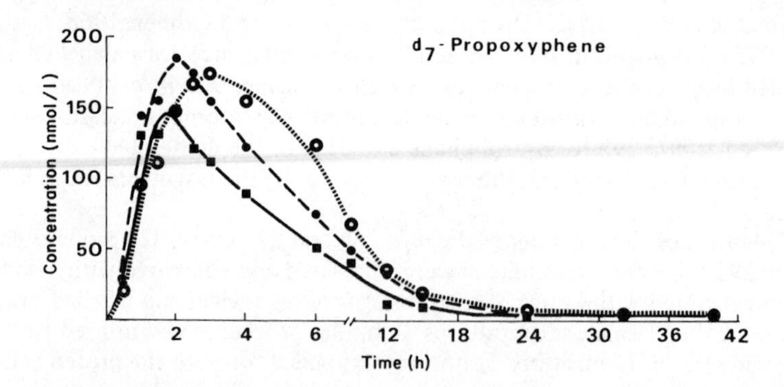

Figure 10.10 The concentration of d_7-D-propoxyphene in plasma. Key: (——■——■——) = 100 mg d_7-dextropropoxyphene; (---●---●---) = 100 mg d_7-dextropropoxyphene plus 100 mg dextropropoxyphene; (....o....o....) = 100 mg d_7-dextropropoxyphene plus 100 mg laevopropoxyphene. All doses were as the napsylate salts. For clarity the early values are plotted at 1/2 h intervals only

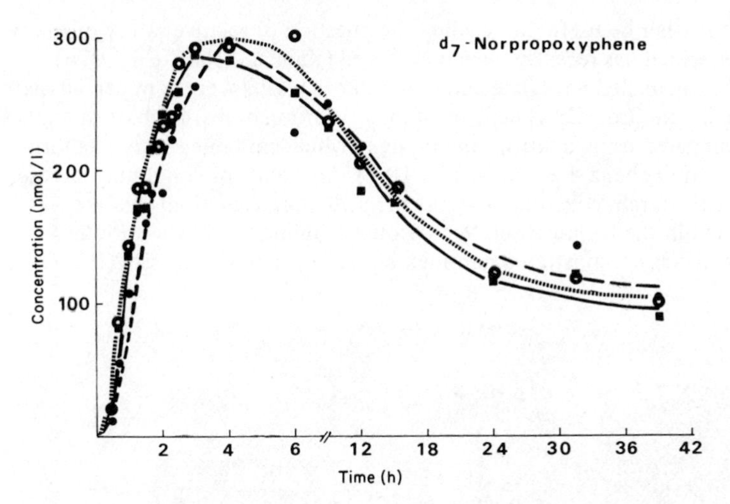

Figure 10.11 The concentration of norpropoxyphene in plasma for which the propoxyphene levels are presented in Figure 10.10. Only the d_7 portion of the administered propoxyphene dose is illustrated. Refer to Figure 10.10 legend for key to treatments

but do not allow of extensive statistical analysis. The concentrations of norpropoxyphene (Figure 10.11) indicate that the differences in plasma levels of the parent drug are not the result of differences in rates of metabolic conversion but rather are the result of modification of the first-pass effect in the liver. These data are also in agreement with the animal data.

The ability to measure various isomeric forms of a given substance in the presence of other isomeric forms of the same compound offers a broad opportunity to study phenomena of metabolism and distribution *in vivo.* Previously this type of study was restricted to techniques such as the use of radiotracers. The stable isotope method eliminates any potential radiologic hazard without unduly reducing one's ability to carry out a successful experiment.

COMPARATIVE ABSORPTION RATES OF THE HYDROCHLORIDE AND NAPSYLATE SALTS OF PROPOXYPHENE

Crossover studies in man indicate that the napsylate salt of dextropropoxyphene is more slowly absorbed than the hydrochloride (Wolen *et al.*, 1971). It was thought that if the two were given as a single dose, some of the difficulties associated with the crossover procedures, particularly the time required for washout and the difficulty of keeping volunteer subjects interested, could be avoided. The study was accomplished by mixing stable-isotope-labelled with unlabelled salt form. The ratio of labelled to unlabelled drug in the plasma was used to measure the relative rates of absorption of the two salts. The drugs were utilised in paired fashion in order to identify any isotope effect which might be incorrectly interpreted as absorption differences. Information on absorption might further be useful in deciding the question of relative safety of the two forms, which has recently been considered (Kaul and Harsfield, 1976).

The unlabelled napsylate and hydrochloride salts were commercial material (Eli Lilly and Co., USA) used for standard formulations. Each of the salts was also prepared using dextropropoxyphene which contained seven deuterium atoms on the benzyl group (Figure 10.7). An excipient containing lactose, cellulose, starch, colloidal SiO_2, stearic acid and magnesium stearate was included in the formulation. Microscopic examination of the various preparations demonstrated considerable variation in crystal size (Figure 10.12).

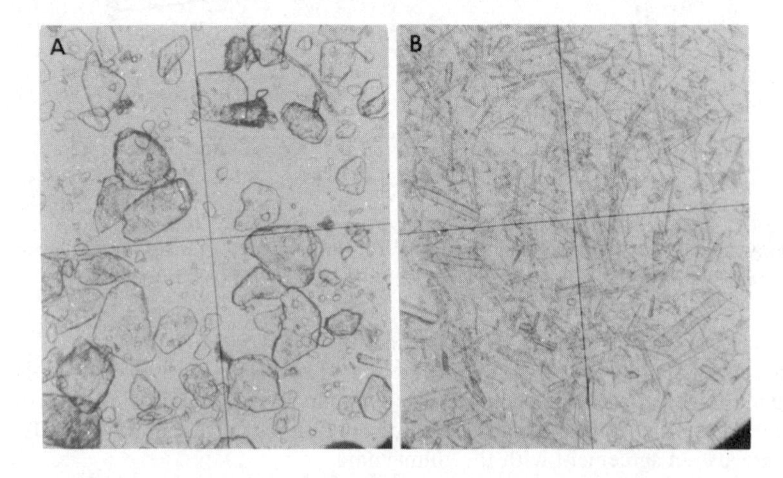

Figure 10.12 Microphotograph of propoxyphene napsylate crystals illustrating the difference in size and form of the commercial material (A) and the deuterium-labelled material (B). Similar differences between the hydrochloride salts were present

Table 10.1 presents the four combinations used in the study. Each drug mixture was ground to a uniform appearance prior to being weighed and added to an equal weight of excipient, then thoroughly mixed and placed in a No. 0 gelatin capsule.

Table 10.1 Dosage formulation: dextropropoxyphene napsylate and hydrochloride absorption study

	mg Drug			
Dose	H-HCl	d_7-HCl	H-Naps.	d_7-Naps.
A	64.4	65.6	–	–
B	–	67.9	97.1	–
C	66.7	–	–	98.3
D	–	–	99.4	100.6

H-HCl = dextropropoxyphene hydrochloride
H-Naps. = dextropropoxyphene napsylate
d_7-HCl = deuterium-labelled dextropropoxyphene hydrochloride
d_7-Naps. = deuterium-labelled dextropropoxyphene napsylate

Each of the four preparations was administered to one volunteer. Blood samples were taken at 0, 0.5, 0.75, 1, 1.25, 1.5, 2, 2.5, 3, 3.5, 4, 8, 12 and 24 h. A 2 ml aliquot of the plasma was analysed by the method previously described (Wolen *et al.*, 1975), with the modification that no internal standard was included. Mass fragmentography utilised the *m/e* 131 and 133 fragments of the unlabelled and deutero forms, respectively, whereas the norpropoxyphene measurements utilised the *m/e* 308 and 315 fragments (Figure 10.9). The choice of the fragments for propoxyphene was to maximise sensitivity, since they are by far the largest fragments present in the chemical ionisation spectra (Figure 10.8). All data were collected on the minicomputer data system and analysed by automated measurement of the area under the mass fragment curves.

The soluble hydrochloride salt of the drug in both the labelled and unlabelled form enters the body at an approximately equivalent rate (Figure 10.13A). Figure 10.13B indicates – by the changing ratio – that the soluble d_7-HCl form enters the circulation more rapidly than does the standard napsylate preparation. The two become relatively constant at 2 h. There is an indication that absorption of the napsylate salt continues over the entire period shown by a gradual increase in the *m/e* 131/133 ratio during the 4–8 h period. Figure 10.13C – representing the d_7-napsylate and soluble hydrochloride forms – presents an interesting contrast to the earlier case of hydrochloride/napsylate comparison. In this instance the napsylate form enters the circulation at a rate approximately equivalent to that of the hydrochloride. Examination of the crystals indicates that the d_7-napsylate material was of smaller crystal size than the unlabelled material (Figure 10.12). The final comparison, Figure 10.13D, is between the d_7- and unlabelled napsylate salts. The earlier observations are confirmed since again the changing ratio indicates that the labelled form (smaller

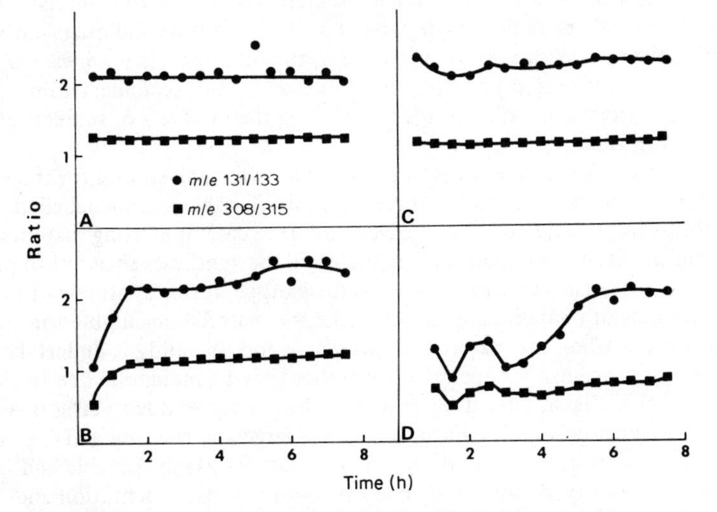

Figure 10.13 Ratios of the deuterium-labelled and unlabelled fragments from propoxyphene (*m/e* 131/133) and norpropoxyphene (*m/e* 308/315) after administration of various equimolar mixtures of labelled and unlabelled hydrochloride and napsylate salts. A = d_7-HCl + H-HCl; B = d_7-HCl + H-Naps.; C = d_7-Naps. + H-HCl; and D = d_7-Naps. + H-Naps.

crystals) enters the circulation more rapidly than does the unlabelled form. The erratic results in this comparison are unexplained.

A pronounced instrumental isotope effect is also demonstrated in this study. The ratio of *m/e* 131/133 in a 1 : 1 by weight mixture is approximately 2. This can only result if the fragmentation of the deuterated form is approximately 1/2 that of the proteo form. Such an effect would not be unreasonable since the phenyl group must be eliminated and the presence of the deuterium may affect this process. In the case of norpropoxyphene, the *m/e* 308/315 ratio of a 1 : 1 by weight mixture is approximately 1, which would be expected if both forms were equally fragmented by dehydration across carbons 2 and 3 of the butyl chain of the amide form.

The conclusions which are drawn from the results are that the napsylate salt form, as commercially produced, is absorbed more slowly during a short initial period with little effect on the overall blood levels when compared with the hydrochloride salt. However, it is also apparent that by varying the nature of the dextropropoxyphene napsylate crystals one can obtain a form of this salt which is comparable in absorption with the soluble hydrochloride salt. If safety in accidental overdose situations is in any way related to the initial absorption rates, it would appear that the currently available napsylate salt would have an advantage.

CONCLUSIONS

These examples present some of the ways stable isotopes may be used in clinical pharmacology. The modern mass spectrometer provides high sensitivity and specificity permitting inferences with small patient samples and decreasing the total discomfort and hazards of human research. The ability to administer two or more forms of a drug to a single subject at the same time and quantitatively determine the fate of each in the presence of the other provides a means of minimising both variation and the subject exposure. As a secondary gain, the techniques permit economic savings by reducing the numbers of subjects and duration of hospitalisation.

One must not ignore the problems associated with such studies for they are many. First, there is the need for good synthesis of stable-isotope-labelled drugs. Often this is impossible because of lack of intermediate or starting materials for the synthesis. Even with good starting material, the syntheses may fail to provide satisfactory material owing to transfer of the isotope during synthesis or owing to the presence of contaminants which make the material unsuitable for human administration. When the materials are available and the study is undertaken, one must guard against isotope effects whether they be metabolic due to the presence of the label at a position of metabolic activity or due to effects such as change in fragmentation. The initial expense is large and the care and maintenance of the instrumentation is no small undertaking. Capable and interested laboratory personnel – operating both the instrumentation and carrying out other necessary procedures – is essential. Carelessness can only result in the generation of marginally useful or useless data at considerable expense of time and effort.

In human studies hundreds of samples may require processing. To avoid delay, it is often necessary to develop instrumentation and procedures which take advantage of technology currently available. This includes systems to automatically accomplish the following: (1) injection of the samples, (2) collection of the data upon command of the injector, (3) a valve to divert materials away from the mass spectrometer source when substances of interest are not present (to minimise the cleaning needs and resultant down-time), (4) periodic calibration procedures to realign and peak the instrument parameters (to assure high sensitivity and specificity at all times), and (5) automated processing of the data (to provide the needed quantitation). All of these are currently available and their application is only limited by each investigator's imaginative approach to his particular problems.

ACKNOWLEDGEMENT

The authors would like to express their appreciation to the volunteers, the nurses and other staff members whose co-operation made the studies possible. The labelled materials synthesised by Mr Richard Booher, Mr William A. White, Dr Frederick J. Marshall and Mr William Turner were essential to the studies. For their effort and willingness to assist, we express our deep appreciation. In addition, Dr Albert Pohland must be recognised for his guidance. Finally, our thanks to Miss Barbara Hatcher, who assisted in preparation of the manuscript.

REFERENCES

Cooper, M. J. and Anders, M. W. (1975). *Life Sci.,* **15**, 1665

Finnigan, R. E., Knight, J. B., Fies, W. F. and DeGragnano, V. L. (1974). *Mass Spectrometry in Biochemistry and Medicine* (ed. A. Frigerio and N. Castagnoli, Jr.), Raven Press, New York

Gordon, A. E. and Frigerio, A. (1972). *J. Chromatog,* **73**, 401

Hammar, C.-G., Holmstedt, B., Lindgren, J.-E. and Tham, R. (1969). *Adv. Pharmacol. Chemother.*, 7

Kaul, A. F. and Harsfield, J. C. (1976). *Drug Intell. Clin. Pharm.,* **10**, 53

Leclercq, P. A. and Desiderio, D. M., Jr. (1971). *Anal. Lett.,* **4**, 305

McCloskey, J. A. (1975). *Selected Approaches to Gas Chromatography-Mass Spectrometry in Laboratory Medicine* (ed. R. S. Melville and V. F. Dobson), US Government Printing Office, Washington, DC

Murphy, P. J., Nickander, R. C., Bellamy, G. M. and Kurtz, W. L. (1976). *J. Pharmacol. Exp. Ther.,* **199**, 415

Nash, J. F., Bennett, I. F., Bopp, R. J., Brunson, M. K. and Sullivan, H. R. (1975). *J. Pharm. Sci.,* **64**, 429

Perrier, D. and Gibaldi, M. (1972). *J. Clin. Pharmacol.,* **12**, 449

Scott, C. D. (1968). *Clin. Chem.,* **14**, 521

Wick, W. E., Preston, D. A., White, W. A. and Gordee, R. S. (1973). *Antimicrob. Agents Chemother.,* **4**, 415

Wolen, R. L. and Gruber, C. M., Jr. (1968). *Analyt. Chem.,* **40**, 1243

Wolen, R. L., Gruber, C. M., Jr., Kiplinger, G. F. and Scholz, N. E. (1971). *Toxic. Appl. Pharmacol.,* **19**, 480

Wolen, R. L., Ziege, E. A. and Gruber, C. M., Jr. (1975). *Clin. Pharmacol. Ther.,* **17**, 15

11
Bioavailability using cold labels: studies on a novel antidiarrhoeal agent SC-27166

N. J. Haskins, G. C. Ford, S. J. W. Grigson and R. F. Palmer (G. D. Searle and Co. Ltd., High Wycombe, Bucks. HP12 4HL, UK

INTRODUCTION

The use of simultaneous administration of a compound to an animal or man by two routes where one route is distinguished by a stable-isotope-labelled derivative has been described in a study of the absolute availability of procainamide in man (Strong *et al.*, 1975; Dutcher *et al.*, 1975). Generally the method has a number of advantages, not least being the necessity to take only one set of blood samples. Also, in the usual crossover study the assumption is made that the kinetics of drug absorption, distribution, metabolism and elimination remain unchanged between doses. However, problems associated with the technique of simultaneous administration could arise if a high circulating plasma concentration of the compound interferes with the absorption of an orally administered dose or if extensive biliary excretion of free drug should take place, thereby making a high percentage of the intravenously administered drug available for intestinal re-absorption.

In order to study the application of a double labelling technique, a study was initiated involving a novel antidiarrhoeal agent SC-27166 (Id_0, Figure 11.1) [2-(3,3-diphenyl-3-(2-methyl-1,3,4-oxadiazol-5-yl)propyl)-2-azabicyclo[2.2.2.] octane]. This compound has been shown to possess antidiarrhoeal activity in a number of animal species (Dajani *et al.*, 1975; Aldelstein *et al.*, 1976). For the present study SC-27166 was prepared with 4, 6 and 9 teuterium atoms incorporated in the molecule. The tetradeutero analogue (d_4-SC-27166, Id_4, Figure 11.1) was labelled in a metabolically inert position. The hexadeutero analogue (d_6- SC-27166, Id_6, Figure 11.1) was used as the internal standard and carrier for the assay, while the nonadeutero analogue (d_9-SC-27166, Id_9, Figure 11.1) was used to establish the fragmentation pathway observed using electron impact ionisation. The d_0-SC-27166 and d_4-SC-27166 were administered simultaneously by intravenous and oral routes to the rat, beagle dog and baboon, and the results obtained are described below.

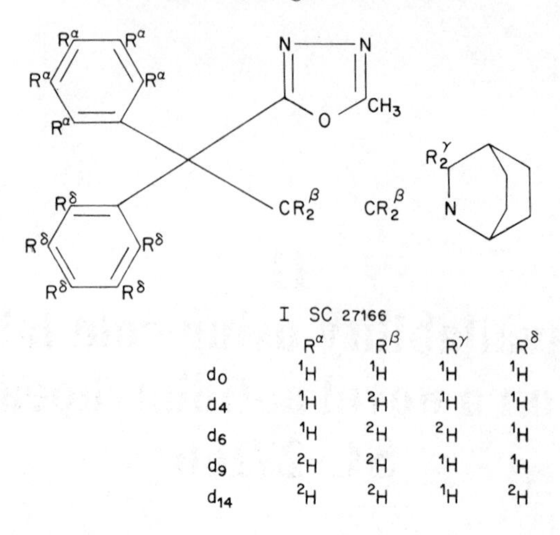

	R^{α}	R^{β}	R^{γ}	R^{δ}
d_0	^1H	^1H	^1H	^1H
d_4	^1H	^2H	^1H	^1H
d_6	^1H	^2H	^2H	^1H
d_9	^2H	^2H	^1H	^1H
d_{14}	^2H	^2H	^1H	^2H

Figure 11.1 Structures of compounds used in this study

EXPERIMENTAL

Materials

SC-27166 was obtained as the pure chemical from G. D. Searle and Co., Skokie, Illinois. The deuterated analogues were prepared in the Chemical Development Laboratory, G. D. Searle and Co. Ltd, High Wycombe. All solvents used were of Analar grade, obtained from Hopkins and Williams Ltd., Chadwell Heath, Essex.

Administration to the rat

Four rats (weights 224–247 g) were treated as two pairs. Rats 1 and 2 received d_0-SC-27166 orally by intubation (10 mg/kg) and d_4-SC-27166 (10 mg/kg) intravenously. Rats 3 and 4 received d_4-SC-27166 orally and d_0-SC-27166 intravenously. The intravenous dosage contained SC-27166 (5 mg/ml) in ethanol/physiological saline (1 : 1, v/v) solution while the oral dosage contained SC-27166 (0.5 mg/ml) in ethanol/water (1 : 19, v/v).

The rats were housed as pairs in two 'metabowls' and urine was collected for 24 h. After 24 h the animals were sacrificed and the blood withdrawn by cardiac puncture.

Administration to the beagle dog

A beagle dog (weight 7.9 kg) was dosed with d_4-SC-27166 (10 mg) intravenously and d_0-SC-27166 (10 mg) orally. A second dog was dosed with d_0-SC-27166 intravenously and d_4-SC-27166 orally. The intravenous injection (1 mg/ml) was made up in a citrate/phosphate buffer and the oral dosage was given in a compressed lactose tablet. Blood samples (5 ml) were withdrawn before dosage and at 10, 20, 30 min and 1, 2, 3, 4, 6, 8 and 24 h after the intravenous injection. The samples were centrifuged immediately after withdrawal and the plasma stored at –20°C until analysed. Urine samples were collected for the first 24 h after dosage, and stored at –20° until analysed.

Administration to the baboon

A baboon, weight 13.24 kg, identified as No. 17, received d_0-SC-27166 (0.25 mg/kg) by gastric intubation followed by water (20 ml). After 15 min the same animal received d_4-SC-27166 (0.25 mg/kg) by intravenous injection. A second baboon, weight 13.88 kg (No. 21), received d_4-SC-27166 orally and d_0-SC-27166 intravenously. The intravenous dosage (1 mg/ml) was made up in a citrate/phosphate buffer. The oral dosage (0.1 mg/ml) was prepared from the solution used for intravenous administration by dilution with distilled water. The baboons were restrained in a primate chair for drug administration and blood withdrawal. Blood samples (4 ml) were withdrawn immediately prior to dosage, then at 10, 20, 30 min and at 1, 2, 3, 4, 6, 8, 12 and 25 h after completion of the intravenous injection. The blood was centrifuged, the plasma collected into tubes and the samples stored at –20° until analysed. Urine was also collected (0–24 h), the volume was measured and it too was stored at –20° until analysed. Both animals were fasted for 16 h prior to drug administration and for 2 h afterwards. Water was provided *ad libitum.*

Extraction procedure for plasma

Plasma (1 ml) was diluted with a solution of d_6-SC-27166 (1 μg) in water (1 ml) and the mixture was allowed to equilibrate. The solution was extracted with chloroform (10 ml) and the chloroform extracts were taken to dryness. The residues were dissolved in ethanol (20 μl) and an aliquot (3–4 μl) taken for analysis by gas chromatography–mass spectrometry.

Extraction procedure for urine

A solution of d_6-SC-27166 (10 μg) in ethanol (1 ml) was added to an aliquot of urine (10 ml) and mixed. The urine was extracted with chloroform (2 x 10 ml) and the combined extracts taken to dryness. The residue was dissolved in ethanol (50 μl) and an aliquot (2–3 μl) was taken for analysis by gas chromatography–mass spectrometry.

Calibration lines

Standard solutions of d_0-SC-27166 and d_4-SC-27166 were prepared in control plasma and urine. Urine contained 2000, 1000, 500, 250, 125, 62.5 and 0 ng/ml of d_0-SC-27166; each sample containing d_0-SC-27166 was further divided into separate samples containing 2000, 1000, 500, 250, 125, 62.5 and 0 ng/ml of d_4-SC-27166. Plasma samples were prepared similarly to contain 100, 50, 25, 12.5 and 0 ng/ml of d_0-SC-27166 and d_4-SC-27166. Samples were extracted as described above and the peak area ratios obtained for d_0/d_6 and d_4/d_6 from selective ion recordings of the $(M + 1)^+$ ions for d_0-, d_4- and d_6-SC-27166 were used to construct calibration lines for the assays.

Gas chromatography–mass spectrometry

Gas chromatography–mass spectrometry was carried out using a Finnigan 3200 E with a chemical ionisation source. Data were analysed using a Finnigan 6000 on-line data system. Gas chromatography was performed on a (1 m x 4 mm) glass column containing 1 per cent Dexsil 300 on gaschrom Q (100–120 mesh) (Phase Separations Ltd, Queensferry, Clwyd). Oven, injector and interface temperatures were kept at 275, 300 and 240°, respectively. Methane was used as the carrier and reactant gas (20 ml/min). Electron impact spectra were obtained using the same instrument but with an electron impact source.

RESULTS

The electron impact and methane chemical ionisation spectra of SC-27166 are shown in Figures 11.2 and 11.3. The fragmentation pattern for SC-27166 and for its deuterated analogues (Figure 11.4) was deduced by the comparison of the electron impact spectra of d_0-, d_4-, d_6- and d_9-SC-27166. By analysis of standard urine samples a straight calibration line was obtained for the range 50–2000 ng/ml for both d_0-SC-27166 and d_4-SC-27166 (Figure 11.5). A similar

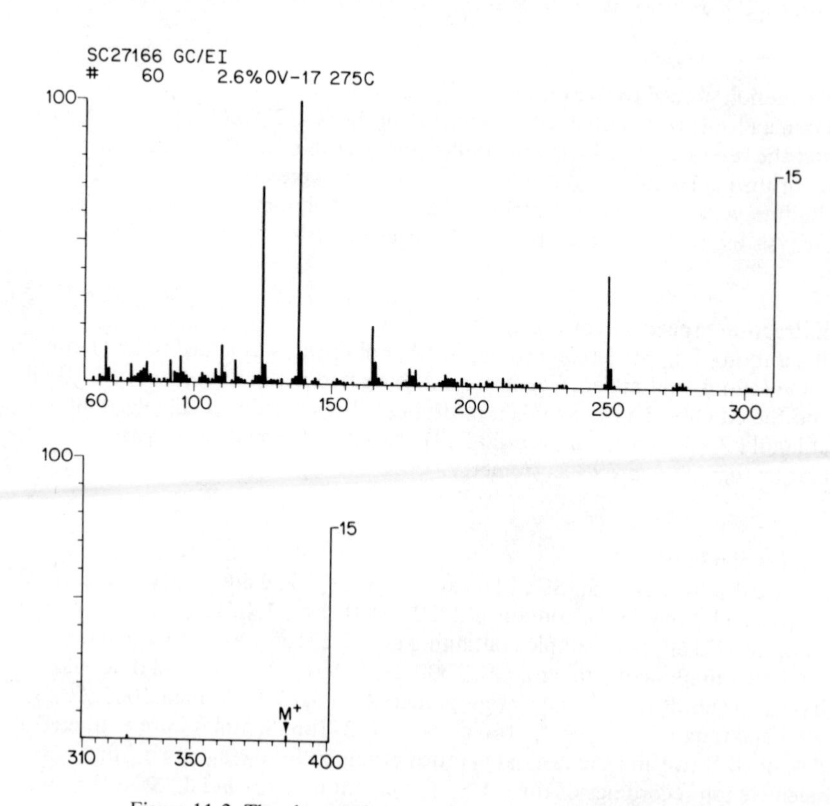

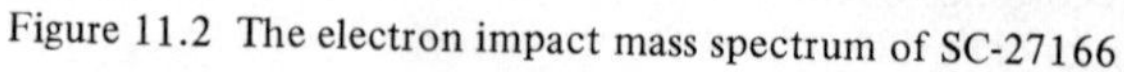
Figure 11.2 The electron impact mass spectrum of SC-27166

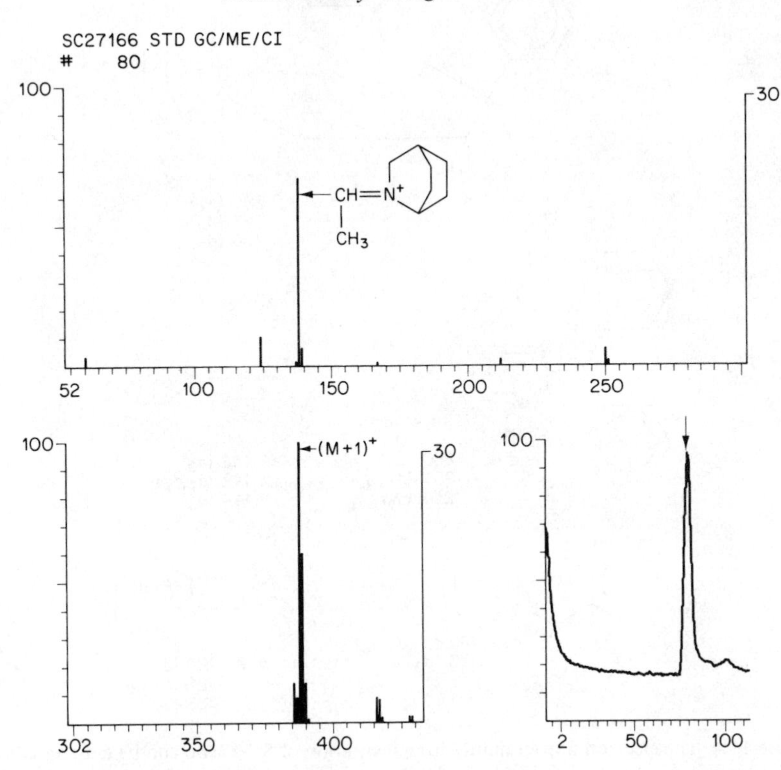

Figure 11.3 The methane chemical ionisation mass spectrum of SC-27166

line was obtained for mixtures of d_0-SC-27166 in plasma over the range 25–2000 ng/ml.

Analysis of the urine obtained from rats dosed simultaneously with d_0-SC-27166 and d_4-SC-27166 showed the ratio d_0/d_4 SC-27166 excreted in the urine to be similar to the ratio d_0/d_4 SC-27166 administered by both routes (Table 11.1). This suggests that absorption of the orally administered dose was essentially complete, although in the rat less than 10 per cent of the dose is excreted in the urine.

Two attempts were made to administer SC-27166 to the beagle dog. On the first occasion d_0-SC-27166 was administered orally followed, 15 min later, by d_4-SC-27166 administered intravenously. After 5 min the dog vomited. On the second occasion the dog was administered d_4-SC-27166 (10 mg) intravenously and allowed to vomit. After 10 min the dog appeared relaxed and 5 min later it was given two lactose tablets containing d_0-SC-27166 (5 mg each) which it swallowed. The analysis of the blood samples is summarised in Table 11.2. The low oral availability of SC-27166 in this case may be due to the intestinal discomfort experienced by the dog, and further experiments with dogs were abandoned. The data obtained (Figure 11.6) were analysed using C. M. Metzler's NONLIN program (Metzler, 1969). The intravenous data fitted a monoexponential function $C_p = Ae^{-\lambda t}$, whereas the oral data fitted a

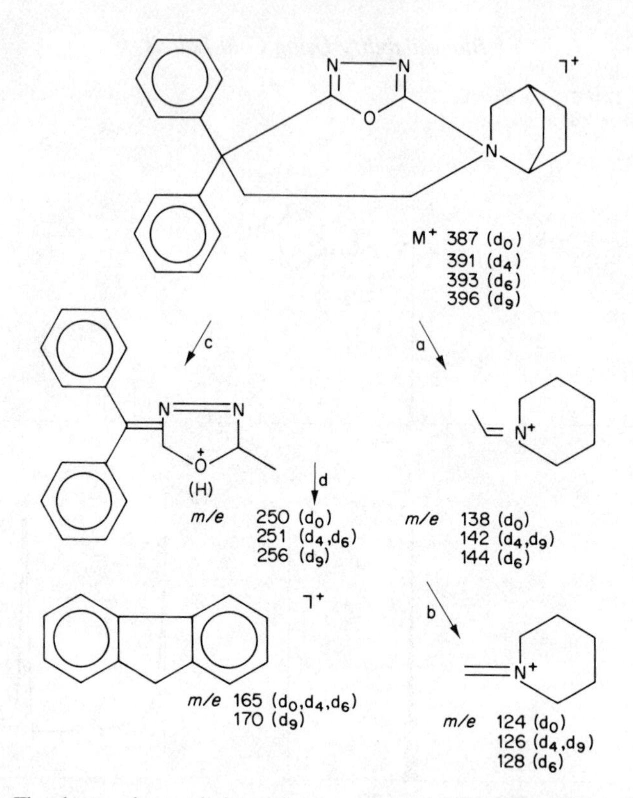

Figure 11.4 The electron impact-induced fragmentation of SC-27166 and its deuterated analogues

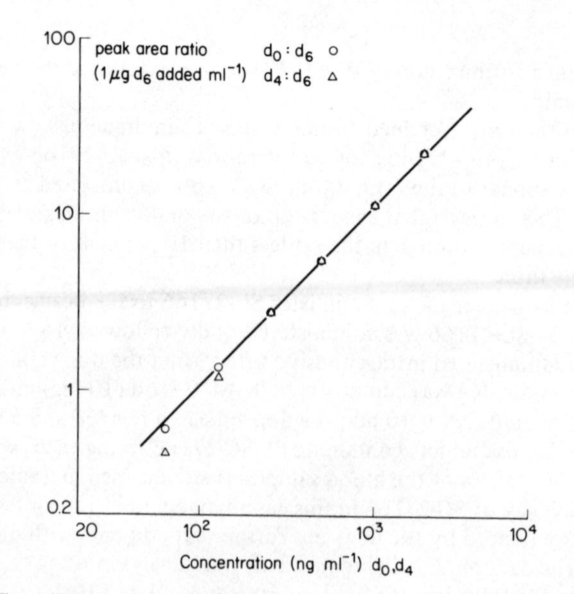

Figure 11.5 The calibration lines obtained for mixtures of d_0-SC-27166 (–○–) and d_4-SC-27166 (–▽–) in plasma using d_6-SC-27166 as internal standard and methane chemical ionisation mass spectrometry

Table 11.1 The ratio of d_0/d_4 SC-27166 administered to rats by oral and intravenous routes, and the ratio of d_0/d_4 observed in SC-27166 extracted from rat urine collected 0–24 h post dose

Rats	p.o. dose	i.v. dose	Ratio d_0/d_4 administered	Ratio d_0/d_4 observed	Availability (%)
1 and 2	2.8 mg d_0	2.4 mg d_4	1.17	1.13	96.6
3 and 4	3.0 mg d_4	2.1 mg d_0	0.70	0.74	94.6

Table 11.2 Pharmacokinetic parameters derived from the plasma concentration–time curves obtained after administering d_0- and d_4-SC-27166 by oral and intravenous routes to a beagle dog. The curves were fitted to a biexponential equation of the form: $C_p = A(e^{-\lambda_1 t} - e^{-\lambda_2 t})$ for oral dosage, and $C_p = Ae^{-\lambda t}$ for intravenous dosage

Parameter	Intravenous	Oral
$A \pm$ SD	220.26 ± 6.05	370.06 ± 82.97
$\lambda_1 \pm$ SD	0.4992 ± 0.0094	0.7662 ± 0.0595
$\lambda_2 \pm$ SD	–	1.0724 ± 0.1641
Correlation	0.989	0.912
$t_{1/2}$ absorption (h)	–	0.65
$t_{1/2}$ elimination (h)	1.39	0.90
fitted AUC (ng h ml^{-1})	499.9	137.9
trapezoidal AUC (ng h ml^{-1})	568.3	151.5
Bioavailability by fitted areas	27.59%	
Bioavailability by trapezoidal areas	26.66%	

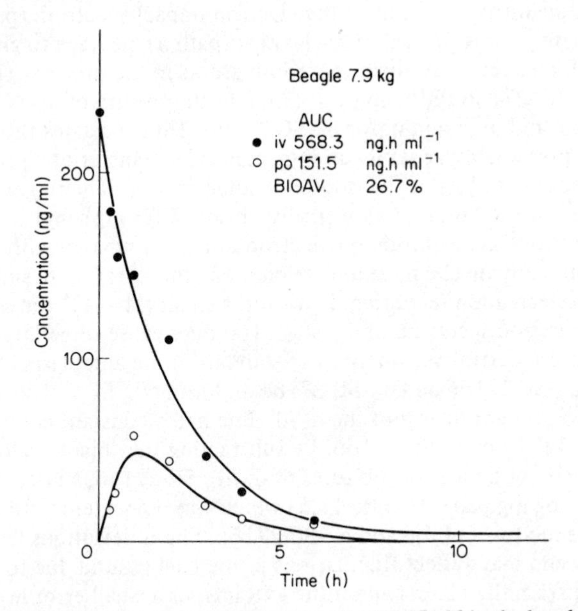

Figure 11.6 The plasma concentration–time curves for SC-27166 in the beagle dog after administration of d_4-SC-27166 (10 mg) intravenously and d_0-SC-27166 (10 mg) orally

biexponential function $C_p = A(e^{-\lambda_1 t} - e^{-\lambda_2 t})$. The areas under the curves were calculated as an integration of the summation of trapezoidal areas. For these analyses zero time was taken as the time of oral administration. The area under the plasma curve following intravenous administration was calculated by back-extrapolation of the fitted curve to −15 min.

No problems were encountered when dosing the baboons and analysis of the plasma samples gave good plasma concentration–time curves fitted to a biexponential equation for the intravenous data, and a triexponential equation for the oral data (Table 11.3). In both cases the 25 h point was omitted from the calculations as this appears to lie on a third elimination phase which is insufficiently defined for analysis. Again the area under the fitted curve was calculated, by back-extrapolation in the case of the oral data, to the actual dosage point.

No free SC-27166 was detected in the urine of the beagle and only low levels in the baboon.

DISCUSSION

Mass spectra

The molecular ion in the electron impact mass spectrum of SC-27166 or its deuterated analogue is small, being less than 1 per cent of the base peak (Figure 11.2), whereas using methane chemical ionisation the pseudomolecular ion $(M + 1)^+$ is the base peak (Figure 11.3). The major fragment ions from the electron impact process are shown in Figure 11.4. Although β-cleavage to the isoquinuclidine nitrogen occurs in the electron impact spectrum (path b, Figure 11.4), the base peak is formed by γ-cleavage (path a), perhaps to give the ion shown. Both ions retain all their deuterium atoms in the process. The ion *m/e* 250 for SC-27166 shifts up by 1 a.m.u. in the spectra of d_4-SC-27166 and d_6-SC-27166, and by 6 a.m.u. for d_9-SC-27166. Thus it retains the biphenyl moiety and presumably arises by an intramolecular transfer of a proton from the ethylene bridge with loss of the isoquinuclidine moiety. The fragment d (*m/e* 165) is the fluorenium ion normally observed for biphenyl compounds.

The extensive fragmentation on electron impact compared with the intense parent ion in methane chemical ionisation made the latter more suitable for an assay using selected ion recording. However, a small $(M - 1)^+$ ion seen in the chemical ionisation spectrum of d_6-SC-27166 does cause some interference with the tetradeutero derivative. Analyses of standard urine and plasma samples containing d_0-SC-27166 and d_4-SC-27166 and using d_6-SC-27166 as an internal standard gave straight lines, but the d_4/d_6 line had a constant contribution due to the $(M - 1)^+$ from d_6-SC-27166. By subtracting this 'blank' value from the peak area ratios obtained for mixtures of d_4/d_6 SC-27166, a linear plot was obtained on log-log paper (Figure 11.5) which was coincident with the plot obtained for mixtures of d_0- and d_6-SC-27166. Slight deviations from linearity at the lower end may reflect fluctuations in the background due to endogenous interference from the plasma and urine extracts, or a small error in the constant term in the equation for the line.

Table 11.3 Pharmacokinetic parameters derived from the plasma concentration–time curves obtained after administering d_0- and d_4-SC-27166 by oral and intravenous routes to two baboons. The curves were fitted to a triexponential equation of the form $C_p = A_1e^{-\lambda_1 t} + A_2e^{-\lambda_2 t} + A_3e^{-\lambda_3 t}$

Parameter	Baboon No. 17		Baboon No. 21	
	Intravenous	Oral	Intravenous	Oral
$A_1 \pm SD$	270.03 ± 138.4	−1369.8 ± 1320.1	423.16 ± 83.06	563.56 ± 64.62
$A_2 \pm SD$	496.68 ± 116.88	322.83 ± 256.80	657.65 ± 49.26	−581.69 ± 39.91
$A_3 \pm SD$	–	259.25 ± 294.46	–	–
$\lambda_1 \pm SD$	0.645 ± 0.089	4.579 ± 2.693	1.448 ± 0.158	0.257 ± 0.011
$\lambda_2 \pm SD$	0.203 ± 0.038	0.366 ± 0.260	0.247 ± 0.010	0.703 ± 0.144
$\lambda_3 \pm SD$	–	0.147 ± 0.074	–	–
Correlation	0.998	0.995	0.996	0.976
$t_{1/2}$ absorption (h)	–	0.15		0.99
$t_{1/2}$ elimination (h)	1.08	1.90	0.48	2.69
$t_{1/2}$ elimination (slow) (h)	3.41	4.73	2.81	–
fitted AUC ngh ml^{-1}	2863	2354	2956	1452
trap AUC ngh ml^{-1}	2921	2369	3023	1637
Bioavailability by fitted areas		82.22%		49.12%
Bioavailability by trapezoidal areas		81.10%		54.15%
Bioavailability from ratio d_0/d_4 in urine		123%		52.90%

Animal experiments

The urinary excretion of intact SC-27166 in the rat is very low. However, the ratio of labelled to unlabelled SC-27166 recovered from the rat urine is not significantly different from the ratio administered by intravenous and oral routes (Table 11.1). The bioavailability in the two pairs of rats is 95.6 ± 1.0 per cent, which suggests that SC-27166 is completely available from an oral solution in the rat.

Studies on the beagle dog were difficult owing to the sensitivity of this animal to intravenous administration of SC-27166. Two factors conspire to reduce availability, namely giving the oral dose as a tablet, thus introducing a dissolution rate factor, and the state of the animal's stomach after vomiting, which probably affects gut motility. The low availability of the oral dose in the beagle therefore is not an accurate estimate of the bioavailability in the normal animal.

The baboon showed no sign of any discomfort after receiving intravenous SC-27166, and this study proceeded normally. The first animal (No. 17) was given d_0-SC-27166 orally and d_4-SC-27166 intravenously. Analysis of the plasma concentration–time curves (Figure 11.7) showed 81 per cent availability of the oral dose. By examining the d_0/d_4 ratio for free SC-27166 in urine a value of 123 per cent was obtained. In this animal SC-27166 is therefore readily

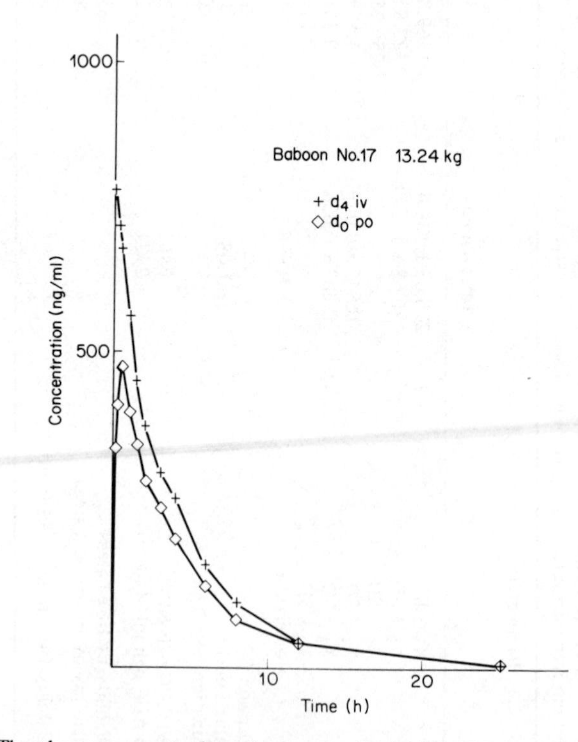

Figure 11.7 The plasma concentration–time curves for SC-27166 in the baboon after administration of d_4-SC-27166 (0.25 mg/kg) intravenously and d_0-SC-27166 (0.25 mg/kg) orally

available after oral administration. In the second animal (No. 21) a better correlation between the bioavailability calculated by comparison of the areas under the plasma concentration–time curve and the urinary d_0/d_4 ratio for free SC-27166 was obtained. In this experiment SC-27166 appeared to be 50–55 per cent available after oral administration.

Communication with Dr A. Karim (G. D. Searle and Co, Skokie, Illinois), who has undertaken a study of SC-27166 availability in the rat, rhesus monkey and man using $[^{14}C]$-SC-27166 (results to be published elsewhere), showed that he had obtained similar figures for bioavailability in the rat and rhesus monkey to those reported here for the rat and the baboon. It therefore appears that the use of simultaneous administration of a compound by intravenous and oral routes can give accurate data concerning the availability of any such compound if one route is differentiated from the other by use of a stable isotope. Although in this study the correlation between the availability calculated from plasma data and the ratio of label excreted in the urine in the baboon is poor, this may reflect inadequate technique in this preliminary study rather than a serious discrepancy in the hypothesis that urinary excretion reflects the ratio of label in plasma.

It seems likely that one could envisage studies in man whereby a compound is administered intravenously in a labelled form and an oral formulation is given simultaneously. By collecting urine for an adequate period to cover excretion of the compound and measuring the ratio of label in the urine, a measure of the availability from a particular formulation can be made. Thus the classical bioavailability study in man involving a crossover design with a long wash-out period between doses, a large number of bleeds in each phase of the study and the analysis of several hundred biological samples may soon be replaced by a study whereby a single intravenous injection and an oral formulation is given, followed by collection of urine only (a relatively painless procedure) with only a few samples for the analyst to contend with. The analytical problem is also reduced inasmuch as for a normal man about 1 litre of urine is available per 24 h, whereas for blood samples 10 ml bleeds are usually the maximum obtainable. Studies to test this procedure are now under way in our laboratories.

ACKNOWLEDGEMENTS

I acknowledge the support given by Dr A. Karim (Searle, Skokie) in this study. Pharmacokinetic analyses were carried out by D. A. Rose and animal experiments by P. N. Spalton and Christine M. Walls. The deuterated analogues were synthesised by T. A. Harrow, I. Fellows, R. Honeyman and P. H. Buckley. The manuscript was typed by Karen Petch and Helen Cuthbert.

REFERENCES

Adelstein, G. W., Yen, C. H., Dajani, E. Z. and Bianchi, R. G. (1976). *J. Med. Chem.*, **19**, 1221

Dajani, E. Z., Bianchi, R. G., Adelstein, G. W. and Yen, C. H. (1975). Studies on the biological properties of SC-27166, a novel antidiarrhoeal agent. Presented at The Philadelphia G.I. Research Forum, Philadelphia, Pa., October 29–30

Dutcher, J. S., Strong, J. M., Lee, W.-K. and Atkinson, A. J. (1975). *Proceedings of the Second International Conference on Stable Isotopes, Oak Brook, Illinois, USA*, National Technical Information Service Document CONF-751027, p. 186

Metzler, C. M. (1969). NONLIN, Technical Report No. 7292/69/7293/005, The Upjohn Co., Kalamazoo

Strong, J. M., Dutcher, J. S., Lee, W.-K. and Atkinson, A. J. (1975). *Clin. Pharmacol. Ther.*, **18**, 613

12
The use of stable isotopes to better define toxic metabolic pathways of *p*-hydroxyacetanilide (acetaminophen) and *p*-ethoxyacetanilide (phenacetin)

S. D. Nelson, W. A. Garland, J. R. Mitchell, Y. Vaishnev, C. N. Statham* and A. R. Buckpitt† (NIH, Bethesda, Maryland and Hoffmann-La Roche, Inc., Nutley, New Jersey, USA)

INTRODUCTION

Phenacetin (acetophenetidine) is a minor analgesic which has been associated with various toxic effects, including methaemoglobinaemia and haemolysis (Beutler, 1969) and renal damage (Abel, 1970). Phenacetin causes both methaemoglobinaemia and mild hepatic necrosis in hamsters. As with the related analgesic, acetaminophen, phenacetin-induced necrosis is potentiated by 3-methylcholanthrene pretreatment. The severity of necrosis parallels the magnitude of the covalent binding of an arylating metabolite of phenacetin to hepatic proteins and depletion of hepatic glutathione by the formation of a glutathione adduct (Mitchell *et al.*, 1975).

Phenacetin is de-ethylated in man and other animals to form acetaminophen (see Garland, Nelson and Sasame, 1976). Studies in rats showed that the de-ethylation of phenacetin is rapid and that the plasma half-life of phenacetin in rats decreases from 28 to 4.5 min after pretreatment with 3-methylcholanthrene (Welch, Hughes and DeAngelis, 1976). Because of this rapid first-pass metabolism of phenacetin, large amounts of acetaminophen are formed. Acetaminophen is known to produce liver necrosis in man and other animals but very little methaemoglobinaemia (Mitchell *et al.*, 1975). Phenacetin-induced methaemoglobin formation has therefore been postulated to require the intact

* Dr C. N. Statham is a staff fellow in the Pharmacology-Toxicology Research Associate Program of the National Institute of General Medical Sciences.
† Dr A. R. Buckpitt was supported by the National Institute of Heart, Lung, and Blood Fellowship No. 1 F32 HL05236 01.

phenetidine structure (Kiese, 1966) and loss of the acetyl moiety by enzymatic hydrolysis (Heymann *et al*., 1969; DiChiara *et al*., 1972; Mitchell *et al*., 1975).

In order to study the effect of phenacetin O-de-ethylation on the toxic metabolic pathways leading to hepatic necrosis and methaemoglobinaemia, we synthesised *p*-[1,1-^{2}H]-ethoxy acetanilide (phenacetin-d_2, Figure 12.1) and *p*-[2,2,2-^{2}H]-ethoxy acetanilide (phenacetin-d_3). With these stable-isotope-labelled analogues we could correlate the effects of deuterium substitution on the de-ethylation reaction, and on the production of hepatic necrosis and methaemoglobinaemia. Covalent binding and glutathione depletion were also monitored after the administration of the stable isotope analogues.

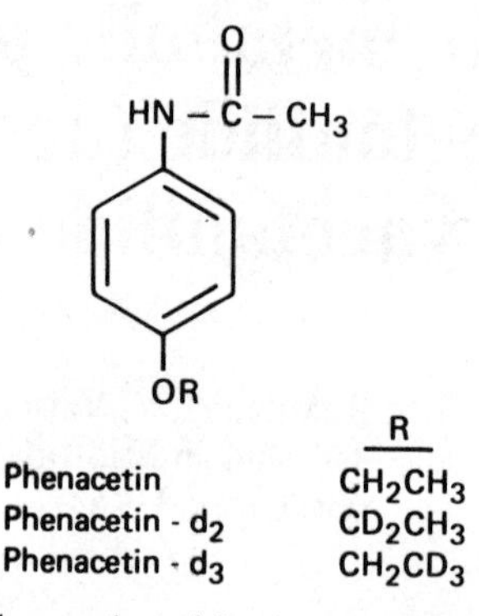

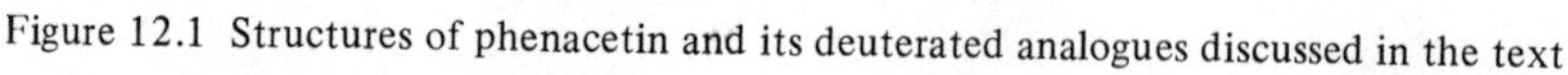
Figure 12.1 Structures of phenacetin and its deuterated analogues discussed in the text

In addition, studies using hamster liver microsomes were carried out to determine the effects of deuterium substitution on the de-ethylation of phenacetin to acetaminophen *in vitro*, and on the covalent binding of radiolabelled analogues to microsomal protein. Results of these experiments indicated that the major pathway for the formation of a hepatotoxic metabolite of phenacetin in hamsters *in vivo* differed from the major pathway for the formation of a reactive metabolite *in vitro* using hamster liver microsomes. Results from experiments carried out previously with hamster liver microsomes and ^{18}O (Hinson, Nelson and Mitchell, 1976, 1977) support the results obtained from the deuterium isotope experiments, and will also be discussed.

MATERIALS AND METHODS

Chemicals

Acetaminophen and phenacetin were purchased from Eastman Organic Chemicals and recrystallised from methanol-water prior to use. Phenacetin-d_2 and phenacetin-d_3 were synthesised as previously described (Garland *et al*., 1976). Ring-labelled [^{3}H]-phenacetin and its deuterated analogues were prepared by hydrolysing *p*-hydroxy acetanilide-[^{3}H] (New England Nuclear) and re-acetylating with acetic anhydride (Aldrich Chemicals), followed by alkylation with the appropriately labelled ethyl bromide. NADP, glucose-6-phosphate and glucose-6-phosphate dehydrogenase were purchased from Sigma Chemical Co.

Hepatic necrosis, covalent binding and glutathione depletion
These parameters were measured as previously described for acetaminophen (Mitchell *et al.*, 1973a, 1973b). Hamsters were all pretreated with 3-methylcholanthrene (20 mg/kg i.p. every 12 h for 3 doses) and administered the appropriately labelled phenacetin (400 mg/kg in Tween 80) 24 h after the last dose of 3-methylcholanthrene.

Methaemoglobin determination
Samples of blood were obtained from the retro-orbital sinus of hamsters and methemoglobin was determined as g/100 g whole blood by the method of Evelyn and Malloy (1938).

Tissue preparation and incubation procedures
Microsomal enzymes were obtained from livers of male Golden Syrian hamsters and microsomal reactions were carried out as previously described (Hinson *et al.*, 1977).

Mass spectrometry
Mass spectrometry was performed on either a Finnigan Model 1015D gas chromatograph–mass spectrometer or a VG MicroMass 16F mass spectrometer coupled to a Varian Aerograph Model 1400 gas chromatograph. Acetaminophen was determined as previously described (Garland *et al.*, 1976), using gas chromatography coupled with chemical ionisation mass spectrometry and selected ion monitoring.

High-pressure liquid chromatography
Acetaminophen and phenacetin, when radiolabelled, were determined by liquid scintillation counting of the eluant from high-pressure liquid chromatography of supernatants from incubations of hamster liver microsomes after the reaction had been terminated by methanol precipitation of the microsomal protein. The same assay was applied to plasma samples from hamsters after hydrolysis of acetaminophen conjugates with β-glucuronidase-arylsulphatase (Calbiochem) and extraction of acetaminophen with ethyl acetate.

RESULTS

In vivo results

Hepatic necrosis
Phenacetin caused extensive necrosis to the centrilobular area of the liver in hamsters pretreated with 3-methylcholanthrene (Table 12.1). Substitution of deuterium for hydrogen on the methylene carbon atom alpha to the phenolic oxygen caused a significant decrease in both the extent and severity of the lesion. Substitution of deuterium for hydrogen on the methyl carbon atom beta to the phenolic oxygen (phenacetin-d_3) did not significantly alter the extent or severity of hepatic necrosis.

Table 12.1 Extent of hepatic necrosis in hamsters pretreated with 3-methylcholanthrene and administered (400 mg/kg, i.p.) phenacetin and its deuterated analogues

Compound administered	Number of animals	Mortality (%)	Extent of necrosis[a,b] (%)			
			0	1+	2+	3+
Phenacetin-d_0	60	45	24	52	18	6
Phenacetin-d_2	60	35	74	23	3	0
Phenacetin-d_3	60	52	41	31	24	4

[a] Extent of hepatic necrosis was scored in survivors by the criteria presented in Mitchell *et al.* (1973a).
[b] The extent of necrosis was significantly different for phenacetin-d_0 and phenacetin-d_2 by χ^2 analysis, $\beta < 0.05$. There was no significant difference between phenacetin-d_0 and phenacetin-d_3.

Covalent binding
Four hours after administration, a time previously determined to be maximal for covalent binding, ring-labelled [^{3}H]-phenacetin and phenacetin-d_3 were bound to hamster hepatic and renal tissue protein at levels significantly higher ($P < 0.02$) than ring-labelled [^{3}H]-phenacetin-d_2 (Table 12.2). For the liver, the ratio of phenacetin to phenacetin-d_2 bound was 1.87, whereas the ratio of phenacetin to phenacetin-d_3 bound was 1.07.

Table 12.2 Covalent binding of radiolabelled phenacetins to hamster tissue protein 4 h after a dose of 400 mg/kg

Drug administered	Liver	Kidney
	(nmol/mg protein)	
Phenacetin-d_0	2.25 ± 0.21	0.32 ± 0.06
Phenacetin-d_2	1.20 ± 0.03	0.10 ± 0.01
Phenacetin-d_3	2.11 ± 0.22	0.19 ± 0.05

Glutathione depletion
The time course for the depletion of glutathione from hamster liver was followed over an 8 h period. Although there were no significant differences in glutathione depletion for the first 2 h after administering phenacetin-d_0, -d_2 or -d_3, phenacetin-d_2 caused significantly less glutathione depletion at times of maximal glutathione depletion (Figure 12.2).

Plasma acetaminophen levels
Plasma acetaminophen levels at this 4 h time period were measured both by a high-pressure liquid chromatographic assay and a gas chromatographic–mass spectral assay as described under Materials and Methods. The plasma concen-

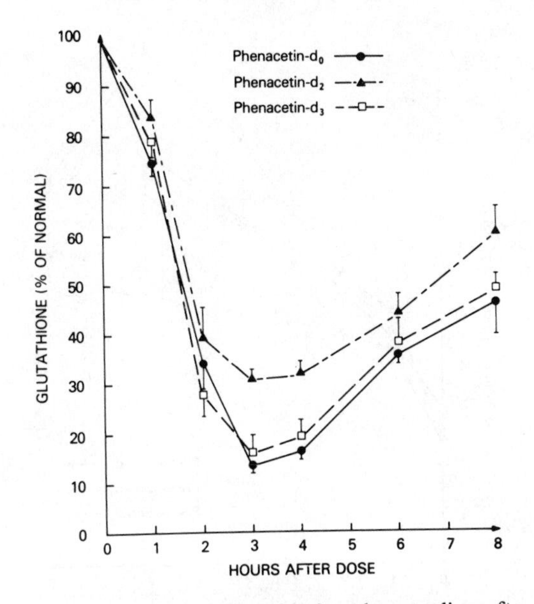

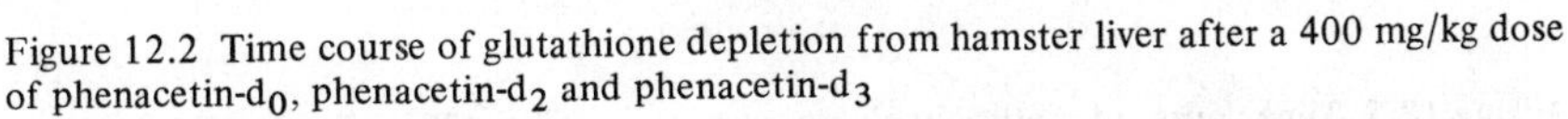
Figure 12.2 Time course of glutathione depletion from hamster liver after a 400 mg/kg dose of phenacetin-d_0, phenacetin-d_2 and phenacetin-d_3

tration of acetaminophen 4 h after the administration of phenacetin was significantly higher ($P < 0.05$) than the acetaminophen concentration after the administration of either phenacetin-d_2 or phenacetin-d_3 (Table 12.3). Calculating from these results, the ratio of the plasma concentration of acetaminophen after phenacetin-d_2 *vs.* phenacetin was 3.15 compared with a ratio of 1.48 for phenacetin-d_3 *vs.* phenacetin.

Table 12.3 Plasma level of acetaminophen (paracetamol) in hamsters 4 h after 400 mg/kg phenacetin

Drug administered (400 mg/kg)	Plasma concentration of acetaminophen (μg/ml)
Phenacetin-d_0	117 ± 2
Phenacetin-d_2	37 ± 8
Phenacetin-d_3	78 ± 17

Methaemoglobin formation

In contrast to the above results, phenacetin-d_2 caused significantly more methaemoglobin formation than did either phenacetin or phenacetin-d_3 (Figure 12.3).

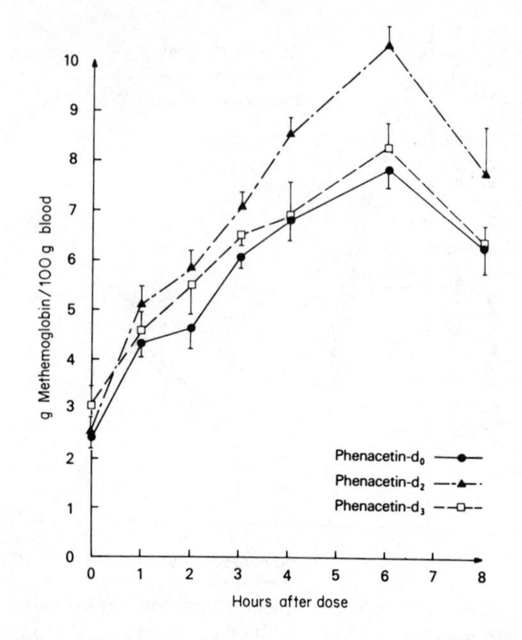

Figure 12.3 Time course of methaemoglobin formation after a 400 mg/kg dose of phenacetin-d_0, phenacetin-d_2 and phenacetin-d_3

In vitro results

Covalent binding

The K_m and V_{max} for the covalent binding of a reactive metabolite of phenacetin to hamster liver microsomes were almost identical for phenacetin and its deuterated analogues (Table 12.4).

Table 12.4 Michaelis constant and maximum velocity for the covalent binding of radiolabelled phenacetins to hamster liver microsomes

Substrate	K_m (mM)	V_{max} (nmol mg^{-1} 10 min^{-1})
Phenacetin-d_0	0.232 ± .015	1.32 ± .076
Phenacetin-d_2	0.204 ± .034	1.32 ± .045
Phenacetin-d_3	0.218 ± .051	1.36 ± .066

O-De-ethylation

The formation of acetaminophen from phenacetin by hamster liver microsomes was determined both by high-pressure liquid chromatography and by gas chromatography–mass spectrometry. The K_m for the de-ethylation of phenacetin and its deuterated analogues to acetaminophen by hamster liver microsomes was virtually the same as for phenacetin and its deuterated analogues (Table 12.5). However, the V_{max} for the O-de-ethylation of phen-

Table 12.5 Oxidative de-ethylation of phenacetins by hamster liver microsomes

Substrate	K_m (mM)	V_{max} (nmol mg^{-1} 10 min^{-1})
Phenacetin-d_0	48 ± 2.4	7.31 ± 0.33
Phenacetin-d_2	44 ± 5.5	3.92 ± 0.42
Phenacetin-d_3	60 ± 7.8	6.52 ± 0.30

acetin-d_2 was significantly slower (k_H/k_{d_2} = 1.86) than that for phenacetin. There was only a slight and insignificant isotope effect (k_H/k_{d_3} = 1.12) for the O-de-ethylation of phenacetin-d_3 when compared with phenacetin. Similar results have been observed for the O-de-ethylation of phenacetin by rabbit liver microsomes (Garland *et al.*, 1976).

DISCUSSION

In order to study the toxic metabolic pathways of phenacetin in detail, we synthesised the stable-isotope-labelled analogues phenacetin-d_2 and phenacetin-d_3. Results of experiments with phenacetin and phenacetin-d_2 showed a significant deuterium isotope effect on the hepatic necrosis caused by phenacetin in hamsters. At a 400 mg/kg dose of phenacetin-d_2, the number of animals with necrosis was decreased by more than 50 per cent when compared with the same dose of phenacetin and the extent of necrosis in the remaining affected animals was significantly reduced. There was also a slight decrease in mortality caused by phenacetin-d_2 *vs.* phenacetin (Table 12.1). On the other hand, no significant difference was found in the incidence or severity of hepatic necrosis between phenacetin-d_3 and phenacetin.

A second parameter, the covalent binding of radiolabelled phenacetin, phenacetin-d_2 and phenacetin-d_3 to hepatic tissue, provided results which paralleled the effects of the isotopes on hepatic necrosis (Table 12.2). Almost twice as much reactive metabolite was bound to hepatic tissue protein after the administration of ring-labelled [^{3}H]-phenacetin or phenacetin-d_3 than after the administration of ring-labelled [^{3}H]-phenacetin-d_2. At this same time period of maximal covalent binding (4 h after the administration of drug), the plasma concentration of the de-ethylated metabolite, acetaminophen, was more than three times higher after the administration of phenacetin than after the administration of phenacetin-d_2 (Table 12.3), when determined by specific assays using either high-pressure liquid chromatography or gas chromatography–mass spectrometry. These results would strongly suggest that in 3-methylcholanthrene-pretreated hamsters the hepatic necrosis caused by phenacetin was mediated by its major metabolite acetaminophen, that is, the de-ethylation of phenacetin to acetaminophen was rate-determining for the hepatic necrosis caused by phenacetin.

This conclusion was strengthened by the time course studies of hepatic glutathione depletion (Figure 12.2), which revealed that phenacetin-d_2 depleted glutathione to a lesser extent than either phenacetin or phenacetin-d_3. Glutathione is known to protect the liver from cellular injury caused by acetaminophen by reacting with an electrophilic metabolite formed by microsomal oxidation of acetaminophen (Mitchell *et al.*, 1973b).

In contrast to these results *in vivo*, no deuterium isotope effect was observed for the covalent binding of radiolabelled phenacetin-d_2 or phenacetin-d_3 *in vitro* when compared with phenacetin, in spite of the fact that there was a significant isotope effect ($k_H/k_d \sim 2$) for the O-de-ethylation of phenacetin to acetaminophen (Tables 12.4 and 12.5). This finding would suggest that the dominant mechanism for the formation of a reactive arylating metabolite of phenacetin *in vitro* is not via the production of acetaminophen and its further oxidation. This is consistent with data from our other studies (Hinson *et al.*, 1976, 1977) which gave the following results *in vitro* using hamster liver microsomes: (1) The maximum velocity of covalent binding for phenacetin exceeds that for acetaminophen, showing that phenacetin is not first de-ethylated to acetaminophen which is then activated. (2) Pretreatment of hamsters with 3-methylcholanthrene increases the rate of covalent binding for acetaminophen but decreases the rate of binding for phenacetin. (3) Phenobarbital pretreatment increases the rate of covalent binding for phenacetin without affecting the rate of binding for acetaminophen. (4) When covalent binding was prevented by trapping of the reactive metabolite with glutathione during incubations carried out under atmospheres of ^{18}O, reduction by Raney nickel of the glutathione conjugates formed from either acetaminophen or phenacetin yielded acetaminophen. However, the acetaminophen conjugate formed during incubations with phenacetin incorporated 50 per cent ^{18}O into the 4 position, as determined by mass spectrometry, whereas the acetaminophen–glutathione conjugate formed during incubations with acetaminophen incorporated no ^{18}O. These results showed that different arylating metabolites were formed *in vitro* after phenacetin and acetaminophen. Thus the dominant mechanisms for the formation of reactive arylating metabolites from phenacetin *in vivo* differ from those with hamster liver microsomes *in vitro*, apparently because phenacetin hepatotoxicity *in vivo* results from its de-ethylation to acetaminophen. Whether this mechanism is important in the genesis of the renal injury caused by phenacetin in man is unknown.

Interestingly, while phenacetin-d_2 is a less potent hepatotoxin in hamsters, it is more toxic with regard to methaemoglobin formation (Figure 12.3). Substitution of deuterium for hydrogen in the methylene carbon atom suppressed the de-ethylation of phenacetin to acetaminophen but augmented methaemoglobin formation. This suggests that the methaemoglobin formation caused by phenacetin requires the intact phenetidine structure (Kiese, 1966) and is in agreement with previous work showing that deacetylation of phenacetin is a prerequisite for methaemoglobinaemia (Heymann *et al.*, 1969; DiChiara *et al.*, 1972; Mitchell *et al.*, 1975).

From our results we can postulate (Figure 12.4) that phenacetin is oxidatively de-ethylated to its major metabolite *in vivo*, acetaminophen, by the formation of a hemiacetal. This unstable intermediate will decompose to acetaldehyde and acetaminophen. Because of the large amounts of acetaminophen formed *in vivo* by this efficient de-alkylation process, reactive metabolite formation and hepatic necrosis results from the further metabolism of acetaminophen. By substituting deuterium for hydrogen at the position being oxidised, this pathway is diminished and the extent of hepatic necrosis is

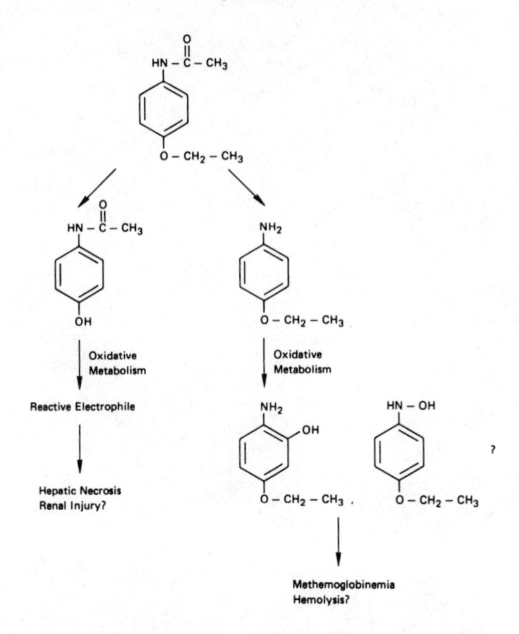

Figure 12.4 Possible pathways for toxic metabolite formation from phenacetin

decreased. However, by blocking this oxidative process, the concentrations of other non-de-ethylated products such as phenetidine may increase, and methaemoglobin formation increases. Thus the use of stable isotopes has better defined the specific toxicological mechanisms underlying the hepatic necrosis and methaemoglobinaemia caused by acetaminophen and phenacetin.

REFERENCES

Abel, J. A. (1970). *Clin. Pharmacol. Ther.*, **12**, 583

Beutler, E. (1969). *Pharmacol. Rev.*, **21**, 73

DiChiara, G., Potter, W. Z., Jollow, D., Mitchell, J. R., Gillette, J. R. and Brodie, B. B. (1972). *Fifth International Congress on Pharmacology* (Abstract), p. 57

Evelyn, K. A. and Malloy, H. T. (1938). *J. Biol. Chem.*, **126**, 655

Garland, W. A., Nelson, S. D. and Sasame, H. A. (1976). *Biochem. Biophys. Res. Commun.*, **72**, 539

Heymann, E., Krisch, K., Buch, H. and Buzello, W. (1969). *Biochem. Pharmacol.*, **18**, 801

Hinson, J. A., Nelson, S. D. and Mitchell, J. R. (1976). *The Pharmacologist* (Abstract), **18**, 24

Hinson, J. A., Nelson, S. D. and Mitchell, J. R. (1977). *Molec. Pharmacol.* (in press)

Kiese, M. (1966). *Pharmacol. Rev.*, **18**, 1091

Mitchell, J. R., Jollow, D. J., Potter, W. Z., Davis, D. C., Gillette, J. R. and Brodie, B. B. (1973a). *J. Pharmacol. Exp. Ther.*, **187**, 185

Mitchell, J. R., Jollow, D. J., Potter, W. Z., Gillette, J. R. and Brodie, B. B. (1973b). *J. Pharmacol. Exp. Ther.*, **187**, 211

Mitchell, J. R., Potter, W. Z., Hinson, J. A., Snodgrass, W. R., Timbrell, J. A. and Gillette, J. R. (1975). *Handbook of Experimental Pharmacology* (ed. O. Eichler, A. Farah, H. Herken and A. D. Welch), Springer Verlag, Berlin, p. 383

Welch, R. M., Hughes, C. R. and DeAngelis, R. L. (1976). *Drug Metab. Dispos.*, **4**, 402

13
A new method for the investigation of inter-individual differences in drug metabolism

J. D. Baty* and P. R. Robinson† (Nuffield Unit of Medical Genetics, Department of Medicine, University of Liverpool, Liverpool, L69 3BX, UK

INTRODUCTION

Inter-individual differences in rates of hepatic oxidation are due to a multiplicity of factors and are probably under polygenic control. The genetic influence in drug oxidation for a particular individual is reflected in their hepatic enzyme activity. We are studying a metabolic pathway which allows us to compare the enzymes involved in aliphatic and aromatic oxidation in man. Acetanilide and phenacetin are both metabolised to paracetamol in man, approximately 80 per cent of each compound being converted to paracetamol, which is found in plasma as the free and conjugated material, and which is excreted in the urine as the sulphate and glucuronide conjugate (Brodie and Axelrod, 1948, 1949). We have studied the C-oxidation of acetanilide-d_5 and phenacetin in man following an equimolar dose of both compounds. Phenacetin is metabolised to paracetamol by oxidative de-alkylation and aromatic hydroxylation of acetanilide-d_5 will produce paracetamol-d_4. The hydroxylated metabolites are excreted as the glucuronide and sulphate (Figure 13.1). Using selected ion monitoring techniques, we can monitor the pharmacokinetics of the two forms of paracetamol and can thus compare the relative amounts of paracetamol produced by different individuals at the same time periods after dosing (Baty and Robinson, 1977). We have studied the biological isotope effects present in this system and in the de-alkylation of phenacetin labelled with five deuterium atoms in the ethyl group.

* Present address: Department of Biochemical Medicine, Ninewells Hospital, Dundee DD2 1UD.
† Present address: Sterling Winthrop Laboratories, Metabolic Studies Section, Fawdon, Newcastle-upon-Tyne NE3 3TT.

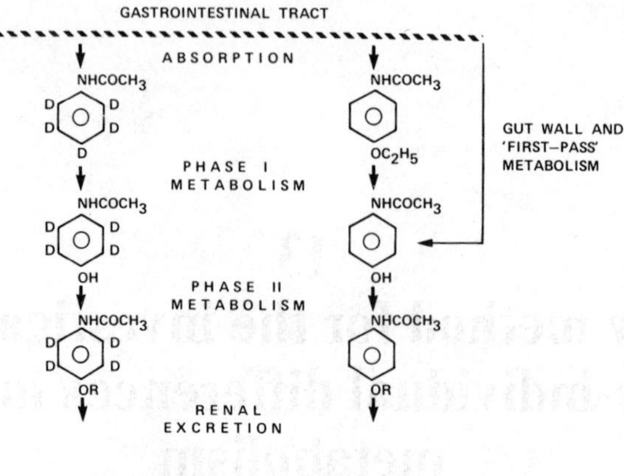

Figure 13.1 The production of paracetamol from phenacetin and acetanilide-d_5

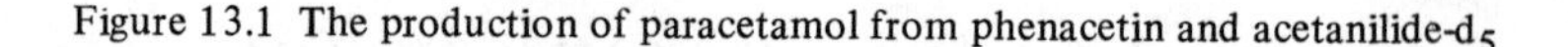

MATERIALS AND METHODS

The deuterium-labelled phenacetin was prepared from paracetamol and ethyl iodide-d_5 in the presence of sodium ethoxide/ethanol. After re-crystallisation from ethanol it had a melting point of 130–133°. The syntheses of the other compounds used in this study have been described in previous communications (Baty, Robinson and Wharton, 1976; Baty and Robinson, 1977). Gas chromatography–mass spectrometry (GC–MS) analyses were carried out using a Pye 104 gas chromatograph coupled via a silicone rubber membrane separator to an MS 12 mass spectrometer (A.E.I., Associated Electrical Industries Ltd, Manchester), operated at 8 kV with a trap current of 300 μA and an ionising voltage of 25 eV. Glass columns (1.5 m x 2 mm), packed with 3 per cent OV-17 on 100/120 mesh Gas Chrom Q, were used in the gas chromatograph.

Selected ion monitoring was carried out using a six-channel multi-peak monitor unit manufactured by A.E.I. The signal from each selected mass value was displayed on a Rikadenki multi-channel chart recorder.

Male volunteers received an equimolar mixture of acetanilide-d_5 and phenacetin, the precise dose being 10.7 mg per kg of metabolically active mass. (MAM = wt (kg) to the power 0.7). The compounds were taken with water after fasting for 12 h. Blood samples (10 ml) were collected in heparinised containers at intervals ranging from 0.5 to 7 h after ingestion of the compounds. Three subjects repeated this experiment after an interval of several weeks when an equivalent equimolar dose of acetanilide-d_5 and phenacetin-d_5 was given. To study a possible *in vivo* biological isotope effect on the aromatic hydroxylation of acetanilide, two subjects ingested an equimolar dose of 400 mg of acetanilide and acetanilide-d_5. Blood samples were taken at 0.5, 1, 2 and 9 h after dosing. All blood samples were centrifuged and the plasma stored at –20° before analysis.

Analysis of plasma samples

The extraction procedure and selected ion monitoring techniques used in the analysis of these samples have been described (Baty and Robinson, 1976). We monitor the intense ions at *m/e* 280 and *m/e* 284, which arise from the protio and deuterio forms of the trimethylsilyl ethers of paracetamol ($M^+ - 15$). The two compounds co-elute on the gas chromatograph. The ratio of the signal at *m/e* 280 over that from *m/e* 284 is then an indication of the amount of paracetamol produced by de-ethylation of phenacetin compared with that produced by hydroxylation of deuterioacetanilide. This ratio is read from a strip chart recording of each mass value. The same technique was used for the acetanilide-d_5/phenacetin-d_5 experiment since the deuterium atoms in the latter compound are lost when it is metabolised to paracetamol.

RESULTS AND DISCUSSION

Figure 13.2 summarises the results of our experiments using acetanilide-d_5 with phenacetin, and Figure 13.3 shows the paracetamol ratios obtained in the experiments involving deuterated phenacetin. The hydroxylation of an equimolar acetanilide/acetanilide-d_5 mixture gave a constant ratio for *m/e* 280/284. This ratio was found to be unity for both the free and total forms of paracetamol. These results indicate that a biological isotope effect does not

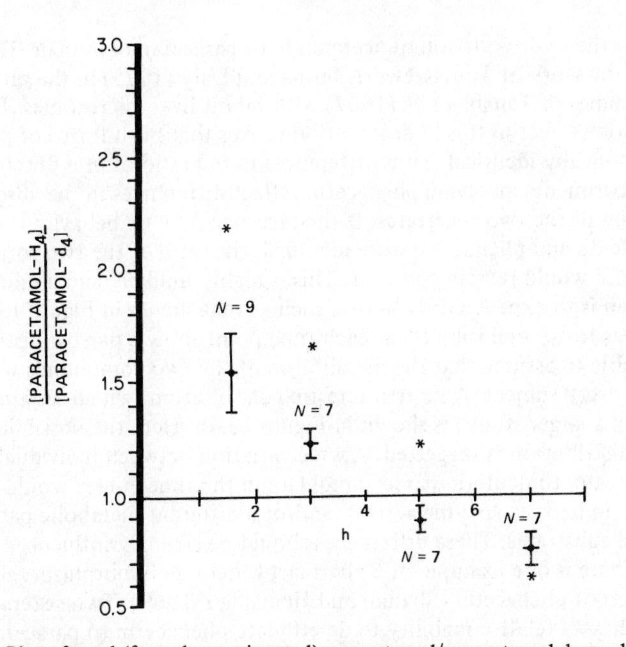

Figure 13.2 Plot of total (free plus conjugated) paracetamol/paracetamol-d_4 peak height ratios (*m/e* 280/284) with time following oral administration of acetanilide-d_5 and phenacetin. Mean value and 95 per cent confidence intervals are shown, together with the number of readings used at each time point. The data points marked * are from a single subject

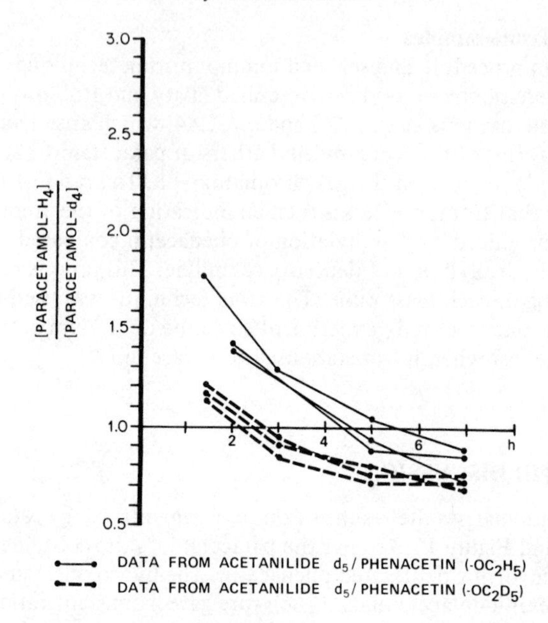

Figure 13.3 An *in vivo* study of a possible deuterium isotope effect. The same three subjects were used in each experiment

operate in the hydroxylation of acetanilide to paracetamol in man. This confirms the work of Tomaszewski, Jerina and Daly (1975) in the rat and the *in vitro* studies of Tanabe *et al.* (1967) with rabbit liver microsomes. The absence of an isotope effect in this hydroxylation proves that both forms of paracetamol are metabolically identical. Thus differences in the ratio of *m/e* 280 to *m/e* 284 in our experiments involving phenacetin reflect differences in the absorption and metabolism of the two substrates. If the pharmacokinetic behaviour of acetanilide-d_5 and phenacetin were identical, the ratio of the two forms of paracetamol would remain constant. This is highly unlikely and a more realistic assumption is to expect a distribution such as that shown in Figure 13.4a. If the ratio values for several subjects at each time point show a narrow distribution, it is reasonable to assume that the metabolism of the two compounds would be similar in every subject. A more interesting distribution (which we are hoping to discover in a larger study) is shown in Figure 13.4b. Here the possibility of a bimodal distribution is suggested. A wide variation between individuals in the ratio of protio- to deuterio-paracetamol late in the time-course would suggest a difference in hepatic enzyme activity and/or a different metabolic pathway for one of the substrates. These differences should be strongly influenced by genetic factors. There is one example of a pharmacogenetic polymorphism related to the metabolism of phenacetin (Shahidi and Hemaidan, 1969). Two sisters were found to have a relative inability to de-ethylate phenacetin to paracetamol. As a consequence they produced relatively large amounts of 2-hydroxy-phenetidine and other metabolites which are known to be effective in producing methaemoglobinaemia. This phenotype is extremely rare but would undoubtedly be detected by our experimental techniques, since the amount of protiopara-

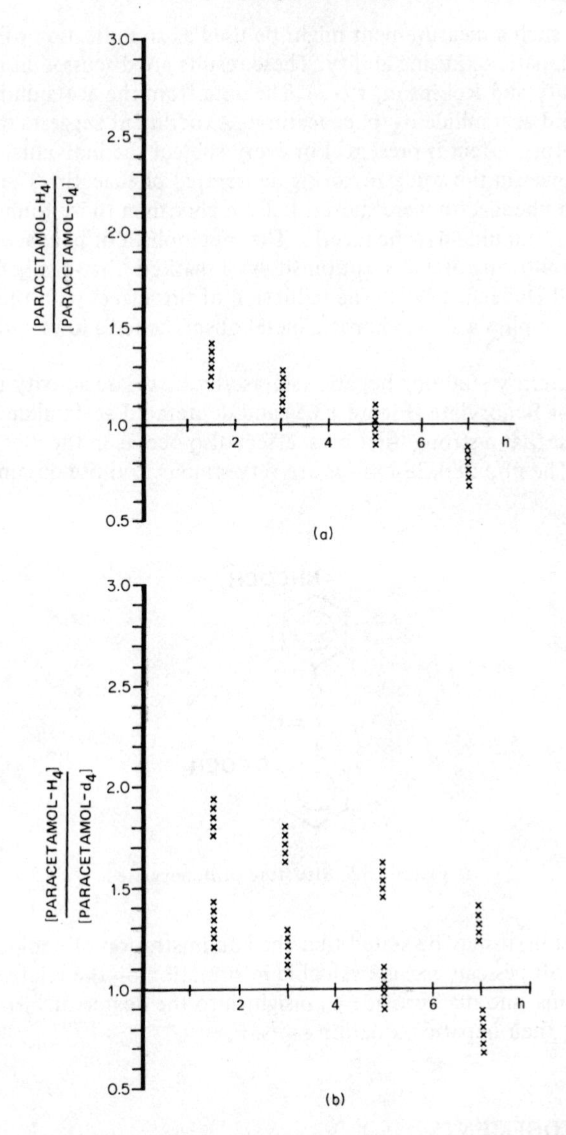

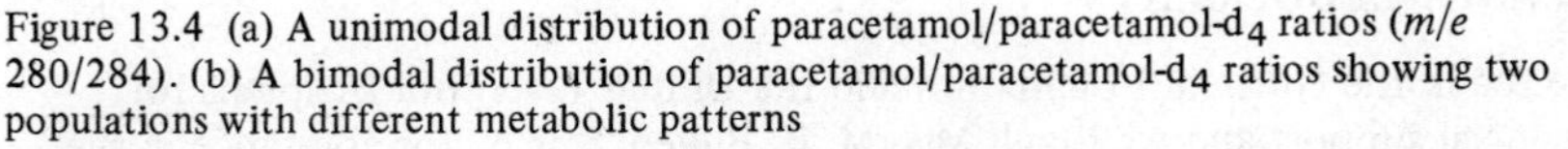

Figure 13.4 (a) A unimodal distribution of paracetamol/paracetamol-d_4 ratios (*m/e* 280/284). (b) A bimodal distribution of paracetamol/paracetamol-d_4 ratios showing two populations with different metabolic patterns

cetamol produced would be small, relative to the amount of deuterioparacetamol.

In the one anomalous result in our initial study (Figure 23.2), the subject was found to have consumed considerable quantities of alcohol prior to fasting. When he repeated the experiment at a later date, his ratio values were in the expected range. Very small intra-individual differences in ratio values were found with three other subjects who repeated the experiment. This is encouraging and suggests that such a measurement might be used as an indication of an individual's hepatic oxidising ability. These results are discussed in more detail elsewhere (Baty and Robinson, 1977). The data from the acetanilide-d_5/phenacetin and acetanilide-d_5/phenacetin-d_5 experiment suggests that an *in vivo* biological isotope effect is present. For every subject the individual paracetamol ratios were lower in the study involving deuterated phenacetin. The plasma levels of deuterated phenacetin were substantially higher than those found after ingestion of acetanilide-d_5/phenacetin. The metabolism of phenacetin at this dosage level following oral absorption shows a marked 'first-pass' effect (Raaflaub and Dubach, 1975). The reduction of this effect with the deuterium-labelled drug implies a slower hepatic metabolism, hence a lower *m/e* 280/284 value.

We are currently studying hepatic esterase/hydroxylase activity in the metabolism of Benorylate (Figure 13.5) and deuterated acetanilide. Preliminary results suggest that a strong 'first-pass' effect also occurs in the metabolism of Benorylate. The *m/e* 280/284 ratios are very similar to those obtained with phenacetin.

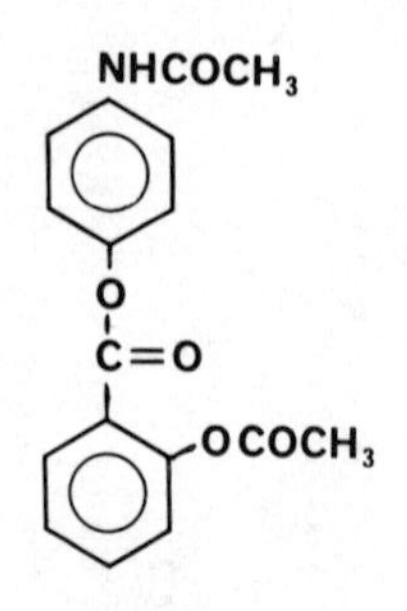

Figure 13.5 Structure of Benorylate

In conclusion, it may be stated that the administration of compounds labelled with stable isotopes can provide valuable information on the relative rates of drug oxidations, and may provide an insight into the degree of variation between individuals in their hepatic oxidation rates.

ACKNOWLEDGEMENTS

We thank the Nuffield Foundation and the United Liverpool Hospitals for financial support and we thank Miss M. F. Bullen, S.R.N. for assistance in the collection of blood samples. P.R.R. was a M.R.C. Research Student.

REFERENCES

Baty, J. D., Robinson, P. R. and Wharton, J. (1976). *Biomed. Mass. Spectrom.*, **3**, 60
Baty, J. D. and Robinson, P. R. (1977). *Clin. Pharmacol. Therap.* (in press)
Brodie, B. B. and Axelrod, J. (1948). *J. Pharm. Exp. Ther.*, **94**, 29

Brodie, B. B. and Axelrod, J. (1949). *J. Pharm. Exp. Ther.*, **97**, 58
Raaflaub, J. and Dubach, U. C. (1975). *Europ. J. Clin. Pharmacol.*, **8**, 261
Shahidi, N. T. and Hemaidan, A. (1969). *J. Lab. Clin. Med.*, **74**, 581
Tanabe, M., Yasuda, D., Tagg, J. and Mitoma, C. (1967). *Biochem. Pharmacol.*, **16**, 2230
Tomaszewski, J. E., Jerina, D. M. and Daly, J. W. (1975). *Biochemistry*, **14**, 2024

14

Discrimination between endogenous and exogenous testosterone in human plasma after oral treatment with testosterone undecanoate using stable-isotope-labelled compounds and mass spectrometric analyses

J. J. de Ridder and P. C. J. M. Koppens (Scientific Development Group, Organon International B.V., Oss, The Netherlands)

INTRODUCTION

In this presentation some possibilities, but also some problems, will be described associated with the use of stable-isotope-labelled compounds in metabolic studies. The examples which will be given are taken from a clinical metabolism experiment and a mass fragmentographic assay method for the development of a new orally active testosterone preparation (Nieschlag *et al.*, 1975): testosterone undecanoate, shown in Figure 14.1.

After administration, this drug is hydrolysed in the body to yield testosterone. We investigated the fate of this exogenous testosterone in a clinical experiment with hypogonadal patients and healthy male volunteers.

Untreated hypogonadal patients show low, but more or less constant, plasma testosterone levels (0–3 ng/ml). Healthy adult males, however, show higher but variable testosterone levels (3–10 ng/ml) with a diurnal variation of approximately 2.5 ng/ml. Because it was expected that a low dose of testosterone undecanoate will cause an increment of plasma testosterone levels of the same order of magnitude as the diurnal variations, it was essential to discriminate between endogenous and exogenous testosterone. This could be achieved by treating the subjects with a preparation containing deuterated testosterone undecanoate. The exogenous testosterone will then be measured as deuterated material and the endogenous testosterone as the non-labelled compound.

A mass fragmentographic assay method was developed for simultaneous

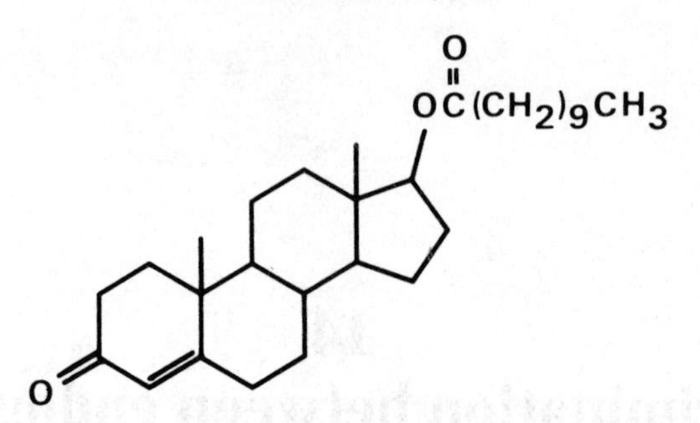

Figure 14.1 Structure of testosterone undecanoate

determination of non-labelled and deuterated testosterone, using tritiated testosterone as internal standard. This assay method will be described first, followed by a discussion of some general problems with stable isotope work. Thereafter, the clinical metabolism study will be discussed.

ASSAY METHOD

The assay method developed is, in some respects, analogous to the method reported by Breuer and Siekmann (1975), although a different clean-up and calibration procedure is used. Moreover, two compounds are now to be determined with the same internal standard. The sample preparation is shown in Figure 14.2.

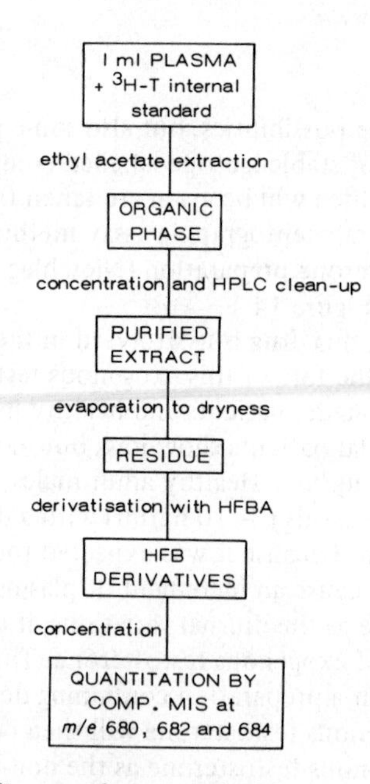

Figure 14.2 Flow diagram of assay method for testosterone in plasma

After addition of the internal standard, the sample is equilibrated and subsequently extracted with ethyl acetate. The extract is concentrated and further purified by a fast and efficient high pressure liquid chromatographic (HPLC) procedure. The collected HPLC fraction is concentrated and the heptafluorobutyryl (HFB) derivatives are made. Subsequently the concentrations are measured by gas chromatography–mass spectrometry (GC–MS). The GC–MS–computer system employed is summarised in Figure 14.3.

GC:	Varian 2740 2 m, 2 mm i.d. 1% OV – 1 on Gas Chrom Q 230°C single-stage WB separator
MS:	Varian MAT 311A combined El/Cl source El mode, 70 eV, 200°
Computer:	Varian MAT Spectro System 100 MS with computer controlled m.i.s. automatic focusing on max. 8 *m/e* channels software controlled acquisition, disc storage integration and ratio calculation

Figure 14.3 Details of GC–MS–computer system employed

The system is composed of a Varian 2740 gas chromatograph, a Varian MAT 311A mass spectrometer and a Spectro System 100 MS. The data system enables an automatic and quasi-simultaneous recording of the ion signals at *m/e* 680, 682 and 684, corresponding to the molecular ions of non-labelled, deuterated and tritiated testosterone derivatives, respectively. Such a recording as monitored on the visual display unit is shown in Figure 14.4.

For calibration, each sample series is accompanied by a series of reference samples containing known concentration ratios of compound to internal standard, yielding calibration graphs as shown in Figure 14.5.

For the calculation of the testosterone concentrations, a three-dimensional calibration curve is, in fact, required. The use of the two two-dimensional calibration curves of Figure 14.5 is also possible, provided the appropriate corrections for isotopes contributions from the third component are made. The assay specifications are summarised in Table 14.1, which shows an accuracy and precision of 3 and 5–6 per cent, respectively, at levels of 1–5 ng/ml. The sensitivity of detection is about 100 pg and for the total assay approximately 200 pg.

PROBLEMS WITH ISOTOPES

It goes without saying that discriminating assay methods as described above depend completely on the use of isotopically labelled compounds. It is therefore unnecessary to explain the usefulness of isotopes in analytical work or in clinical

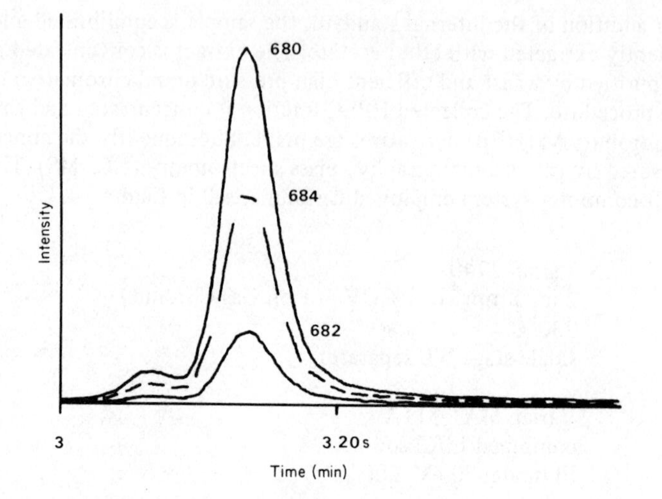

Figure 14.4 Multiple ion recording of the traces with *m/e* 680, 682 and 684 of a standard containing 1.4 ng [^{1}H]-T, 0.10 ng [^{2}H]-T and 1.0 ng [^{3}H]-T

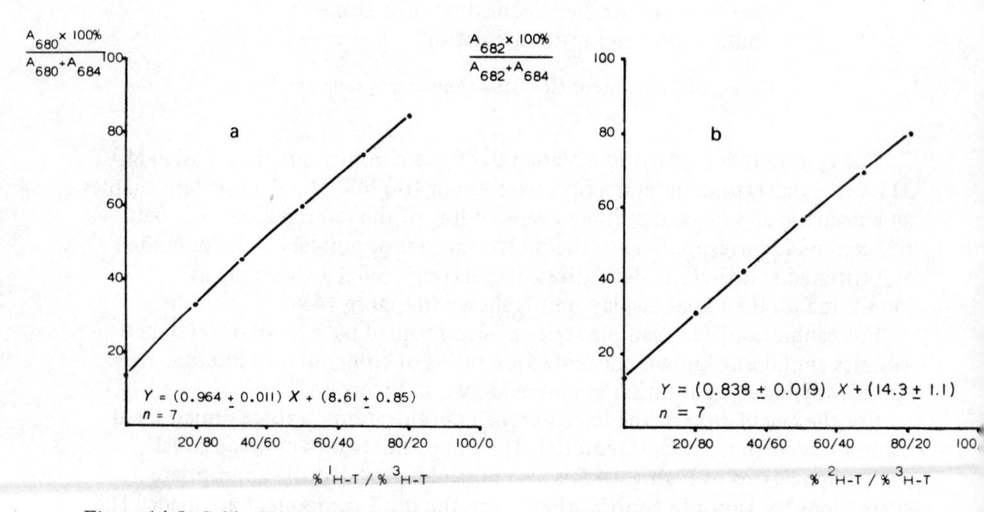

Figure 14.5 Calibration graphs for the quantification of (a) testosterone and (b) deuterated testosterone, both with tritiated testosterone as internal standard

Table 14.1 Assay specifications

Accuracy	3% for 1–5 ng/ml
Precision	5–6% for 1–5 ng/ml (n = 14)
Sensitivity	Detection: 100 pg; ± 8% (n = 10) Assay: 200 pg

experiments. Perhaps it is more interesting to focus our attention on those cases where isotope labels cause problems.

Most problems with (stable) isotopes arise from two different sources of error: loss of label and isotope effects.

Loss of label

Loss of label affects both the results of the biological experiment as well as the accuracy of the assay method. In the case of our testosterone undecanoate, for example, we used a preparation which was deuterated and tritiated at C-1 and C-2, which is the most common position for labelling of testosterone. The label appeared to be stable in refluxing methanol and was only partly lost in alkaline methanol. The pharmaceutical formulation of this drug is a soft-shell gelatine capsule containing a solution in oleic acid. Comparison of the deuterium contents before and after capsulation revealed the data shown in Table 14.2, indicating a loss of label of approximately 30 per cent. In the same way, the

Table 14.2 Loss of label from [1,2-^{2}H]-testosterone undecanoate after capsulation in oleic acid and storage for about 1 year at room temperature

%	Before capsulation in oleic acid	Extracted from capsules
d_0	0	0
d_1	17.3	75
d_2	82.7	24

tritium label is partly lost. This can, however, hardly be detected in the standard quality control tests for radioactive formulations, but might have serious consequences for all published work with 1,2-tritiated testosterone*. For assay methods loss of label will yield erroneous results because of the uncertain amount of internal standard. As an example, Table 14.3 shows loss of a deuterium label from a steroid during GC–MS analysis.

It cannot be assumed that loss of label is automatically corrected via the calibration function, because these losses are hardly reproducible.

Table 14.3 Loss of label of Org 6001 : 3α-amino-2β-hydroxy-5α-androstan-17-one hydrochloride during GC analysis

Deuterium content 16,16[$^{2}H_2$]-Org 6001 (TFA, TMS derivative) (%)		
Direct probe:	d_0	3.2
	d_1	14.5
	d_2	80.4
	d_4	1.1
GC–MS:	d_0	17.1
	d_1	44.4
	d_2	36.8
	d_3	0.9
	d_4	0.8

* Note added in proof: Additional experiments indicated that partial loss of label occurred only in acidic media. In the arachis oil commonly used, loss of label was not observed.

Consequently, any possible loss of label should be adequately investigated for all labelled formulations and all internal standards. The labelled compounds may only be used when the stability of the label has been proven. This proof, however, is, in general, more easy to obtain when dealing with stable isotopes than in case of radioactive isotopes.

Isotope effects

Although isotope effects are always neglected, they do occur and in many cases they are responsible for assay results being some factors in error. Isotope effects in biological experiments are well known, but let us consider only the effects on the assay method. When, for example, introduction of the label influences the p*K* of an acidic or basic drug, resulting in different partition coefficients, the extractions of the compound and internal standard are not identical, as required. Also, during the clean-up procedure, particularly with HPLC, discrimination between labelled and non-labelled compound may occur. The most serious example we met was of a basic tetracyclic anti-migraine drug which was completely separated from its tetradeuterated analogue on HPLC as shown in Figure 14.6 (de Ridder and Van Hal, 1976).

A pseudo or apparent isotope effect, encountered in the gas chromatograph, probably is responsible for the most serious error which can occur in quantitative mass fragmentography. This is the phenomenon of adsorption and desorption of material on the column. When previously adsorbed material is desorbed from

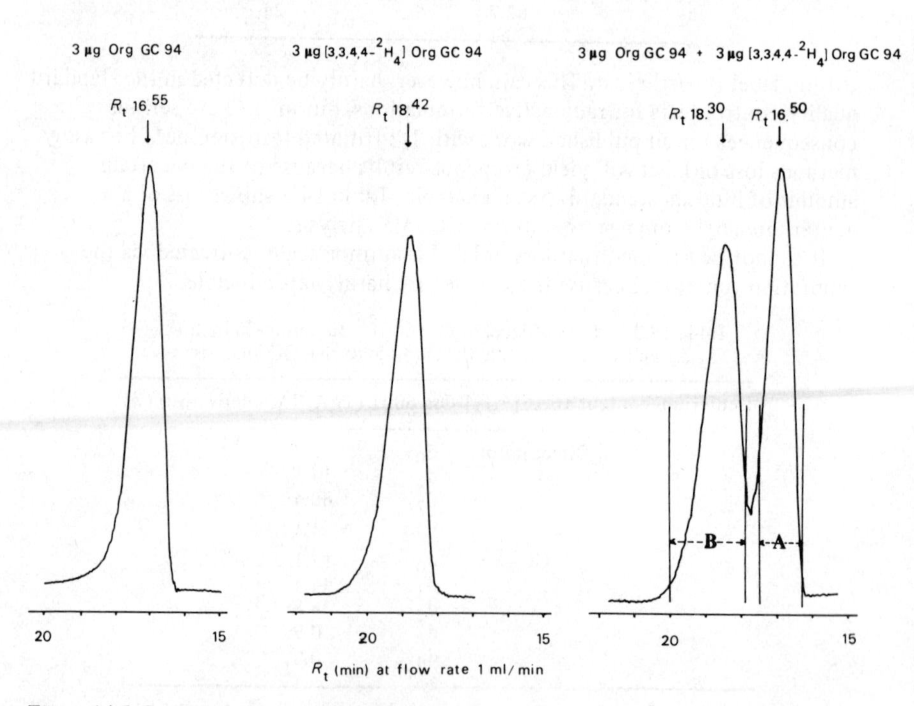

Figure 14.6 Separation of the tetracyclic anti-migraine drug Org GC 94 from its tetra-deuterated analogue in high-pressure liquid chromatography

the column during the analysis of a sample which contains a different amount of standard from the sample injected, serious errors are the result, as illustrated in Table 14.4.

These are only some of the sources of errors in quantitative mass fragmentography. They are mentioned because too often the erroneous opinion is encountered that once the internal standard has been added no further effects can be detrimental to the analysis. We should realise that although the labelled compounds look identical, as a matter of fact they remain different and they will behave differently.

Table 14.4 Analyses of a test sample containing 100 ng of the steroid Org 6001 per ml of plasma. The analyses are performed with 100 ng [9,11,16-$^2H_3$]-Org 6001 as internal standard and using the TFA, TMS derivatives. Owing to absorption and desorption phenomena, different results were obtained after treating the GC column with standard solutions containing different ratios of non-labelled *vs.* labelled Org 6001

After injection of standard mixture Cpd: int. st.	Result (ng/ml)
1 : 1	98.4; 99.9; 99.7; 101.3
1 : 2	68.9; 74.1; 93.0
2 : 1	172; 126
	Mean 103.7 ng/ml; SD 29%

CLINICAL EXPERIMENT

Let us now return to the clinical experiment. The subjects were treated in a stable therapy with three 40 mg capsules of non-labelled testosterone undecanoate per day. After 4 weeks of treatment, one dose of non-labelled drug was replaced by a capsule containing deuterated and tritiated testosterone undecanoate. The volunteers were also treated with a single dose of labelled testosterone undecanoate. Because of the loss of label mentioned above, the exact amount of labelled material administered is somewhat uncertain. There are, however, good reasons to assume an amount of 6.7 mg of deuterated and 61 μCi of tritiated testosterone undecanoate.

RESULTS

The results of the plasma level measurements are shown in Figure 14.7 for a single dose to a healthy volunteer showing a peak level of deuterated testosterone of approximately 1 ng/ml occurring 3–4 h following dosing and, at the same time, a decrease of 2.5 ng/ml of the endogenous testosterone level. The dotted line represents the plasma level of total radioactivity. From this figure it is clear that the amount of exogenous testosterone is small in comparison with the variation in the amount of endogenous testosterone. Therefore it can be

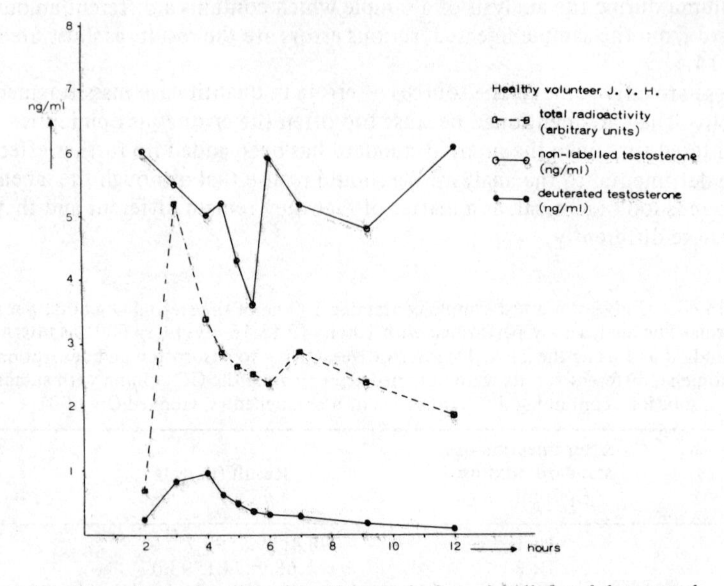

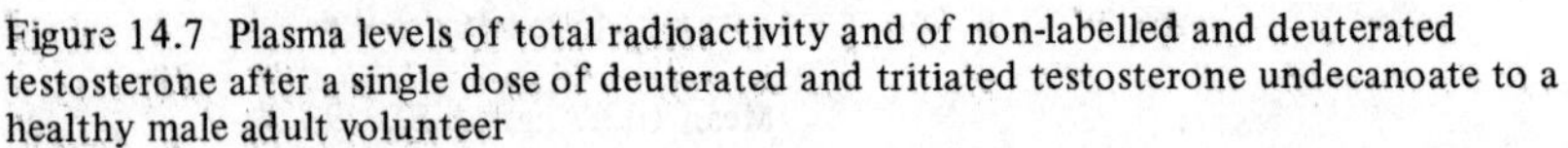
Figure 14.7 Plasma levels of total radioactivity and of non-labelled and deuterated testosterone after a single dose of deuterated and tritiated testosterone undecanoate to a healthy male adult volunteer

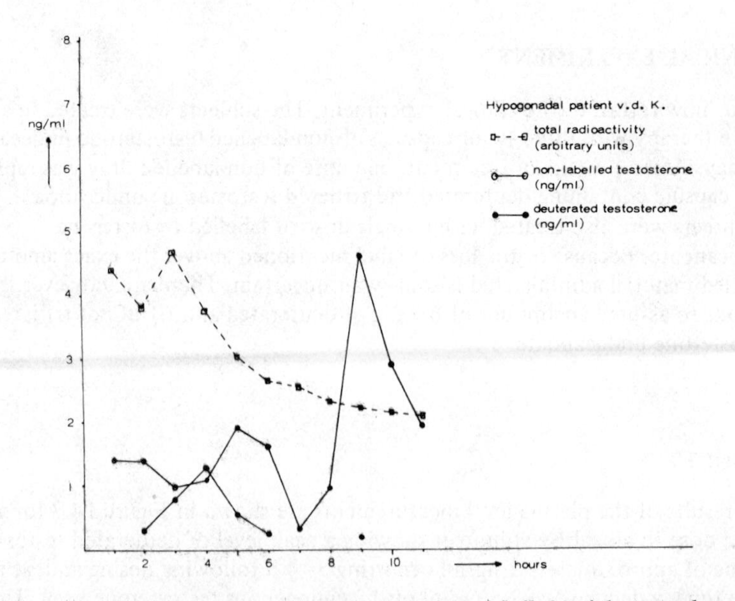

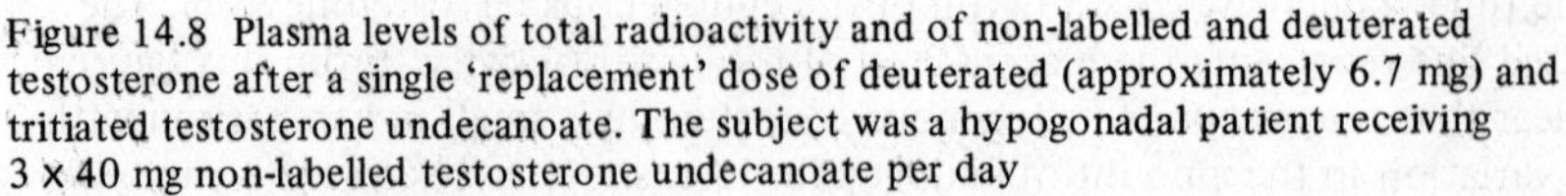
Figure 14.8 Plasma levels of total radioactivity and of non-labelled and deuterated testosterone after a single 'replacement' dose of deuterated (approximately 6.7 mg) and tritiated testosterone undecanoate. The subject was a hypogonadal patient receiving 3 x 40 mg non-labelled testosterone undecanoate per day

concluded that, at least at this dose level, a discriminating method of analysis was essential to monitor the disposition of the exogenous material.

An example of the plasma levels in a patient during a stable therapy is shown in Figure 14.8, demonstrating the initially low levels of non-labelled testosterone which increase after intake of a non-labelled capsule 7 h after the labelled dose. The levels of deuterated testosterone again remain very low and do not exceed 1.3 ng/ml. Also in this case it is obvious that the discrimination between non-labelled and deuterated testosterone was required to monitor the exogenous testosterone, albeit for a limited period of time.

In conclusion, we may state that the use of the deuterium label in combination with the mass fragmentographic analysis in this case gave information which could not be obtained otherwise. On the other hand, it was demonstrated that phenomena such as loss of label or isotope effects may hamper the stable isotope work seriously. The use of radioactive isotopes in metabolism studies cannot, however, be eliminated because of the requirements for excretion data and in order to facilitate the isolation and identification of metabolites.

REFERENCES

Breuer, H. and Siekmann, L. (1975). *J. Steroid Biochem.* **6**, 685
de Ridder, J. J. and Van Hal, H. J. M. (1976). *J. Chromatog.*, **121**, 96
Nieschlag, E., Mauss, J., Coert, A. and Kićović, P. (1975). *Acta Endocrinol.*, **79**, 366

Section 3
Applications in clinical research

15
Application and potential of coenzyme labelling

T. Cronholm, T. Curstedt and J. Sjövall (Department of Chemistry, Karolinska Institutet, Stockholm, Sweden)

INTRODUCTION

Reaction rates and equilibria in intermediary metabolism are regulated by several factors such as substrate concentrations, allosteric effects and feed-back control of enzyme synthesis. Another factor that may be of importance is the interaction between reactions that utilise the same coenzyme. A typical example is the decreased conversion of lactate to pyruvate during ethanol oxidation, mediated by the increased ratio between the concentrations of free (i.e. not protein-bound) NADH and NAD^+ in the liver cells. Compartmentation of redox reactions and coenzymes may determine the degree of interaction. For example, the impermeability of the mitochondrial membrane to pyridine nucleotides results in the well-known existence of cytosolic and mitochondrial pools of NAD with widely different redox states. The diffusion rate in the cytosol is generally considered to be sufficiently high to prevent the formation of several coenzyme pools in this compartment. However, the equilibration of NAD bound to enzymes may be limited by slow dissociation rates. Another possibility for compartmentation of coenzymes when whole organs are studied is the existence of functionally different cell types in the organ.

Interactions mediated by common pyridine nucleotide pools are usually studied by measurements of the influence of the redox state of one reaction on that of another, using whole organisms, organs or isolated cells. Another method is to determine the transfer of hydrogen atoms from an administered oxidisable donor compound to a reducible endogenous or exogenous acceptor. The latter method may give information regarding compartmentalised coupling of an oxidation with specific reductions. Both tritium and deuterium have been employed in such studies. The advantage of using tritium is the possibility to trace all of the hydrogen removed in the oxidation step, and the comparatively simple instrumentation needed. However, in most cases only a minute fraction of the isotope appears in reduced metabolites and large doses of tritium have to be used to permit determinations of specific activity. This will prevent studies on humans.

Although only a small fraction of administered isotope may appear in reduced metabolites, a large fraction of the hydrogen atoms in specific positions may be derived from the isotope donor. This means that the isotope excess in these positions is high and deuterium excess can be measured by gas chromatography–mass spectrometry even when small amounts of metabolite are present. Thus, for studies of metabolic interactions via common coenzyme pools, stable isotopes are often better suited than radioactive ones. Additional advantages associated with the use of stable isotopes are the possibilities of direct determination of the site of labelling and of analysing the distribution of molecules containing different numbers of heavy atoms.

The principle of coenzyme labelling was first used by Hoberman (1958) in experiments where deuterated hydroxy acids were used to provide deuterium-labelled NADH *in vivo*. A marked specificity in the transfer from the different acids into liver glycogen was noted, suggesting that intracellular NAD is kinetically inhomogeneous. Several studies have dealt with the origin of the reduced coenzyme for fatty acid synthesis (see Rous, 1971), and the principle of measuring isotope transfer via coenzymes has been used to study the shuttle systems which transfer reducing equivalents over the mitochondrial membrane (e.g. Hassinen, 1967; Carnicero, Moore and Hoberman, 1972; Rognstad and Katz, 1970). The problems caused by the presence of two transferable hydrogens (4-*pro-R* and 4-*pro-S*) in the pyridine nucleotides ('intramolecular compartmentation') have been observed and discussed (Hung and Hoberman, 1972; Rognstad and Clark, 1974). With the exception of Hoberman's early experiments, all these studies have been made using tritium-labelled hydrogen donors.

STUDIES WITH DEUTERIUM-LABELLED ETHANOL

Ethanol is well suited as a donor of isotope in studies of compartmentation of NADH and NADPH in the liver. It is oxidised almost exclusively in the liver and the rate of oxidation is constant within a wide concentration range. When [1-$^2H_2$]-ethanol is administered to rats, a constant level of labelling of coenzyme pools in the liver is rapidly obtained. Reducible compounds can then be administered as probes for studies of the labelling of coenzyme pools participating in the reduction of these compounds. Alternatively, one can analyse the labelling of normal metabolites formed in equilibrium reactions, or of endogenous compounds formed in irreversible reductions or by reduction followed by a second irreversible metabolic reaction (e.g. esterification or conjugation of a hydroxy compound). By comparing the labelling of different compounds in the liver it might be possible to define reactions which are particularly susceptible to the production of NADH during ethanol oxidation.

When cyclohexanone is administered to rats oxidising [1-$^2H_2$]-ethanol highly labelled cyclohexanol is produced in a reaction catalysed by the alcohol dehydrogenase-NADH complex (Cronholm, 1974). This shows that the labelled bound NADH equilibrates only to a small extent with free NADH before being used in the reduction of cyclohexanone. This is similar to results obtained in *in vitro* studies of tritium transfer during reduction of 17-ketosteroids in the presence of 17β-hydroxysteroids labelled with ^3H in the 17α position. Thus, the mutual interaction in oxidoreductions of oestradiol/oestrone and 17β-hydroxy-/17-keto-C_{19}-steroids (Mützel and Wenzel, 1971; Wenzel, Pitzel and Riesselmann, 1975) has been explained by a slow dissociation of the enzyme–NADH complex (Wenzel, Bollert and Ahlers, 1974). On the other hand, the specific transfer of hydrogen from the 17α position of oestradiol to C-20 of progesterone in human placenta is considered to be due to compartmentation of the reduced coenzyme (Pollow *et al.*, 1974).

When [1-$^2H_2$]-ethanol was used to study oxidoreductions of steroids in man, several C_{19} steroid sulphates with a 17β-hydroxy group were found to be highly labelled, whereas a 20α-hydroxysteroid was much less labelled (Cronholm and Sjövall, 1970). This shows that reductive reactions in steroid metabolism may be influenced by ethanol oxidation to different extents. The possible significance of this finding for the development of hormonal disturbances in alcoholics is not known. This study and subsequent studies in rats have shown that several coenzyme pools labelled to different extents from [1-$^2H_2$]-ethanol are involved in steroid reductions (Cronholm, 1972; Cronholm, Makino and Sjövall, 1972a,b). For example, different pools of NADPH are used in the reduction of the 4,5 double bond of intermediates in bile acid biosynthesis and the 4,5 double bond of 3-keto-4-cholenoic acid (Cronholm *et al.*, 1972a,b). Possibly, these reductions occur in different cell types (Nabors, Berliner and Dougherty, 1967). The transfer of deuterium from NADH (primarily formed during ethanol oxidation) to $NADP^+$ in these experiments may be explained by labelling of malate and citrate (Cronholm *et al.*, 1974b) which can subsequently be oxidised using $NADP^+$ as coenzyme.

The steady state labelling of coenzymes with [1-$^2H_2$]-ethanol may be used to study compartmentation of compounds synthesised in the liver. Thus, if a compound is synthesised with the use of labelled coenzyme and then serves as the precursor of another compound, a time course study of the labelling of the two compounds will give information about a possible heterogeneity of the precursor compound pool. Using this approach, it was found that bile acids produced by bile fistula rats during metabolism of [1-$^2H_2$]-ethanol were formed from a cholesterol that had a higher deuterium content than the cholesterol excreted in bile (Cronholm *et al.*, 1972a; Cronholm, Burlingame and Sjövall, 1974a). Further studies using [2-$^2H_3$]-ethanol to label the acetyl-CoA pool showed that the bile acids were labelled at a much faster rate than cholesterol, indicating heterogeneity of the hepatic cholesterol pool. Thus, in the bile fistula rat, bile acids are formed from newly synthesised cholesterol molecules which have not become fully equilibrated with total hepatic cholesterol.

Another example of the use of coenzyme labelling to study compartmentation and coupling of reactions is the investigation of phosphatidylcholine biosynthesis during metabolism of [1-$^2H_2$]-ethanol in the bile fistula rat (Curstedt and Sjövall, 1974; Custedt, 1974a,b). Newly formed phosphatidylcholines incorporate deuterium via NADH at C-1, C-2 and C-3 of the glycerol moiety. The rate of incorporation depends on the turnover rate of the individual molecular species of phosphatidylcholines. From the positional distribution of deuterium and the isotopic composition (i.e. percentage of mono-, di- and trideuterated molecules) of the glycerol moiety, it is possible to calculate the deuterium excess at different positions of the glycerol moiety of newly formed phosphatidylcholines synthesised *de novo* (Curstedt and Sjövall,

1974). The results obtained for two different molecular species are shown in Figure 15.1. It is seen that the degree of labelling and the distribution of isotope is the same for both species and constant throughout the experiment. Analysis of total hepatic glycerol-3-phosphate shows that the deuterium excess of this compound is 2–3 times lower than that of the glycerol moiety of newly formed phosphatidylcholine molecules. This shows that phosphatidylcholines are formed from a special pool of glycerol-3-phosphate which is apparently acylated at a rate that is high enough to prevent complete mixing with other glycerol-3-phosphate pools in the liver. This finding may be related to the

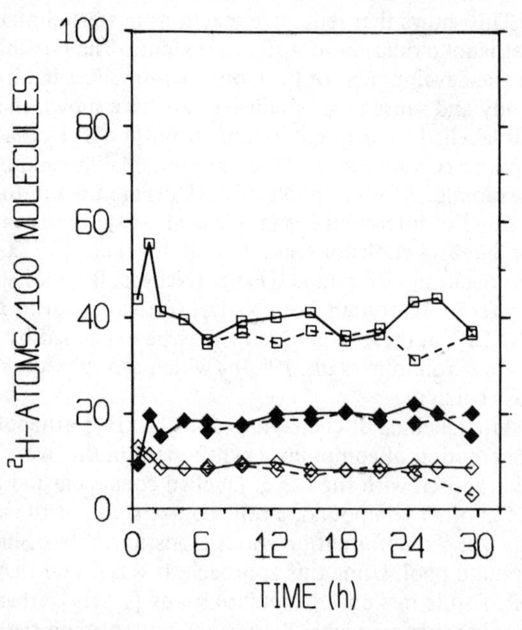

Figure 15.1 Calculated deuterium excess at C-1 (◇), C-2 (□) and C-3 (◆) of the glycerol moiety of 1-palmitoyl-2-oleoyl (dotted lines) and 1-palmitoyl-2-linoleoyl (solid lines) phosphatidylcholines synthesised *de novo* and excreted in the bile of a rat given [1-$^2H_2$]-ethanol for 24 h

suggested existence of separate glycolytic and gluconeogenic compartments in the liver (Threlfall and Heath, 1968; Zehner *et al.*, 1973). It is of interest that the high labelling at C-3 of the glycerol moiety of phosphatidylcholines indicates that the major part of glycerol-3-phosphate used in the biosynthesis of these glycerolipids is of gluconeogenetic origin (malate is labelled to about the same extent (Cronholm *et al.*, 1974a)). This conclusion is supported by the low labelling of plasma glucose (Curstedt, 1974b).

The conversion of ethanol to acetic acid in the liver involves two oxidations. The 1-*pro-R* hydrogen is transferred to NADH in the alcohol dehydrogenase reaction and the 1-*pro-S* hydrogen in the aldehyde dehydrogenase reaction. By using stereospecifically labelled ethanol the coenzyme pools labelled in the two reactions can be studied separately. This has been done with ethanols labelled with tritium (Rognstad and Clark, 1974) or deuterium (Cronholm and Fors,

1976). The results of the latter study in whole animals show that the alcohol and aldehyde dehydrogenases do not share a common pool of NAD, and that NADH formed during acetaldehyde oxidation is utilised for reductions in the cytosol to a much smaller extent than NADH formed in the alcohol dehydrogenase reaction. This supports the concept (based on cell fractionation procedures) that aldehyde oxidation is mainly an intramitochondrial process. It is of interest that some NADPH-requiring reactions in the cytosol to some extent utilise hydrogen that was originally derived from the 1-*pro-S* position of ethanol, i.e. hydrogen atoms which were transferred to NADH in the aldehyde dehydrogenase reaction. This transfer may make it possible to study changes in the compartmentation of coenzymes used for synthesis of compounds that are formed slowly and not in equilibrium reactions, such as cholesterol. Thus the ratio of the transfer from the 1-*pro-R* and 1-*pro-S* positions of ethanol is independent of any dilution by cholesterol formed prior to ethanol administration (Cronholm and Curstedt, 1977).

TURNOVER STUDIES BY COENZYME LABELLING

Labelling of coenzymes *in vivo* or *in vitro* by isotope transfer from a labelled substrate may also be used to study the turnover and production of metabolites which are formed through one or several reductions. This principle is being used in many laboratories to study fatty acid and cholesterol synthesis with the aid of ^3H_2O (Jungas, 1968) or ^2H_2O (Wadke *et al.*, 1973). In this case labelling of the lipid occurs both directly and via pyridine nucleotides. The possibility of using [1-$^2H_2$]-ethanol to provide labelled coenzymes without significant labelling of the water has been demonstrated with fatty acids (Curstedt and Sjövall, 1974), cholesterol and bile acids (Cronholm *et al.*, 1974a). However, the latter study indicated that labelling of the acetyl-CoA pool by use of [2-$^2H_3$]-ethanol gave more reliable turnover rate measurements for the steroid skeleton. In the studies of phosphatidylcholine biosynthesis during oxidation of [1-$^2H_2$]-ethanol in bile fistula rats (Curstedt and Sjövall, 1974), differences in half-life times for different molecular species could be determined from the rate of labelling of the glycerol moiety. In addition, palmitic acid residues became labelled via NADPH. This labelling increased markedly during administration of glucose or fructose, indicating an increased synthesis of palmitic acid. The rate of labelling of the palmitoyl residues was the same for all molecular species of phosphatidylcholines but different from the rate of labelling of the glycerol moieties. This indicates that labelled palmitoyl residues in different molecular species are derived from the same palmitic acid pool.

These experiments illustrate the potential of coenzyme labelling with deuterium followed by analysis of metabolites by gas chromatography–mass spectrometry. They show that detailed information can be obtained regarding the origin and turnover rates of different parts of the molecules studied. For example, a comparison of the labelling of molecular species of phosphatidylcholines in liver and bile showed that both half-life times and labelling of the glycerol moiety were the same for corresponding molecular species from the two sources. This was interpreted to indicate that the major phosphatidylcholines in liver and bile are derived from the same pool of newly synthesised molecules (Curstedt, 1074a).

CONCLUSIONS

Labelling of coenzymes by administration of a compound labelled with a stable isotope is a useful technique for studies of compartmentation or coupling of reactions in whole animals, perfused organs or cell preparations. While this brief review has dealt almost exclusively with deuterium-labelling of pyridine nucleotides, the principle may be applicable to other situations where a coenzyme acts as the carrier for atoms or molecules. Labelling of acetyl-CoA by administration of [2-$^2H_3$]-ethanol was mentioned above. It seems possible that coenzyme labelling with stable isotopes could be used for clinical studies of organ function under normal conditions and in disease. Simultaneous administration of a donor and an acceptor of the stable isotope, both being metabolised in the same organ, could reveal changes in normal metabolic relationships in this organ. Ethanol is only one example of a suitable donor compound. It is possible that interactions of drugs with normal metabolism could be detected by this method. A major practical problem at present is that new and faster methods have to be developed to permit accurate determinations of excess and position of heavy atoms in a wide variety of naturally occurring metabolites in a large number of samples.

ACKNOWLEDGEMENTS

This work was supported by grants from the Swedish Medical Research Council (grant No. 25X-2189) and from Karolinska Institutet.

REFERENCES

Carcinero, H. H., Moore, C. L. and Hoberman, H. D. (1972). *J. Biol. Chem.*, **247**, 418
Cronholm, T. (1972). *Europ. J. Biochem.*, **27**, 10
Cronholm, T. (1974). *Europ. J. Biochem.*, **43**, 189
Cronholm, T., Burlingame, A. L. and Sjövall, J. (1974a). *Europ. J. Biochem.*, **49**, 497
Cronholm, T. and Curstedt, T. (1977). In preparation
Cronholm, T. and Fors, C. (1976). *Europ. J. Biochem.*, **70**, 83
Cronholm, T., Makino, I. and Sjövall, J. (1972a). *Europ. J. Biochem.*, **24**, 507
Cronholm, T., Makino, I. and Sjövall, J. (1972b). *Europ. J. Biochem.*, **26**, 251
Cronholm, T., Matern, H., Matern, S. and Sjövall, J. (1974b). *Europ. J. Biochem.*, **48**, 71
Cronholm, T. and Sjövall, J. (1970). *Europ. J. Biochem.*, **13**, 124
Curstedt, T. (1974a). *Biochim. Biophys. Acta*, **369**, 196
Curstedt, T. (1974b). *Eur. J. Biochem.*, **49**, 355
Curstedt, T. and Sjövall, J. (1974). *Biochim. Biophys. Acta*, **369**, 173
Hassinen, I. (1967). *Ann. Med. Exp. Fenn.*, **45**, 35
Hoberman, H. D. (1958). *J. Biol. Chem.*, **232**, 9
Hung, H. C. and Hoberman, H. D. (1972). *Biochem. Biophys. Res. Commun.*, **46**, 399
Jungas, R. L. (1968). *Biochemistry*, **7**, 3708
Mützel, W. and Wenzel, M. (1971). *Hoppe-Seyler's Z. Physiol. Chem.*, **352**, 304
Nabors, C. J., Berliner, D. L. and Dougherty, T. F. (1967). *J. Reticuloend. Soc.*, **4**, 237
Pollow, K., Sokolowski, G., Schmalbeck, J. and Pollow, B. (1974). *Hoppe-Seyler's Z. Physiol. Chem.*, **355**, 515
Rognstad, R. and Clark, D. G. (1974). *Europ. J. Biochem.*, **42**, 51
Rognstad, R. and Katz, J. (1970). *Biochem. J.*, **116**, 483
Rous, S. (1971). *Adv. Lipid Res.*, **9**, 73

Threlfall, C. J. and Heath, D. F. (1968). *Biochem. J.*, **110**, 303
Wadke, M., Brunengraber, H., Lowenstein, J. M., Dolhun, J. J. and Arsenault, G. P. (1973). *Biochemistry,* **12**, 2619
Wenzel, M., Bollert, B. and Ahlers, B. (1974). *Hoppe-Seyler's Z. Physiol. Chem.*, **355**, 969
Wenzel, M., Pitzel, L. and Riesselmann, B. (1975). *Hoppe-Seyler's Z. Physiol. Chem.*, **356**, 459
Zehner, J., Loy, E., Müllhofer, G. and Bücher, T. (1973). *Europ. J. Biochem.*, **34**, 248

16
Pathways of steroid metabolism *in vivo* studied by deuterium labelling techniques

T. A. Baillie*, R. A. Anderson†, M. Axelson, K. Sjövall and J. Sjövall (Department of Chemistry, Karolinska Institute, and Department of Obstetrics and Gynaecology, Karolinska Hospital, Stockholm, Sweden)

INTRODUCTION

Studies on the biosynthesis of progesterone in human pregnancy have indicated that the placenta has a very high capacity for the production of this steroid, and that rates of synthesis of up to 300 mg/day may be attained towards term (Solomon and Fuchs, 1971). However, it is not known why such large amounts of progesterone are required and what role, if any, its metabolites play in regulating hormonal activity.

Progesterone is metabolised mainly by reduction of the Δ^4 double bond and of the oxo groups at C-3 and C-20 to give a series of isomeric pregnanolones and pregnanediols. The glucuronide conjugates of 3α-hydroxy-5β-pregnan-20-one and 5β-pregnane-3α,20α-diol are the quantitatively major metabolites present in urine, although their combined excretion rates only account for some 10–30 per cent of the total progesterone elimination (Solomon and Fuchs, 1971). Recently, a series of sulphoconjugated pregnane derivatives has been identified in the plasma of pregnant women (Sjövall, Sjövall and Vihko, 1968), the structures of the major components of which are shown in Figure 16.1. In contrast to most of the urinary metabolites of progesterone, these sulphate conjugates in plasma possess a 5α-H configuration; all of the steroids are sulphated at the C-3 position, while the pregnanediols are also present as their 3,20-disulphates. The total concentration of these steroid sulphates has been shown to increase steadily throughout gestation and to reach values as high as 10 μg/ml at term (Sjövall, 1970), which is very much higher than that of progesterone itself and related

*Present address: Department of Clinical Pharmacology, Royal Postgraduate Medical School, Ducane Road, London W12 0HS, UK.
†Present address: Department of Forensic Medicine, University of Glasgow, Glasgow G12 8QQ, UK.

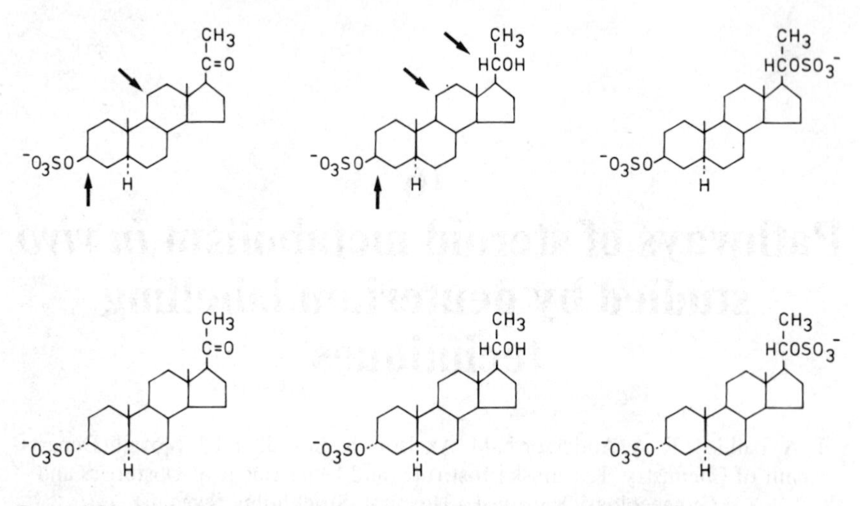

Figure 16.1 Structures of the major sulphoconjugated pregnane derivatives in human pregnancy plasma. Positions at which deuterium atoms were incorporated in the labelled analogues used as tracers are indicated by arrows for representative pregnanolone and pregnanediol sulphates

unconjugated derivatives (Axelson and Sjövall, 1974). While steroid sulphates are known to be actively involved as intermediates in certain biosynthetic pathways, e.g. in the conversion of dehydroepiandrosterone to oestrogens in the feto-placental unit, their role in the metabolism of progesterone is not understood. The object of the present investigation, therefore, was to determine the pool sizes, turnover rates and metabolic inter-relationships of the major sulphoconjugated pregnane derivatives in late pregnancy plasma and thereby to assess the quantitative importance of the 'sulphate' pathway of metabolism in human pregnancy.

On ethical grounds, radioactive tracers cannot normally be administered to healthy pregnant women. Steroids labelled with stable isotopes, on the other hand, are ideally suited to investigations in obstetric subjects (Pinkus, Charles and Chattoraj, 1971) and we have confirmed the usefulness of such compounds as tracers for *in vivo* metabolic studies (Baillie *et al.*, 1975a). The administration of steroid sulphates labelled specifically with deuterium, coupled with sample analysis by gas chromatography–mass spectrometry (GC–MS), was thus adopted as the method of choice for the present investigation.

MATERIALS AND METHODS

Deuterium-labelled steroid sulphates

The requirements for deuterium-labelling for the purposes of this study were twofold. Firstly, it was considered essential to be able to follow the turnover of the steroid skeleton in each of the injected compounds, and deuterium atoms were therefore required at a metabolically 'stable' position; C-11 was chosen as the site for such labelling. Secondly, information was required on whether sulphates of 3β- and 3α-hydroxysteroids could be interconverted by the processes of hydrolysis, oxido-reduction and resulphation. Deuterium substitution at C-3 was chosen as the means by which this reaction sequence could be followed. Similarly, introduction of a deuterium atom at C-20 in the pregnanediol sulphates was required to monitor side-chain metabolism in these compounds. Analogues of each of the compounds shown in Figure 16.1 were synthesised accordingly to afford pregnanolone sulphates labelled with a single atom of deuterium at C-3 and two deuterium atoms at C-11, while the pregnanediol sulphates contained, in addition, a fourth deuterium label in the 20β position. The deuterium excess at both C-3 and C-20 was in the order of 95 atoms per cent, while the isotopic purity at C-11 was over 90 per cent $^2H_2$. A detailed account of the synthesis of these labelled steroids has been published (Baillie *et al.*, 1975b; Baillie, Sjövall and Herz, 1975c).

Clinical studies and plasma steroid analysis

Healthy women in the last trimester of an uncomplicated pregnancy were given 1.5–3.0 mg of one of the labelled steroid sulphates, injected intravenously as the potassium salt dissolved in isotonic glucose solution (10 ml). A blood sample (20 ml) was collected in a heparinised tube immediately before, and at six intervals over a period of 15–24 h after injection. The samples were immediately centrifuged and the plasma stored at –20° until analysed. Steroid sulphates were extracted from plasma, isolated by chromatography on Sephadex LH-20 and solvolysed in acidified ethyl acetate solution (Sjövall, 1970). The unconjugated steroids so obtained were converted into either trimethylsilyl (TMS) ethers (3β-hydroxysteroids) or *O*-methyloxime trimethylsilyl (MO-TMS) derivatives (3α-hydroxysteroids) for analysis by GC–MS.

Gas chromatography–mass spectrometry (GC–MS)

GC–MS was carried out on a modified LKB 9000 instrument (Reimendal and Sjövall, 1972). 3β-Hydroxysteroid derivatives were analysed using a glass column (3 m x 3.5 mm i.d.) packed with 1.5 per cent SE-30 and operated at 210°, while for derivatives of 3α-hydroxysteroids, a 25 m open-tubular glass capillary column coated with OV-1 was employed. The mass spectrometer was operated at an electron energy of 22.5 eV and with a basic accelerating voltage of 3.5 kV. Full mass spectra (0–700 a.m.u.) were recorded by repetitive magnetic scanning and data were recorded on magnetic tape using the incremental mode of recording (Jansson *et al.*, 1970). Partial mass spectra (5–10 a.m.u. in 2–5 s) were obtained with repetitive accelerating voltage scanning and a data sampling rate of 10 kHz. Intensity readings were bunched to give 10–15 values/a.m.u., which were recorded on magnetic tape. While repetitive magnetic scanning was used for screening of isotopic composition of ions, the accelerating voltage scan mode was employed in carrying out accurate determinations of isotope content. Data were evaluated by off-line processing on an IBM 1800 computer (Axelson *et al.*, 1974).

RESULTS

Metabolic reactions

Computer-based analysis of the data obtained from repetitive magnetic scanning of plasma extracts indicated which of the steroid sulphate pools had become labelled following the injection of deuterated analogues of each of the compounds shown in Figure 16.1. From this information, a number of general metabolic reactions could be discerned and the inter-relationships between metabolites were established. It was anticipated that the sulphates of steroids with a 3β-hydroxyl group would be metabolically related, while a similar finding was expected for the 3α-hydroxy series. This was, in fact, found to be the case and the results of the metabolic studies are discussed below accordingly. Pathways of metabolism were as follows.

(i) *Oxido-reduction at C-20* All of the steroid monosulphates studied underwent rapid oxido-reduction at C-20, which was the major metabolic reaction in both the 3α and the 3β series. Thus both of the $^{2}H_{3}$-pregnanolone sulphates were reduced to the corresponding $^{2}H_{3}$-pregnanediol monosulphates, while the injected pregnanediol monosulphates, labelled with four atoms of deuterium, underwent oxidation with the concomitant loss of label at C-20 to afford the $^{2}H_{3}$-pregnanolone derivatives.

(ii) *Sulphoconjugation at C-20* Conversion of the pregnanediol 3-monosulphates into the corresponding 3,20-disulphate conjugates was also a quantitatively important process, although this reaction occurred somewhat more slowly than did oxido-reduction at C-20. When the $^{2}H_{4}$-pregnanediol disulphates were injected, no deuterium appeared in any other steroid, thus indicating that sulphoconjugation at C-20 is an irreversible reaction.

(iii) *16α-Hydroxylation* A minor part (approximately 5 per cent) of each of the injected steroid monosulphates was converted into the corresponding 16α-hydroxylated derivative. The products of this reaction in both the 3α and 3β series, i.e. the 16α-hydroxypregnanolone and pregnane-3,16α,20α-triol 3-monosulphates, were also found to be interconverted by rapid oxido-reduction at C-20, in a similar fashion to the pregnanolone/pregnanediol pairs.

(iv) *21-Hydroxylation* Following injection of the labelled analogue of either 3α-hydroxy-5α-pregnan-20-one sulphate or 5α-pregnane-3α,20α-diol monosulphate, deuterium appeared in 5α-pregnane-3α,20α,21-triol monosulphate. The corresponding 20-oxosteroid, 3α,21-dihydroxy-5α-pregnan-20-one monosulphate, has been tentatively identified as a minor component in late pregnancy plasma (Baillie *et al.*, 1976), but its low concentration precluded measurements of deuterium incorporation into this steroid in the present studies. Although it is possible that the monosulphates of the pregnane-3,20,21-triol and 3,21-dihydroxypregnan-20-one are interconverted by oxido-reduction at C-20, the rate of this process must be very slow in comparison with the oxido-reductions described under (i) and (ii) above. Thus $^{2}H_{4}$-pregnane-3,20,21-triol monosulphate was detectable in plasma some 14 h after injection of the corresponding $^{2}H_{4}$-pregnanediol monosulphate, a finding which is in marked contrast to that obtained for the 16α-hydroxylated steroids, which undergo rapid oxido-reduction at C-20 with concomitant loss of label at this position.

In none of the experiments was any evidence obtained for the interconversion of sulphates of the 3α- and 3β-hydroxysteroids. Furthermore, no loss of label occurred from either the C-3 or C-11 position of any of the injected compounds and all metabolites contained either three or four atoms of deuterium. The metabolic inter-relationships between the sulphated pregnane derivatives in pregnancy plasma are summarised in Figure 16.2.

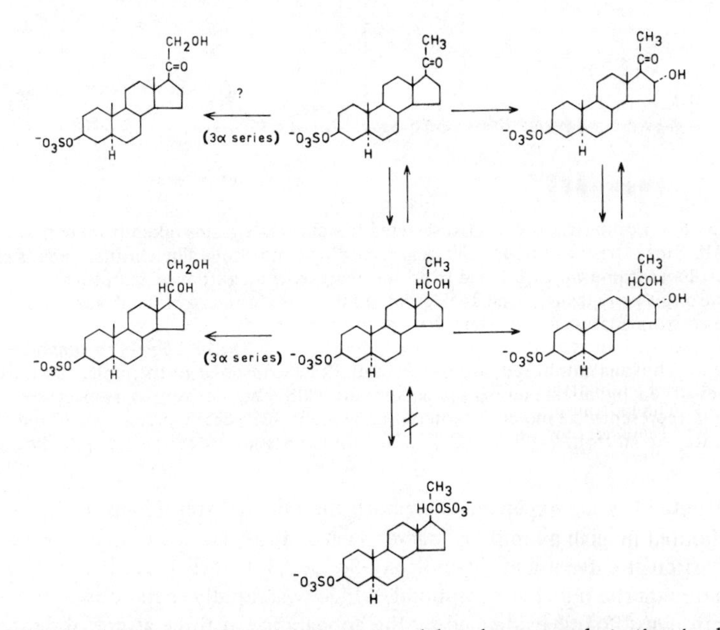

Figure 16.2 Metabolic inter-relationships between sulphated pregnane derivatives in plasma from pregnant women. Pathways of metabolism are common to both 3β- and 3α-hydroxysteroid derivatives, although 21-hydroxylation is quantitatively important only in the 3α series. Adapted from Baillie *et al.* (1977), with permission from Spectrum Publications Inc.

Pharmacokinetic studies

3β-Hydroxy-5α-pregnane derivatives

Steroids in this structural group were analysed by GC–MS as their TMS ether derivatives, using packed columns with SE-30 as the stationary phase. Values for the endogenous pool size, production rate and elimination half-life for each of the three major 3β-hydroxysteroid sulphates were obtained from analysis of the relative amounts of labelled and unlabelled molecular species in plasma samples taken at timed intervals following administration of the appropriate tracer compound. Measurements of the extent to which steroid sulphate pools were labelled at the times of sampling were based on the areas under peaks in reconstructed selected ion chromatograms (Figure 16.3) and were obtained by off-line data processing using programs ION4X and ISOTN (Axelson *et al.*, 1974). Semi-log plots of percentage excess labelled species against time were

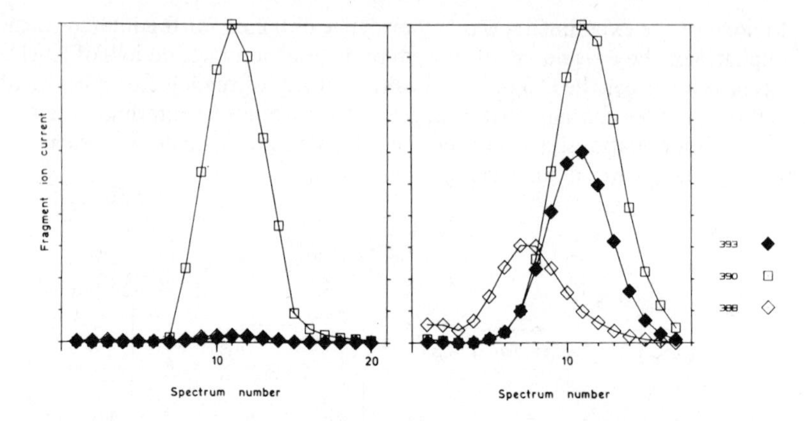

Figure 16.3 Computer-constructed selected ion chromatograms taken from consecutive GC–MS runs carried out under identical recording conditions. The chromatograms shown in the left-hand frame were obtained from the analysis of an authentic sample of 3β-hydroxy-5α-pregnan-20-one TMS ether, while those in the right-hand frame were obtained from analysis of the steroid monosulphate fraction of a plasma sample, taken from a volunteer 15 min after the injection of 3β-hydroxy-5α-[3α,11,11-$^2H_3$] pregnan-20-one sulphate. The ions monitored, at *m/e* 390 and 393, correspond to the molecular ions of the unlabelled (endogenous) and $^2H_3$-pregnanolone TMS ether derivatives, respectively, while *m/e* 388 represents the molecular ion of endogenous 3β-hydroxy-pregn-5-en-20-one TMS ether. Reprinted from Baillie *et al.* (1977), with permission from Spectrum Publications Inc.

constructed in each experiment for both the injected steroid and for its metabolites, to give a family of curves such as those shown in Figure 16.4. In this particular experiment, 3β-hydroxy-5α[3α,11,11-$^2H_3$] pregnan-20-one sulphate was the injected compound, which was rapidly metabolised by reduction at C-20, as evidenced by the appearance of three atoms of deuterium in the corresponding pregnanediol monosulphate. Label then appeared in the 16α-hydroxypregnanolone monosulphate, pregnane-3,16,20-triol monosulphate and pregnane-3,20-diol disulphate, in that order. The function for disappearance of the $^2H_3$-pregnanolone sulphate intersected those for the labelled monosulphates of the pregnanediol and the two 16α-hydroxysteroids at their maxima, which indicates that the injected compound served as the exclusive precursor of those monosulphates which became labelled.

Pharmacokinetic data were obtained by use of a further computer program which incorporated a routine to correct measured values of isotope content for errors introduced by expansion of endogenous pools, i.e. to compensate for the perturbation from steady state which results from the administration of semi-tracer doses of labelled compound (Nystedt, 1977). The decay curve for the $^2H_3$-pregnanolone sulphate (Figure 16.4) could be resolved by a curve-peeling procedure into two exponential components. However, the rate at which deuterium was incorporated into the pregnanediol monosulphate pool was more rapid than that which would result if the pregnanolone sulphate and the pregnanediol monosulphate formed a simple two-pool system. Assuming that the volume of distribution for the pregnanolone sulphate was the same as those for its metabolites, pool sizes for these metabolites could be calculated on the basis of their plasma concentrations. The pharmacokinetic data obtained from the single experiment illustrated in Figure 16.4 are given in Table 16.1.

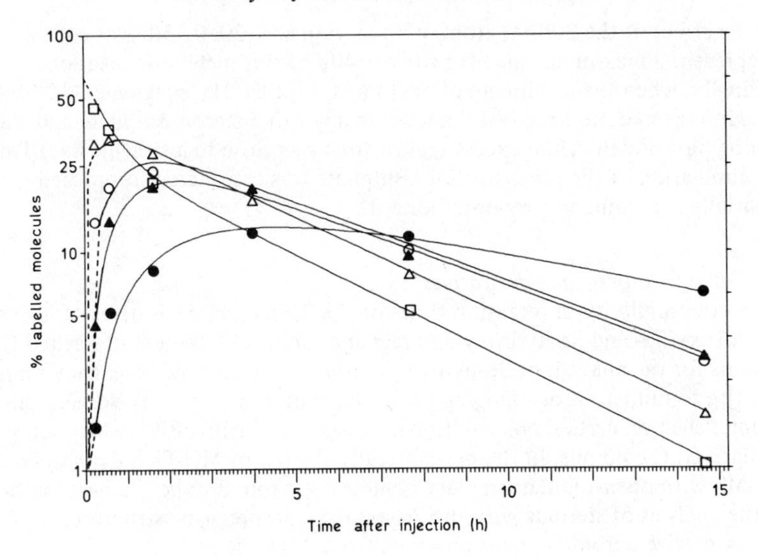

Figure 16.4 Semilog plots of percentage excess trideuterated molecules against time for several steroid sulphates in plasma which became labelled following the injection of 3β-hydroxy-5α-[3α,11,11-$^2H_3$] pregnan-20-one sulphate. These steroids, together with the conjugate fractions in which they appeared and the ions used to obtain values of isotope excess, are as follows:
□ 3β-hydroxy-5α-pregnan-20-one, monosulphate, *m/e* 390 and 393; △ 5α-pregnane-3β,20α-diol, monosulphate, *m/e* 269 and 272; • 5α-pregnane-3β,20α-diol, disulphate, *m/e* 269 and 272; ○ 3β,16α-dihydroxy-5α-pregnan-20-one, monosulphate, *m/e* 388 and 391; ▲ 5α-pregnane-3β,16α,20α-triol, monosulphate, *m/e* 462 and 465

In experiments where 5α-[3α,11,11,20β-$^2H_4$] pregnane-3β,20α-diol monosulphate was injected, deuterium was incorporated into the same steroids as when the pregnanolone sulphate was administered. Oxidation to 3β-hydroxy-5α-pregnan-20-one sulphate was found to follow first-order kinetics with an apparent half-life of approximately 0.75 h, but in view of the potential primary

Table 16.1 Pharmacokinetic data for 3β-hydroxy-5α-pregnan-20-one sulphate and its major metabolites obtained by computer-based analysis of the family of curves shown in Figure 16.4

Steroid	Pool size (μmol)	Elimination half-life (h)	Production rate (μmol/day)
3β-Hydroxy-5α-pregnan-20-one sulphate	2.9	2.7	} 73
5α-Pregnane-3β,20α-diol monosulphate	9.0[a]	2.7	
5α-Pregnane-3β,20α-diol disulphate	17.0[a]	7.4	38
3β,16α-Dihydroxy-5α-pregnan-20-one monosulphate	ND	4.1	ND
5α-Pregnane-3β,16α,20α-triol monosulphate	ND	4.1	ND

[a]Calculated on the assumption that pool size is proportional to concentration of the steroid in plasma.
ND = not determined.

isotope effect of the 20β-^{2}H atom on oxidation at C-20, 0.75 h was considered to represent a maximum value for the half-life of this metabolic reaction.

Finally, when the disulphate of 5α-[3α,11,11,20β-$^2H_4$] pregnane-3β,20α-diol was administered, no label was detected in any other steroid sulphate and the semilog plot of deuterium excess against time was close to a straight line. Thus, the elimination of the pregnanediol disulphate was considered to proceed essentially according to first-order kinetics.

3α-Hydroxy-5α-pregnane derivatives

Due to the similarity in retention times of the TMS ether derivatives of isomeric 3α-hydroxy-5α- and 3α-hydroxy-5β-pregnanes on SE-30, the use of packed GLC columns for the analysis of 3α-hydroxysteroids in extracts of pregnancy plasma gives rise to numerous overlapping peaks. Resolution of 5α/5β-H isomers can be accomplished on certain polar stationary phases, e.g. HiEff 8BP, or on non-polar capillary GLC columns. In the present study the use of MO-TMS derivatives and GC–MS with open-tubular capillary columns was found to be the best method for the analysis of steroids with the 3α-hydroxy-5α-pregnane structure. Representative chromatograms obtained from the same plasma extract

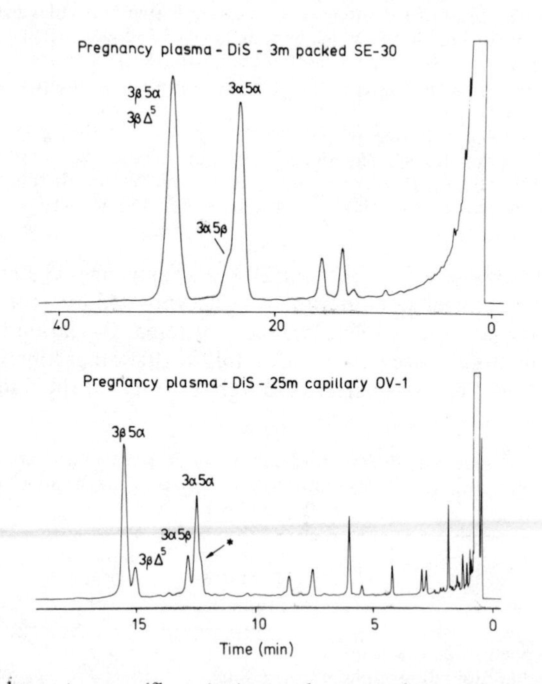

Figure 16.5 Gas chromatograms (flame ionisation detector) obtained from the analysis of a disulphate fraction of human pregnancy plasma. The upper trace was obtained on a 3 m packed column of SE-30 (isothermal, 210°), while the lower trace was obtained on a 25 m open-tubular capillary column, coated with OV-1 (isothermal, 250°). Peaks correspond to the TMS ethers of the following steroids:
'3α5α', 5α-pregnane-3α,20α-diol; '3α5β', 5β-pregnane-3α,20α-diol; '3β5α', 5α-pregnane-3β,20α-diol; '3βΔ^5', pregn-5-ene-3β,20α-diol. The identity of the unresolved component in the '3α5α' peak, marked with an asterisk, has not yet been established.

chromatographed on a 3 m packed column of SE-30 and on a 25 m open-tubular capillary coated with OV-1 are shown in Figure 16.5.

The procedure by which the degree of isotope incorporation into the various 3α-hydroxysteroid sulphate pools was determined paralleled that employed in studies on the 3β series. However, despite the high resolving power of the capillary columns used, GC–MS analysis of one steroid, 5α-pregnane-3α,20α-diol TMS ether, was still subject to interference from an unidentified endogenous compound. The nature of this interference is illustrated in Figure 16.6, which shows that both the endogenous pregnanediols and the unidentified component

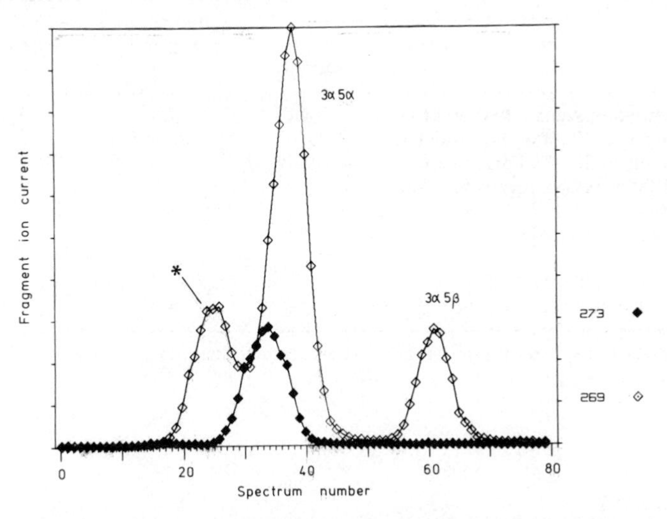

Figure 16.6 Computer-constructed selected ion chromatograms obtained from the analysis of a disulphate fraction of plasma taken from a subject who had been injected with 5α-[3β,11,11,20β-$^{2}H_4$] pregnane-3α,20α-diol disulphate. The ions monitored, at *m/e* 269 and 273, correspond to the [M-2 x 90-15]$^+$ ion in the spectra of unlabelled and $^{2}H_4$-pregnandiol TMS ether derivatives, respectively. Peaks are labelled with the same notation as used in Figure 16.5. The analysis was carried out by temperature programming from 210 to 260° at 0.6°/min

give rise to a fragment ion at *m/e* 269. In fact, the interfering compound afforded ions at all other *m/e* values characteristic of pregnanediol TMS ethers and was thus tentatively identified as a pregnanediol isomer. Since accurate values for the area under the *m/e* 269 chromatogram could not be obtained for the 5α-pregnane-3α,20α-diol peak, a modified program, ISOTL, was used (Curstedt, Reimendal and Sjövall, 1976). In this program, the investigator specifies the number of consecutive data points, taken across the peak centre, to be used for integration; this *same* number of data points is then used in each subsequent ion chromatogram, so that the *same proportion* of each peak is taken for isotope excess measurements. In the sample shown in Figure 16.6, the number of data points used was 12 (spectra Nos. 31–42 in the chromatogram for *m/e* 269 and Nos. 28–39 for *m/e* 273). By this approach, interferences of the type described above could be effectively eliminated.

Semilog plots, such as those illustrated in Figure 16.4, were constructed from data obtained from administration of each of the 3α-hydroxysteroid sulphates and the corresponding pharmacokinetic parameters calculated as before. Representative values, obtained from an experiment in which 3α-hydroxy-5α-[3β,11,11-$^2H_3$] pregnan-20-one sulphate was injected, are given in Table 16.2.

Table 16.2 Pharmacokinetic data for the major sulphated 3α-hydroxy-5α-pregnane derivatives in pregnancy plasma. The figures were obtained from a single experiment in which the subject was injected with 3α-hydroxy-5α-[3β,11,11-$^2H_3$] pregnan-20-one sulphate

Steroid	Pool size (μmol)	Elimination half-life (h)	Production rate (μmol/day)
3α-Hydroxy-5α-pregnan-20-one sulphate	4.1	2.8	} 89
5α-Pregnane-3α,20α-diol monosulphate	11.0[a]	2.8	
5α-Pregnane-3α,20α-diol disulphate	9.4[a]	2.8	56
3α,16α-Dihydroxy-5α-pregnan-20-one monosulphate	0.3[a]	3.3	1.5
5α-Pregnane-3α,16α,20α-triol monosulphate	0.4[a]	3.3	2.0
5α-Pregnane-3α,20α,21-triol monosulphate	7.4[a]	4.3	29

[a]Calculated on the assumption that pool size is proportional to concentration of the steroid in plasma.

DISCUSSION

The present investigation has shown that the C_{21} steroid monosulphates found in plasma of pregnant women are not merely 'dead' end-products of progesterone metabolism, but that they actively undergo interconversion and further metabolism *in vivo*. Rapid oxido-reduction at C-20 is the major metabolic reaction for these compounds, while conversion of pregnanediol monosulphates into the corresponding 3,20-disulphates is also a quantitatively important, and essentially irreversible, process. In addition, each of the monosulphates studied underwent 16α-hydroxylation, while only compounds with a 3α-sulphate group were hydroxylated at the 21 position; metabolism by these routes appeared to account for some 30 per cent of the turnover of the parent $C_{21}O_2$ steroid sulphates. These metabolic reactions are summarised in Figure 16.2.

Following injection of a labelled steroid with the 3β-sulpho-oxy-5α-pregnane structure, no deuterium appeared in the isomeric 3α-sulpho-oxy derivatives, and vice versa. Furthermore, in none of the experiments could dideuterated steroids be detected. These two observations together support the contention that the sulphate moiety at C-3 remained intact, since a free hydroxyl group at the 3 position in all probability would undergo oxido-reduction with concomitant loss of deuterium, as is the case in the rat (Cronholm and Makino, 1972). Thus it seems that the enzyme systems involved in the above metabolic reactions utilised the intact steroid sulphates as substrates. Human fetal liver tissue is known to contain 16α-hydroxylase and 20α-hydroxysteroid oxido-reductase systems active on pregnanolone, sulphate (Huhtaniemi, 1974; Ingelman-Sundberg, Rane and Gustafsson, 1975) although the same substrate has recently been shown to undergo direct 16α-hydroxylation in microsomal preparations from adult human liver (Einarsson *et al.*, 1976). Since placental tissue has a high capacity to hydrolyse steroid 3-sulphates (Diczfalusy, 1969), and a high 3β-hydroxysteroid oxido-reductase activity (Palmer *et al.*, 1966), retention of deuterium at C-3 in all steroids investigated would point to the maternal, rather than the fetal, compartment as the site for metabolic conversions observed.

Values for the combined production rates of the six $C_{21}O_2$ steroid sulphates studied in this investigation were in the order of 80 mg/day (approximately 250 μmol/day), which indicates that the conversion of progesterone into sulphoconjugated pregnane derivatives must be a metabolic pathway of considerable quantitative importance in late pregnancy. In contrast to urinary steroid profiles, where glucuronide conjugates of 5β-reduced pregnane derivatives predominate, bile from pregnant women has been shown to contain high concentrations of sulphate esters of many 5α-reduced C_{21} steroids (Laatikainen and Karjalainen, 1972). It seems likely, therefore, that excretion into bile is a major route for the clearance from plasma of sulphated progesterone metabolites.

In the present investigation we have shown that GC–MS, coupled with the use of tracers specifically labelled at several positions with deuterium, is a powerful technique for metabolic studies in humans. Thus labelling at metabolically 'stable' positions may be used to determine the fate of the carbon skeleton of the administered compound, while deuterium atoms introduced at potentially labile positions, e.g. at sites of hydroxyl group substitution, serve as markers for specific metabolic reactions. With the present analytical limitations, relatively large quantities of the tracer compound may have to be administered if accurate pharmacokinetic measurements are required; in the present studies, amounts comparable with the endogenous pool size had to be injected in order to follow the disappearance of the tracer over a period equivalent to at least three half-lives. However, corrections can be made in such cases for the resultant perturbations in endogenous pool size and clearance rate. Furthermore, in view of the current trends in development of GC–MS and computer hardware for quantitative analyses and measurements of low levels of isotope enrichment, such limitations may be significantly reduced in the near future.

ACKNOWLEDGEMENTS

We wish to thank Dr Tore Curstedt and Mr Robert Reimendal for their help with the computer programs and Mrs Kerstin Robertsson and Miss Irene Ferdman for skilful technical assistance. This work was supported by grants from the Swedish Medical Research Council (Project No. 13X-219) and from the World Health Organization. T. A. Baillie was the holder of a Postdoctoral Fellowship from the Royal Society (London), under the European Science Exchange Programme.

REFERENCES

Axelson, M., Cronholm, T., Curstedt, T. Reimendal, R. and Sjövall, J. (1974). *Chromatographia*, 7, 502

Axelson, M. and Sjövall, J. (1974). *J. Steroid Biochem.*, **5**, 733

Baillie, T. A., Anderson, R. A., Sjövall, K. and Sjövall, J. (1976). *J. Steroid Biochem.*, 7, 203

Baillie, T. A., Eriksson, H., Herz, J. E. and Sjövall, J. (1975a). *Europ. J. Biochem.*, **55**, 157

Baillie, T. A., Herz, J. E., Anderson, R. A. and Sjövall, J. (1975b). *Proceedings of the Second International Conference on Stable Isotopes* (ed. E. R. Klein and P. D. Klein), Oak Brook, Illinois, USA, p. 358, National Technical Information Service Document CONF-751027

Baillie, T. A., Sjövall, J. and Herz, J. E. (1975c). *Steroids*, **26**, 438

Baillie, T. A., Sjövall, K., Herz, J. E. and Sjövall, J. (1977). *Advances in Mass Spectrometry in Biochemistry and Medicine*, Vol. 2, (ed. A. Frigerio), Spectrum Publications, New York, Chapter 12 (in press)

Cronholm, T. and Makino, I. (1972). *Europ. J. Biochem.*, **26**, 251

Curstedt, T., Reimendal, R. and Sjövall, J. (1976). Unpublished results

Diczfalusy, E. (1969). *The Foetoplacental Unit* (ed. A. Pecile and C. Finizi), Excerpta Medica Foundation (ICS 183), Amsterdam, p. 65

Einarsson, K., Gustafsson, J.-Å., Ihre, T. and Ingelman-Sundberg, M. (1976). *J. Clin. Endocrinol. Metab.*, **43**, 56

Huhtaniemi, I. (1974). *Acta Endocrinol. (Kbh.)*, **76**, 525

Ingelman-Sundberg, M., Rane, A. and Gustafsson, J.-Å. (1975). *Biochemistry,* **14**, 429

Jansson, P.-Å., Melkersson, S., Ryhage, R. and Wikström., S. (1970). *Arkiv Kemi*, **31**, 565

Laatikainen, T. and Karjalainen, O. (1972). *Acta Endocrinol. (Kbh.),* **69**, 775

Nystedt, L. (1977). To be published

Palmer, R., Eriksson, G., Wiqvist, N. and Diczfalusy, E. (1966). *Acta Endocrinol. (Kbh.*), **52**, 598

Pinkus, J. L., Charles, D. and Chattoraj, S. C. (1971). *J. Biol. Chem.*, **246**, 633

Reimendal, R. and Sjövall, J. (1972). *Anal. Chem.*, **44**, 21

Sjövall, J., Sjövall, K. and Vihko, R. (1968). *Steroids*, **11**, 703

Sjövall, K. (1970). *Ann. Clin. Res.*, **2**, 393

Solomon, S. and Fuchs, F. (1971). *Endocrinology of Pregnancy* (ed. F. Fuchs and A. Klopper), Harper and Row, New York, p. 66

17
Characterisation of bile acid metabolism in man using bile acids labelled with stable isotopes

A. F. Hofmann, and P. D. Klein (Gastroenterology Research Unit, Mayo Clinic and Mayo Foundation, Rochester, Minnesota, and Division of Biological and Medical Research, Argonne National Laboratory, Argonne, Illinois, USA)

BILE ACID PHYSIOLOGY

Bile acids are acidic steroids formed in the liver from cholesterol and secreted into bile as their *N*-glycine or *N*-taurine conjugates. Their biological and clinical aspects have been thoroughly reviewed (Heaton, 1972; Dietschy, 1972; Nair and Kritchevsky, 1973; Paumgartner, 1977). Bile acids are amphipathic molecules which can disperse lecithin in micellar form, and bile is a concentrated micellar solution containing mixed micelles of bile acids, lecithin and cholesterol. When this mixed bile acid micelle reaches the small intestinal lumen, its composition changes and now the major lipid constituting the mixed micelle is fatty acid (plus some monoglyceride), which derives from pancreatic lipolysis of dietary triglyceride. Micellar solubilisation of the relatively insoluble fatty acid and monoglyceride greatly increases their aqueous concentration and thus accelerates the diffusive flux through the unstirred layer. Most of the bile acids are reabsorbed in the small intestine, especially the terminal ileum, where there is an active transport site. The bile acids return in the portal blood to the liver, where they are efficiently extracted and rapidly secreted into bile. This circulation of bile acids from liver to intestine and back to liver is termed the enterohepatic circulation, and it occurs about twice per meal. A small amount of bile acids are lost from the enterohepatic circulation and are excreted in faeces; the bile acid pool is maintained by hepatic synthesis. There is no urinary loss of bile acids (Figure 17.1). The bile acid conjugates are mostly absorbed in conjugated form from the small intestine, but a small amount of the bile acid conjugates is hydrolysed by bacterial deconjugases, releasing the unconjugated bile acid moiety. Most of this is reabsorbed to return to the liver, where it is reconjugated (Hepner *et al.*, 1973).

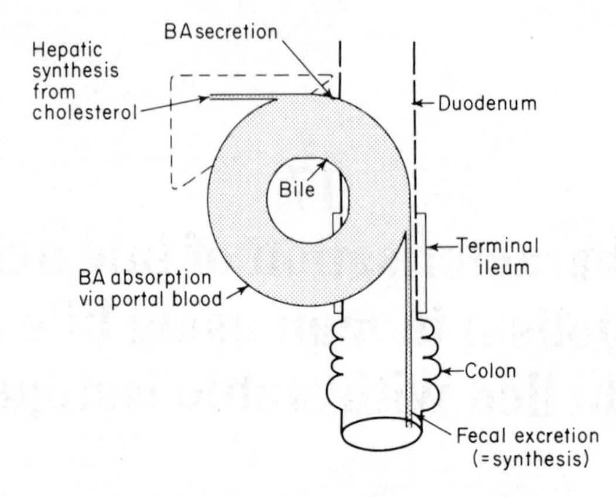

Figure 17.1 Schematic depiction of the enterohepatic circulation of bile acids in man under steady state conditions. The diagram indicates that the only input to the bile acid pool is hepatic synthesis from cholesterol, and that bile acid loss is only by faecal excretion. The anatomical location of the bile acid pool is a consequence of active transport systems in the terminal ileum and liver

The chemical constituents of the bile acid pool are the major primary bile acids, cholic and chenodeoxycholic acid, the major secondary bile acids, deoxycholic and lithocholic acid, which are formed by bacterial 7α-dehydroxylation of these primary bile acids, and the tertiary bile acid, ursodeoxycholic acid, which is thought to be formed in the liver from 7-keto-lithocholic acid formed in the small intestinal lumen by bacterial dehydrogenation of chenodeoxycholic acid (Figures 17.2 and 17.3).

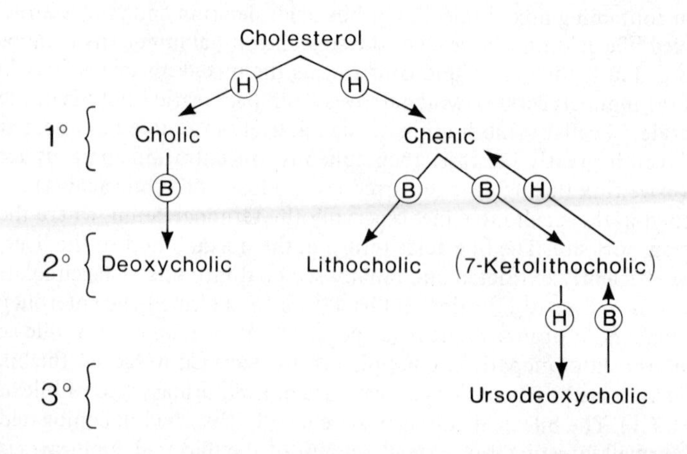

Figure 17.2 Major primary and secondary bile acids in the enterohepatic circulation in man. 7-Keto-lithocholic is present in the small intestinal lumen, but at present is considered to be reduced during hepatic passage so that it is lacking in bile. Abbreviations: H = hepatic; B = bacterial

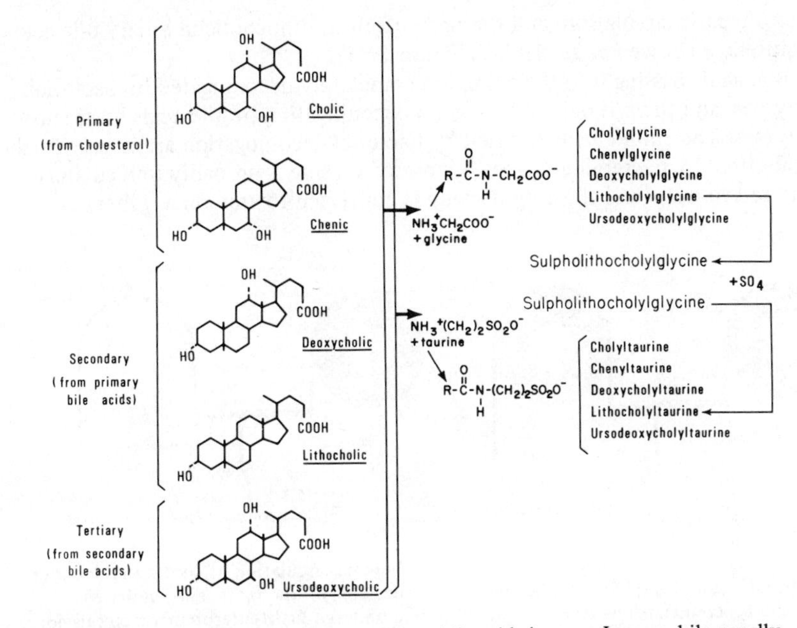

Figure 17.3 Chemical structure of major biliary bile acids in man. In man, bile usually contains about equal parts of cholyl and chenyl conjugates (30–40 per cent) and less deoxycholyl conjugates (10–30 per cent). The proportion of ursodeoxycholyl and lithocholyl conjugates is usually less than 5 per cent; most of the lithocholyl conjugates are sulphated

Current views of bile acid formation may be summarised as follows. Cholic and chenodeoxycholic acid are formed in the liver from cholesterol. About twice as much cholic as chenodeoxycholic is synthesised daily, but because intestinal conservation of chenodeoxycholic is more efficient, bile contains equal proportions of these two primary bile acids. Cholic, which is lost from the enterohepatic circulation, passes into the colon, where it is converted to deoxycholic by bacterial 7-dehydroxylation. About one-third to one-half of deoxycholic which is formed is reabsorbed; it passes to the liver, where it is conjugated with glycine or taurine, and the subsequent metabolism of the deoxycholyl conjugates resembles that of the primary bile acid conjugates (Hepner, Hofmann and Thomas, 1972a,b). Chenodeoxycholic, which is lost from the enterohepatic circulation, has two fates: some is dehydrogenated by bacterial enzymes to 7-keto-lithocholic in the small intestine, which in turn is absorbed and reduced to chenodeoxycholic and ursodeoxycholic in the liver. Some chenodeoxycholic is lost from the enterohepatic circulation and passes into the colon, where it is 7-dehydroxylated by bacteria to form lithocholic. About one-fifth of the lithocholic newly formed in the colon is absorbed and passes to the liver (Allan, Thistle and Hofmann, 1976). Here it is not only conjugated with glycine and taurine, but also mostly sulphated to form two new conjugates, sulpholithocholylglycine and sulpholithocholyltaurine. These are poorly reabsorbed from the intestine, as sulphation decreases both passive and active absorption. As a consequence, lithocholic is rapidly lost from the enterohepatic circulation, and the proportion of lithocholic in biliary bile acids is quite low (Cowen *et al.*, 1975) (Figure 17.4).

Bile acids passing into the colon are completely deconjugated by bacterial enzymes, and there is believed to be no deconjugation of bile acids by tissue enzymes. The amino acids released by bacterial deconjugation are largely further catabolised by bacterial enzymes. Liberated glycine is probably immediately oxidised to CO_2, and the amino group is converted to ammonia. Liberated

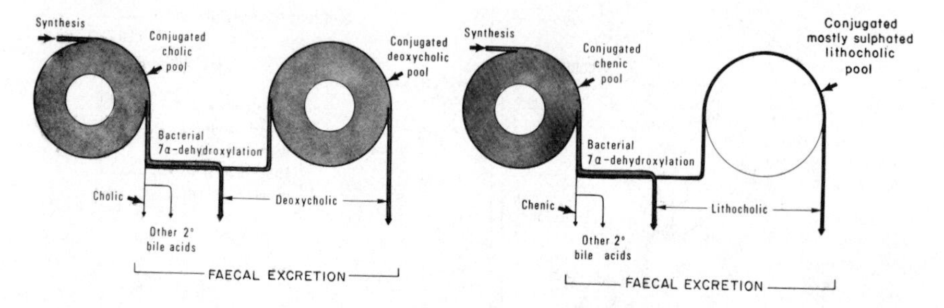

Figure 17.4 Schematic depiction of the enterohepatic circulation of cholic and deoxycholic acids, left, and chenodeoxycholic and lithocholic acids on the right. The sulphated lithocholyl conjugates which are secreted in bile undergo little enterohepatic circulation, and as a consequence, lithocholic acid is rapidly eliminated, and its concentration in bile is quite low

taurine is catabolised similarly, although sulphate is also formed (Hepner *et al.*, 1973) (Figure 17.5). There is essentially no re-utilisation of the liberated glycine or taurine as such or any of their constituent atoms for bile acid conjugation.

Considered dynamically, the bile acid pool circulates rapidly during digestion and circulates slowly during fasting (Figure 17.6). Hepatic obstruction is always extremely efficient, so that peripheral bile acid levels are quite low. Considerable evidence suggests that the first-pass clearance of bile acids remains relatively

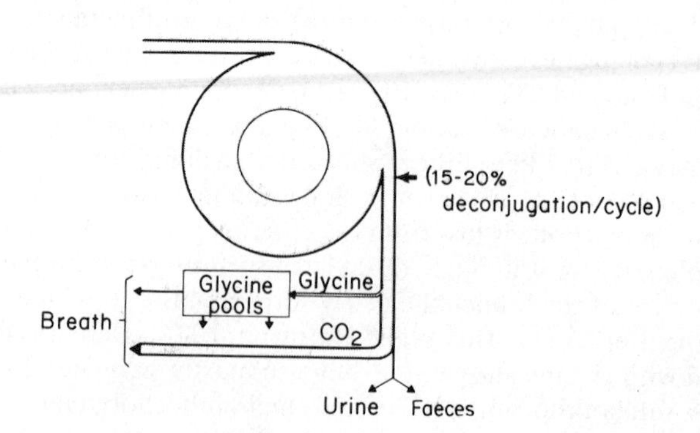

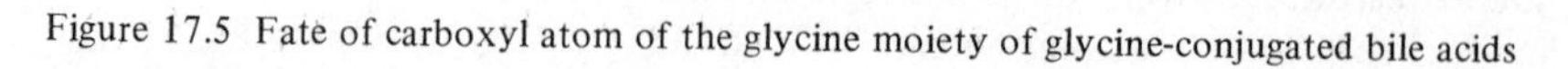
Figure 17.5 Fate of carboxyl atom of the glycine moiety of glycine-conjugated bile acids

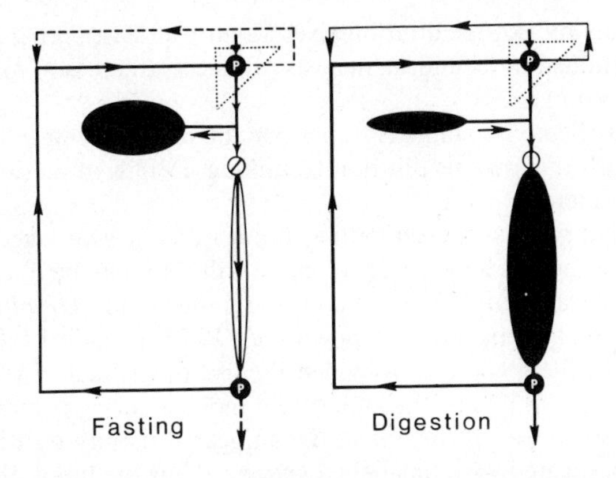

Figure 17.6 Schematic depiction of the enterohepatic circulation of bile acids in man, emphasising dynamic aspects. The ellipsoids indicate the gall bladder and the small intestine. P represents the chemical pumps in the terminal ileum and liver, and the upper horizontal line represents the plasma compartment. The diagram indicates that capacity for storage of the bile acid pool is located only in the gall bladder and small intestine. Note also the sphincter of Oddi, indicated as a valve. The source of plasma bile acids in the healthy individual is hepatic spillover of bile acids returning from the intestine

constant, so that the increased return of bile acids to the liver which occurs during digestion is signalled by an increase in the level of bile acids in peripheral blood. However, the magnitude of this increase is small, and the postprandial increase in serum bile acid levels can only be detected easily by extremely sensitive analytical methods, such as radioimmunoassay (Simmonds *et al.*, 1973).

CHARACTERISATION OF BILE ACID METABOLISM IN MAN WITH ISOTOPICALLY LABELLED BILE ACIDS

Exchangeable pools and inputs

Lindstedt (1957) injected labelled cholic as tracer, sampled duodenal bile daily, isolated the cholic moiety by chromatography after alkaline hydrolysis, and found that the specific activity declined mono-exponentially. He concluded that the metabolism of cholic could be described by a single pool model, and this was subsequently shown to be true for chenodeoxycholic and deoxycholic. Lithocholic was originally reported to conform to a single pool model, but when the specific activity decay curve was followed at more frequent time intervals in its early portion, it was found to be bi-exponential (Allan *et al.*, 1976). With this technique of isotope dilution, one may estimate pool size by extrapolating the specific activity decay curve to zero, and thus determining the degree to which injected tracer would be diluted, if mixing were instantaneous. If the specific activity decay curve is plotted using the natural logarithm of the specific activity values, then the slope is equivalent to the fractional turnover rate, and the product of the fractional turnover rate times the pool size equals daily input, which, in the absence of exogenous bile acid, should be equal to synthesis (Lindstedt, 1957; Hofmann and Hoffman, 1974). Curiously, bile acid input when determined by isotope dilution gives a figure at least 50 per cent larger than that determined by chemical analysis of faeces; the reason for this discrepancy is not clear.

Bile acids labelled with stable isotopes may be used for determination of bile acid kinetics with this isotope dilution technique. Details of methodology are discussed subsequently.

If the bile acid pool is labelled with a conjugated bile acid tagged in the amino acid moiety, the specific activity decay curve will generally be linear or nearly linear, but will decay more rapidly than if the same conjugated bile acid labelled in the steroid moiety is injected (Hepner *et al.*, 1973) (Figure 17.7). The more rapid turnover of the specific activity decay curve of the amino acid is explained by deconjugation with loss of the amino acid moiety during enterohepatic cycling. As noted above, the liberated, unconjugated moiety returns to the liver, where it is reconjugated with unlabelled glycine. Thus the input of amino acid to the conjugated bile acid pool is usually greater than that of the steroid moiety because of deconjugation and reconjugation during enterohepatic cycling.

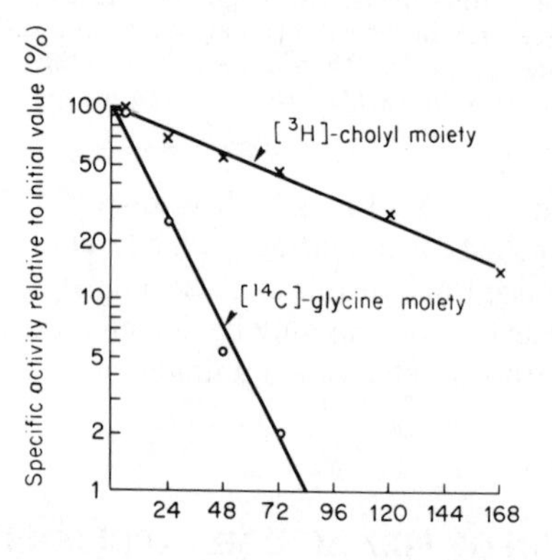

Figure 17.7 Specific activity decay curves of ^{3}H-cholylglycine-1-^{14}C. The more rapid turnover of the amino acid moiety is explained by intestinal deconjugation and hepatic reconjugation of the steroid moiety. The ^{14}C-glycine moiety is lost after deconjugation, being rapidly degraded to CO_2 and ammonia

Biotransformations: product–precursor relationships

In man, cholic is 7-dehydroxylated by bacteria to form deoxycholic; chenodeoxycholic is similarly dehydroxylated to form lithocholic. Ursodeoxycholic derives from the 7-keto-lithocholic compound which is formed from either chenodeoxycholic or ursodeoxycholic in the small intestine by bacterial 7-dehydrogenation (Salen *et al.*, 1974). Precursor–product relationships can be clarified using bile acids labelled with radioactive or stable isotopes.

Concentrations of bile acids and biological fluids

The concentration of bile acids in biological fluids varies remarkably from 50 000 μM in bile to 5 μM in plasma. Measurement of bile acids in fluids in which their concentration is high is not difficult. It may be carried out using a specific hydroxy steroid dehydrogenase which oxidises the 3- (or 7-) hydroxy group to a keto group, the oxidation being coupled to the reduction of DPN, which can be quantified by ultraviolet spectroscopy (Palmer, 1969). For serum bile acids, the sensitivity of the method is increased by quantifying the DPN reduction by fluorimetry. In a recent modification of this method, the hydrogen is transferred to a second fluor resazurin, for still greater sensitivity (Osuga *et al.*, 1977). Bile acids may also be quantified by gas–liquid chromatography using trimethylsilyl ether or trifluoroacetate (or acetate) derivatives. Successful gas chromatography requires tedious work-up procedures, since bile acids must be deconjugated, esterified and derivatised.

However, for accurate quantification of plasma levels, neither of the above methods is sufficiently sensitive. At the moment, only two sensitive methods have been described. The first is conventional radioimmunoassay, using antibodies which are specific for the number and position of hydroxylic substituents in the steroid moiety (Simmonds *et al.*, 1973; Murphy *et al.*, 1974). The second is inverse isotope dilution, using deuterium- or ^{13}C-labelled bile acids. These are added to the plasma sample, and then the isotope ratio is determined using gas–liquid chromatography–mass spectrometry with accelerating voltage alternation (Klein, Haumann and Hachey, 1975).

Successful inverse isotope dilution measurements of serum bile acids has now been reported in abstract form (Mede, Tserng and Klein, 1976).

PREPARATION OF BILE ACID LABELLED WITH STABLE ISOTOPES

Deuterium

For satisfactory deuterium labelling of bile acids, one wishes to use a method which gives a product of high isotopic content and high isotopic purity. Further, the site of labelling should not influence biological behaviour, and the label should not be removed by tissue or bacterial enzymes.

These requirements exclude any non-specific labelling methods such as the Wilzbach procedure. It is possible to oxidise any or all of the hydroxy groups on the bile acid molecule and then reduce these with borodeuteride, but this gives a mixture of epimers, and although the desired epimer can be isolated chromatographically, bile acids containing epimeric deuterium are unsatisfactory for metabolic studies, since hydrogenation is a common bacterial biotransformation in the small intestine. However, such bile acids can be used for inverse isotope dilution studies, provided a satisfactory dilution curve can be obtained.

Some years ago we reported successful exchange labelling with tritium at the 2 and 4 positions, and this technique can also be used for deuterium (Hofmann, Szcepanik and Klein, 1968). However, it is difficult to obtain a product with high isotopic purity, and indeed, bile acids prepared in this manner are quite heterogeneous with respect to number and position of label. The ideal route to bile acids tagged with deuterium should involve reductive deuteration of an olefinic precursor. The olefinic bile acids may be prepared by elimination of a hydroxyl group, and we recently described in some detail the synthesis of Δ^{11}-chenodeoxycholic and lithocholic (Cowen *et al.*, 1975) (Figure 17.8). These

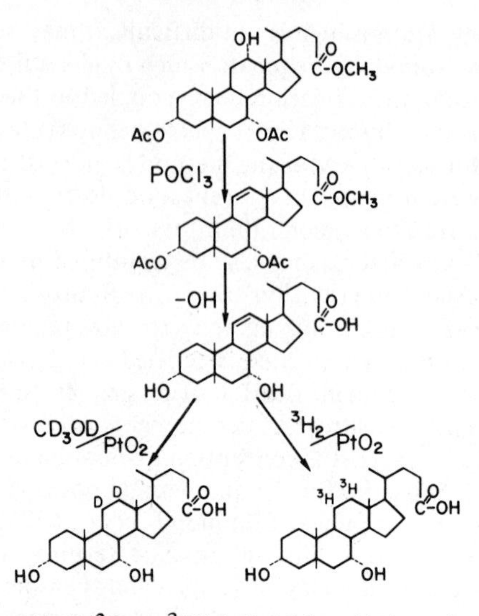

Figure 17.8 Preparation of 11,12-^{2}H- or ^{3}H-labelled bile acids by reductive tritiation of a Δ^{11} precursor prepared from cholic acid

olefinic bile acids appear to be excellent precursors for 11,12-dideutero bile acids. It should also be possible to prepare or to insert a double bond into the isopentanoic acid side chain of bile acids, but such has not been reported.

For conjugated bile acids labelled in the amino acid moiety, deuterium-labelled glycine is available commercially, and bile acids may be conjugated using the mixed carboxylic–carbonic anhydride method first detailed by Norman (1955), or by a method employing one of the newer bi-functioning coupling agents (Lack *et al.*, 1973; Tserng, Hachey and Klein, 1977). Deuterium-labelled taurine has not been described in the literature.

Carbon-13

Bile acids labelled with ^{13}C in the carboxyl moiety may be prepared by the conventional decarboxylation–recarboxylation procedure, described some years ago for the preparation of 24-^{14}C-bile acids by Bergström, Rottenberg and Voltz (1953), and more recently modified for the preparation of ^{13}C-bile acids (Hachey *et al.*, 1974; Tserng and Klein, 1977).

VALIDATION OF ISOTOPE STABILITY *in vivo* FOR BILE ACIDS LABELLED WITH STABLE ISOTOPES

Necessity of validation

The deuterium label on a bile acid may not be stable during enterohepatic cycling because of removal by tissue or bacterial enzymes. Our group has argued that no bile acid labelled with deuterium (or tritium) should be used for isotope dilution or biotransformation studies in man, until the biological stability of that particular labelled bile acid has been defined. To do this, one must administer the deuterium-labelled bile acid together with a 24-^{14}C-bile acid and compare the specific activity decay curves for the two isotopes of that bile acid, after its isolation from bile. Since the label may not be lost during enterohepatic cycling, but may be lost during colonic passage, it is desirable to isolate the bile acid from faecal samples and also determine the isotope ratio. If there is loss of the deuterium relative to ^{14}C, then it indicates attack of the deuterium label during colonic passage.

Results of validation studies

2,2′,4,4′-^{2}H-*Labelled bile acids*

In rather limited studies, we showed that this label is stable during enterohepatic cycling in man (Balistreri *et al.,* 1975). No information is available on the those obtained with 24-^{14}C-labelled bile acids (LaRusso, Hoffman and Hofmann, 1974). The subsequent fate of the deuterium label was not followed, but for 2,2′,4,4′-^{3}H-bile acids we found evidence for substantial loss of ^{3}H during colonic passage. The biotransformation product of cholic, deoxycholic, contained much less ^{3}H than its precursor. Thus it seems likely that bile acids labelled with deuterium at the 2 and 4 positions, prepared by exchange labelling, are satisfactory for isotope dilution studies, but not for detailed metabolic studies; in addition, as noted, this exchange-labelling technique gives a product of low isotopic purity.

11,12-^{2}H-Chenodeoxycholic acid appears to be stable during enterohepatic cycling in man (Balistreri *et al.*, 1975). No information is available on the stability of this label during colonic passage. When similar experiments were carried out with 11,12-^{3}H-bile acids, we found that there was an immediate loss of about 10 per cent of the ^{3}H label (Ng, Allan and Hofmann, 1977). We attributed this to the presence of tritium in a position other than at C-11 or C-12 since our compound was prepared with carrier-free tritium gas; but we are not certain. We also observed additional loss of ^{3}H during intestinal passage, since the ^{3}H/^{14}C ratio of stool was lower than that of bile. When these experiments were carried out with lithocholic, loss of tritium was considerable during enterohepatic cycling, suggesting that 11,12-^{2}H- or 11,12-^{3}H-lithocholic cannot be used in man for any biological studies. Lithocholic has a unique enterohepatic circulation among bile acids, since it is excreted in bile mostly as the sulphated conjugate, and since reabsorption of the steroid moiety can probably occur only after desulphation and deconjugation, which may well only occur in the colon. Accordingly, lithocholic is exposed to more bacteria during its enterohepatic cycling, and this may explain the lability of the label.

USE OF BILE ACIDS TAGGED WITH STABLE ISOTOPES TO MEASURE BILE ACID KINETICS

Steroid moiety

The Lindstedt technique for isotope dilution has already been discussed. Labelled bile acid is given intravenously on the evening preceding the first bile collection. Bile is collected daily for 5 days or every other day for 9 days by means of an oro-duodenal tube with intravenous cholecystokinin to induce gall bladder contraction. The bile sample is hydrolysed with a deconjugating enzyme or strong base and the liberated bile acids are extracted, esterified and derivatised for coupled gas chromatography–mass spectrometry. The time course of the change in atoms per cent excess is determined, permitting calculation of pool size and fractional turnover rate. With stable isotopes it is possible to determine the specific activity decay curve for all major bile acids simultaneously, since they are separated by gas chromatography prior to isotope measurement. The simultaneous determination of specific activity decay curves for the major bile acids has seldom been carried out using radioactive isotopes, because of the difficulty of separating chenodeoxycholic and deoxycholic by thin-layer chromatography, which was nearly essential, since radioactivity was determined before gas–liquid chromatographic separation and quantification, and since cholic radioactivity soon appeared in deoxycholic. Description of a new solvent system giving superb resolution of deoxycholic and chenodeoxycholic (Chavez and Krone, 1976) should now permit simultaneous measurement of bile acid kinetics for all major bile acids in man using radioactive isotopes.

In the isotope dilution technique in man bile has been sampled most frequently because the concentration of bile acids in it is so high. In some studies intestinal content has also been used (Bennion and Grundy, 1975). Plasma samples have not been used, because the very large pool of bile acids dilutes injected tracer, and the very low concentration of bile acids in serum means that the amount of radioactivity in any plasma sample is too low to measure. Stable isotopes give promise of permitting plasma samples to be used for isotope dilution studies in man. For example, if every tenth molecule in the bile acid pool could be labelled, using stable isotopes, then it should be possible to determine the specific activity decay curve on plasma samples. However, it has not yet been shown that plasma bile acids and biliary bile acids are in fact in isotopic equilibrium. Studies to test this point are in progress in a number of laboratories.

Since the specific activity of bile acids in bile and proximal small intestinal contents should be identical, it should be possible to sample the bile acid pool in the small intestine, rather than bile, and this could be done non-invasively if a suitable technique were developed. We have described the construction of capsules made of dialysis tubing containing cholestyramine, and which bind bile acids during intestinal passage (Hoffman, LaRusso and Hofmann, 1976) (Figure 17.9). Available data suggest the precision of the method is less than the sampling bile, but that the method probably is of sufficient accuracy and precision for screening purposes (Figure 17.10).

Amino acid moiety

As discussed, conjugated bile acids labelled in the amino acid moiety may be used for isotope dilution studies, and the specific activity decay reflects the amount of deconjugation occurring during enterohepatic cycling. To date, such studies have not been carried out with conjugated bile acids whose glycine or taurine moiety was labelled with deuterium or ^{13}C. Methodologically, such experiments would involve isolation and quantification of the amino acid moiety after alkaline hydrolysis of the conjugated bile acid.

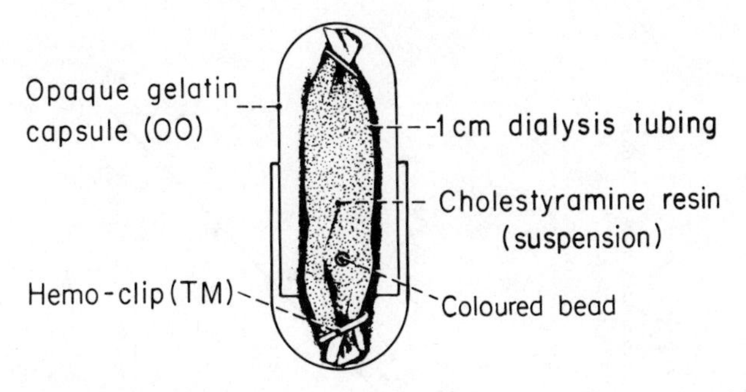

Figure 17.9 Construction of a cholestyramine-containing capsule for trapping bile acids in the small intestinal lumen

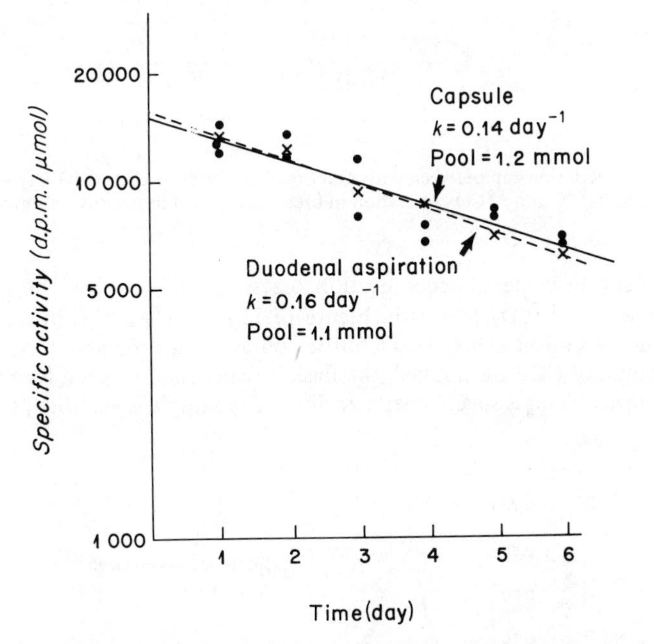

Figure 17.10 Specific activity decay curve of labelled chenodeoxycholic acid isolated from bile in samples obtained either by the cholestyramine-containing capsule or by direct duodenal aspiration

If the conjugated bile acid labelled in the amino acid moiety with ^{13}C is given, the ^{13}C will appear in breath as $^{13}CO_2$ almost instantaneously after the amino acid is released by bacterial enzymes (Hofmann *et al.*, 1972; Fromm and Hofmann, 1971; Solomon *et al.*, 1977). The rate of appearance of $^{13}CO_2$ is linearly correlated with the fractional turnover rate of the amino acid moiety of the conjugated bile acid (Hepner *et al.*, 1972a,b), indicating that $^{13}CO_2$ excretion can be used to quantify deconjugation (Figure 17.11). Thus, Solomon *et al.* (1977) have reported the use of cholylglycine-1,2-^{13}C for the

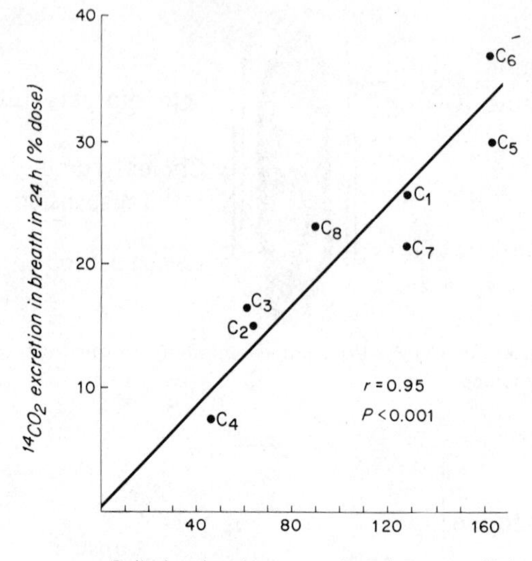

Figure 17.11 Relationship between daily fractional turnover of glycine moiety of cholylglycine-1-^{14}C and $^{14}CO_2$ excretion in breath over 24 h in healthy volunteers

measurement of bacterial deconjugation, assessing release of the ^{13}C-glycine by measurement of $^{13}CO_2$ in breath (Figure 17.12). $^{13}CO_2$ in breath is collected by having the subject exhale into a bottle containing a strong base. After several mmol of CO_2 are trapped, the flask is sealed, and shipped for remote processing at the mass spectrometry facility. The sample is acidified, the CO_2

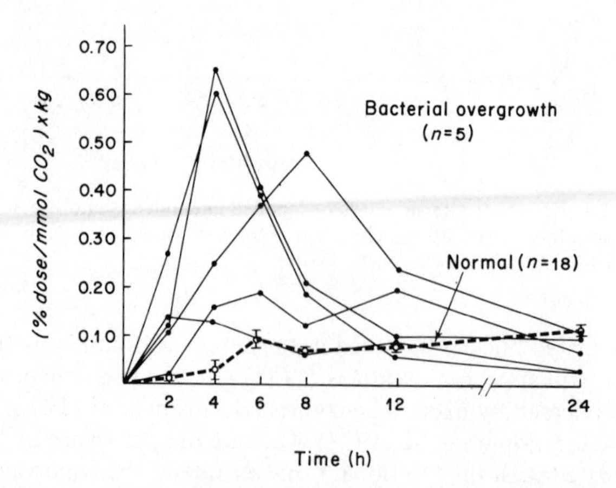

Figure 17.12 Increased $^{14}CO_2$ excretion in breath after administration of cholylglycine-1-^{14}C in a test meal to five patients with bacterial overgrowth (stagnant loop syndrome). Data from 18 healthy control subjects are shown

trapped, and its atoms per cent excess determined by mass spectrometry, usually using a twin-cup instrument. A critical analysis of methodological problems has recently been published by Schoeller *et al*. (1977).

USE OF BILE ACIDS LABELLED WITH STABLE ISOTOPES FOR INVERSE ISOTOPE QUANTIFICATION OF SERUM BILE ACIDS

As discussed above, radioimmunoassay and inverse isotope dilution are the only two methods possessing adequate sensitivity for quantifying the diurnal variation in serum bile acid levels in healthy subjects. The methodology here is entirely conventional, except that the ^{2}H- or ^{13}C-labelled bile acids should be added in the same chemical form as the bile acid to be measured exists in plasma. It is necessary to synthesise a family of ^{2}H- or ^{13}C-labelled bile acids which will simulate the spectrum of bile acid species present in plasma. Only one group has reported its results to date, and only in abstract form (Mede *et al.*, 1976).

STUDIES OF BILE ACID METABOLISM IN MAN USING BILE ACIDS LABELLED WITH STABLE ISOTOPES

Infants

Watkins *et al.* (1973, 1975a) have shown that the bile acid pool is markedly reduced in both premature and newborn infants compared with adults, when pool size is expressed on a mg/kg basis. They have also shown (Watkins *et al.*, 1975b) that in children with cystic fibrosis the bile acid turnover rate is somewhat increased, in agreement with other studies showing increased faecal bile acids in patients with cystic fibrosis.

Isotope dilution studies

To date, bile acids labelled with stable isotopes have not been used to characterise bile acid metabolism in pregnancy, despite its importance.

Biotransformation

To date, bile acids labelled with stable isotopes have not been used for any detailed characterisation of biotransformation studies in man. Figure 17.13 shows the conversion of 24-^{13}C-chenodeoxycholic into lithocholic and ursodeoxycholic. Figure 17.14 shows the conversion of 24-^{13}C-cholic into deoxycholic. These studies should be extended.

Inverse isotope dilution measurement of plasma bile acid concentrations

As noted, Mede *et al.* (1976) have reported in abstract form preliminary measurements of serum bile acid levels using inverse isotope dilution.

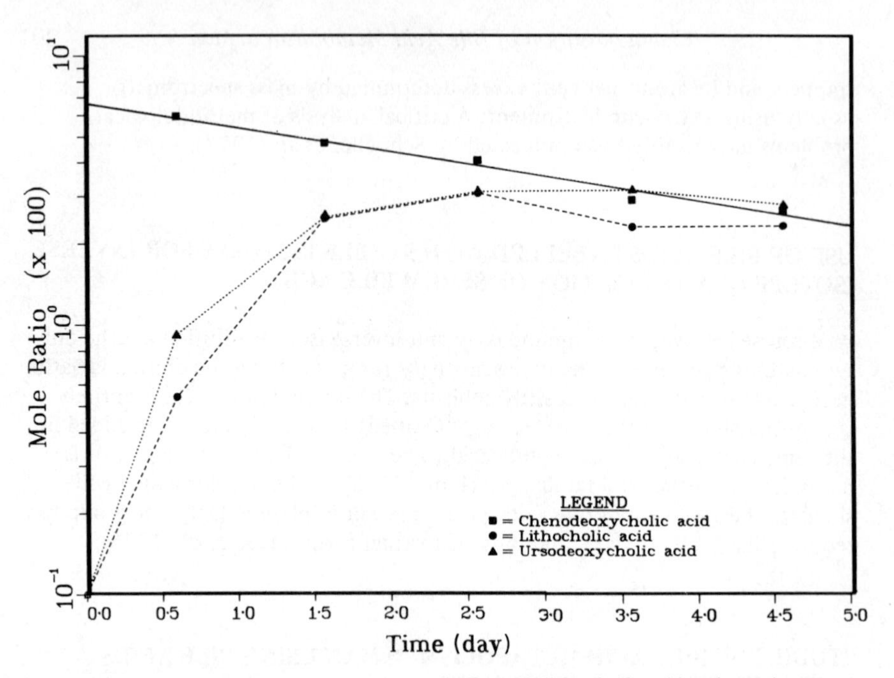

Figure 17.13 Isotope content in chenodeoxycholic acid, lithocholic acid and ursodeoxycholic acid in human bile samples obtained in a healthy volunteer after administration of 24-^{13}C-chenodeoxycholic acid

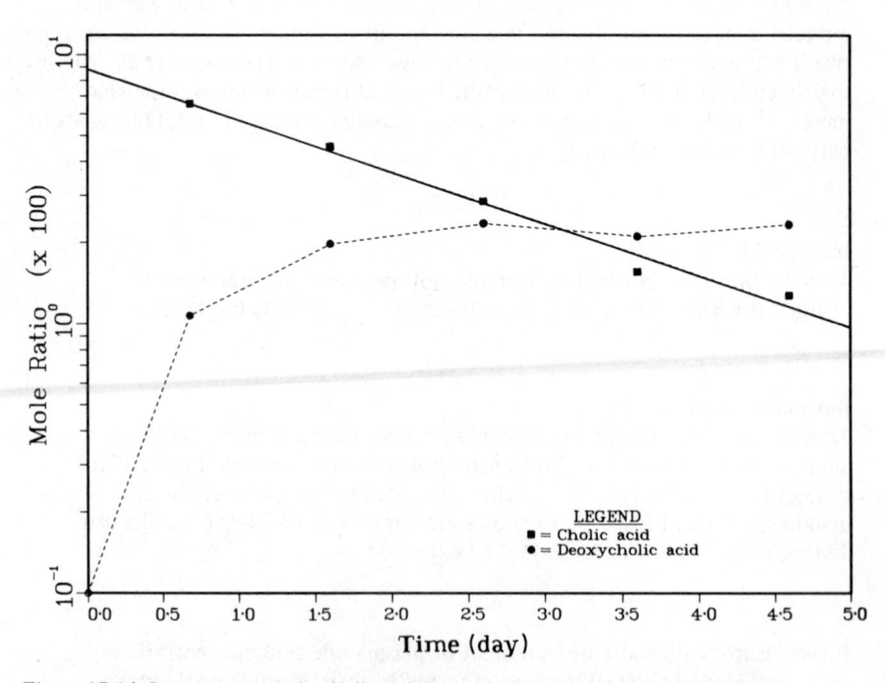

Figure 17.14 Isotope content in cholic and deoxycholic acids in human bile samples obtained in a healthy volunteer after administration of 24-^{13}C-cholic acid

PERSPECTIVE

Bile acids labelled with radioactive isotopes are far simpler to use, and the clinical investigator is usually much happier if he can carry out the analyses in his own laboratory. Stable isotopes often involve remote processing. For studies in children and pregnant women, bile acids labelled with stable isotopes are essential and their use will continue. The pace of the research will relate to whether or not bile acids are involved in important physiological or pathological events in children or pregnant women.

If plasma and biliary bile acids are shown to be in isotopic equilibrium, and if it can be shown that bile acid kinetics can be determined with precision and accuracy using plasma samples, after the administration of a bile acid labelled with stable isotopes, then isotope dilution measurements could be extended to large population groups, assuming that mass spectrometry facilities were available for the analyses. Since patients with cholelithiasis have a decreased bile acid pool (Vlahcevic *et al.*, 1970), it could be that the finding of a decreased bile acid pool might be used to define a population at risk which could have therapeutic implications, since cholesterol gallstones can now be treated medically (Thistle and Hofmann, 1973). Prospective studies are needed.

The use of the inverse isotope dilution technique to measure serum bile acid levels could be important in permitting validation of radioimmunoassays. The technique of radioimmunoassay is evolving continuously and successful enzyme immunoassays for bile acids are likely to be present soon. It seems unlikely that inverse isotope dilutions will compete in convenience with radioimmunoassay.

If bile acids labelled with stable isotopes were relatively inexpensive, and if measurement of isotope ratios could be greatly simplified by the development of dedicated, turn-key instruments which could be present in the investigator's laboratory, then the majority of experiments carried out with radioactive bile acids would probably be carried out with bile acids tagged with stable isotopes.

For the amino acid moiety bile acids tagged with stable isotopes appear to have a solid, continuing future, if simpler and less expensive instrumentation for measuring breath CO_2 can be developed. It seems likely that several breath tests (aminopyrine (Lauterburg and Birchner, 1976; Hepner and Vessel, 1974), triolein (Newcomer, Thomas and Hofmann, 1976), and cholylglycine) will become standard diagnostic tests in clinical gastroenterology. As the use of breath tests proliferates, then it seems likely that the use of ^{13}C substrates will be mandated to obviate the release of large amounts of $^{14}CO_2$ into the atmosphere.

ACKNOWLEDGEMENTS

The work described here was supported at Mayo Clinic by Mayo Foundation and NIH Grant AM 16770. Collaborative work at Argonne National Laboratory was supported by ERDA. The experiments at Mayo Clinic were carried out by a number of associates, including Neville E. Hoffman, Rudy G. Danzinger, Peter Y. Ng, Alistair E. Cowen and Paul J. Thomas.

REFERENCES

Allan, R. N., Thistle, J. L. and Hofmann, A. F. (1976). *Gut*, **17**, 413

Balistreri, W. F., Cowen, A. E., Hofmann, A. F., Szczepanik, P. A. and Klein, P. D. (1975). *J. Ped. Res.*, **9**, 757

Bennion, L. J. and Grundy, S. M. (1975). *J. Clin. Invest.*, **56**, 996

Bergström, S., Rottenberg, M. and Voltz, J. (1953). *Acta Chem. Scand.*, 7, 481

Chavez, M. N. and Krone, C. L. (1976). *J. Lipid Res.*, **17**, 545

Cowen, A. E., Korman, M. G., Hofmann, A. F., Cass, O. W. and Coffin, S. B. (1975). *Gastroenterology*, **69**, 67

Dietschy, J. M. (1972). *Arch. Intern. Med.*, **130**, 473

Fromm, H. and Hofmann, A. F. (1971). *Lancet*, **2**, 621

Hachey, D. L., Szczepanik, P. A., Berngruber, O. W. and Klein, P. D. (1974). *J. Labelled Comp.*, **9**, 703

Heaton, K. W. (1972). *Bile Salts in Health and Disease*, Williams and Wilkins, Baltimore, p. 252

Hepner, G. W., Hofmann, A. F. and Thomas, P. J. (1972a). *J. Clin. Invest.*, **51**, 1889

Hepner, G. W., Hofmann, A. F. and Thomas, P. J. (1972b). *J. Clin. Invest.*, **51**, 1898

Hepner, G. W., Sturman, J. A., Hofmann, A. F. and Thomas, P. J. (1973). *J. Clin. Invest.*, **52**, 433

Hepner, G. W. and Vessel, E. S. (1974). *N. Engl. J. Med.*, **291**, 1384

Hoffman, N. E., LaRusso, N. F. and Hofmann, A. F. (1976). *Mayo Clin. Proc.*, **51**, 171

Hofmann, A. F., Szczepanik, P. A. and Klein, P. D. (1968). *J. Lipid Res.*, **9**, 707

Hofmann, A. F., Danzinger, R. G., Hoffman, N. E., Klein, P. D., Berngruber, O. W. and Szczepanik, P. A. (1972). *Proc. Seminar on the Use of Stable Isotopes in Clinical Pharmacology* (ed. P. D. Klein and L. J. Roth), p. 37. National Technical Information Services Document CONF-711115

Hofmann, A. F. and Hoffman, N. E. (1974). *Gastroenterology*, **67**, 341

Klein, P. D., Haumann, J. R. and Hachey, D. L. (1975). *Clin. Chem.*, **21**, 1253

Lack, L., Dorrity, F. O., Jr., Walker, T. and Singletary, G. D. (1973). *J. Lipid Res.*, **14**, 367

LaRusso, N. F., Hoffman, N. E. and Hofmann, A. F. (1974). *J. Lab. Clin. Med.*, **84**, 759

Lauterburg, B. H. and Birchner, J. (1976). *J. Pharmacol. Exp. Ther.*, **196**, 501

Lindstedt, S. (1957). *Acta Physiol. Scand.*, **40**, 1

Mede, K. A., Tserng, K. Y. and Klein, P. D. (1976). *Gastroenterology*, **71**, 921 (abstract)

Murphy, G. M., Edkins, S. M., Williams, J. W. and Catty, D. (1974). *Clin. Chim. Acta*, **54**, 81

Nair, P. P. and Kritchevsky, D. (1973). *The Bile Acids, Chemistry, Physiology, and Metabolism*. Vol. II, *Physiology and Metabolism*, Plenum Press, New York

Newcomer, A. D., Thomas, P. J. and Hofmann, A. F. (1976). *Gastroenterology*, **70**, 923 (abstract)

Ng, P. Y., Allan, R. N. and Hofmann, A. F. (1977). *J. Lipid Res.* (in press)

Norman, A. (1955). *Arkiv Kemi*, **8**, 331

Osuga, T., Mitamura, K., Mashige, F. and Imai, K. (1977). *Clin. Chim. Acta*, **75**, 81

Palmer, R. H. (1969). *Methods in Enzymology*, Vol. XV, *Steroids and Terpenoids* (ed. S. P. Colowick and N. O. Kaplan), Academic Press, New York, p. 280

Paumgartner, G. (1977). *Clinics in Gastroenterology*, Vol. 6, Saunders, London

Salen, G., Tint, G. S., Eliav, B., Deering, N. and Mosbach, E. H. (1974). *J. Clin. Invest.*, **53**, 612

Schoeller, D. A., Schneider, J. F., Solomons, N. W., Watkins, J. B. and Klein, P. D. (1977). *J. Lab. Clin. Med.* (in press)

Simmonds, W. J., Korman, M. G., Go, V. L. W. and Hofmann, A. F. (1973). *Gastroenterology*, **65**, 705

Solomon, N. W., Schoeller, D. A., Wagonfeld, J. B., Ott, D., Rosenberg, I. H. and Klein, P. D. (1977). *J. Lab. Clin. Med.* (in press)

Thistle, J. L. and Hofmann, A. F. (1973). *N. Engl. J. Med.*, **289**, 655

Tserng, K. Y., Hachey, D. L. and Klein, P. D. (1977). *J. Lipid Res.* (in press)

Tserng, K. Y. and Klein, P. D. (1977). *J. Lipid Res.* (in press)

Vlahcevic, Z. R., Bell, C. C., Jr., Buhac, I., Farrar, J. T. and Swell, L. (1970). *Gastroenterology*, **59**, 165

Watkins, J. B., Ingall, D., Szczepanik, P. A., Klein, P. D. and Lester, R. (1973). *N. Engl. J. Med.*, **288**, 431

Watkins, J. B., Szczepanik, P. A., Gould, J., Klein, P. D. and Lester, R. (1975a). *Gastroenterology*, **69**,706

Watkins, J. B., Tercyak, A. M., Szczepanik, P. A. and Klein, P. D. (1975b). *Gastroenterology*, **68**, 1087 (abstract)

18

Detection and quantification of stable isotope incorporation by high-resolution mass spectrometry and $^{13}C\{^2H, ^1H\}$ nuclear magnetic resonance difference spectroscopy. Utilisation of carbon and hydrogen atoms of ethanol in the biosynthesis of bile acids

D. M. Wilson and A. L. Burlingame (Space Sciences Laboratory, University of California, Berkeley, California, USA), S. Evans (AEI Scientific Apparatus Ltd, Manchester, UK) and T. Cronholm and J. Sjövall (Kemiska Institutionen, Karolinska Institutet, Stockholm, Sweden)

INTRODUCTION

We have been interested for some time in developing reliable and sensitive methods of detecting stable isotope incorporation in steroids. The focus of this interest has been the transfer of deuterium and ^{13}C atoms from labelled ethanol to bile acids, which can be isolated in rather large amounts using bile fistula rats. Initially, low-resolution GC–MS–computer techniques were used to calculate heavy isotope excess at certain characteristic *m/e*s (Cronholm, Burlingame and Sjövall, 1974). For example, Figure 18.1 shows the differently labelled molecular species of cholesterol and chenodeoxycholic acid that result from continuous administration of $[2,2,2\text{-}^2H_3]$-ethanol. The filled bars show the distributions of labelled species actually found. However, the sites and extents of deuterium incorporation along the steroid skeleton cannot be directly determined by GC–MS. Instead, using a number of assumptions and known incorporation sites of the atoms transferred from ethanol to the steroid, one can construct a mathematical model to generate theoretical distributions, which tend to support or disprove the initial premises. The open bars in Figure 18.1 show a theoretical distribution, arrived at by such an approach. However, the method is

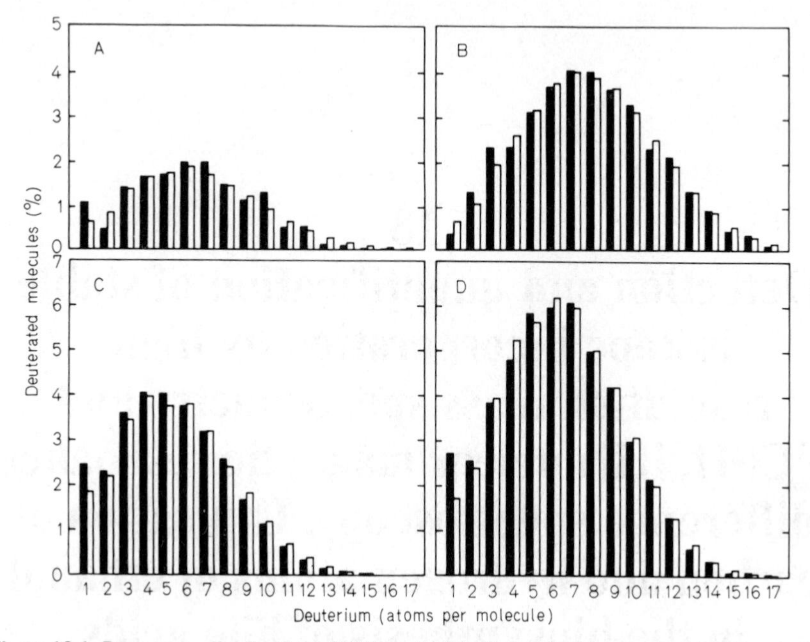

Figure 18.1 Percentage distribution of differently labelled molecular species of cholesterol (A,B) and chenodeoxycholic acid (C,D) in bile collected during the 4th (A,C) and 24th (B,D) hour of administration of [2-$^2H_3$]-ethanol to a bile fistula rat. Filled bars: distribution found by low-resolution GC–MS; open bars: theoretical distribution calculated using assumptions of biosynthetic mechanisms. From Cronholm *et al.* (1974), reproduced by permission from *European Journal of Biochemistry*

severely underdetermined, particularly in the case of [1,1-$^2H_2$]-ethanol metabolism. In this case, deuterium is lost immediately and can be incorporated only via NADPH-dependent reductions later on in steroid biosynthesis. There may, for example, be different NADPH pools for mevalonate synthesis and for later reductions in the completed steroid skeleton, leading to different levels of incorporation per site.

For these reasons other spectroscopic means have been sought which might provide more information. One of these, $^{13}C\{^2H,^1H\}$ NMR difference spectroscopy, has evolved from some initial NMR results which showed promise for isotope incorporation site specificity (Wilson *et al.*, 1974a). The other is very-high-resolution ($m/\Delta m \sim 100\,000$) mass spectrometry (HRMS), which is similar to the above GC–MS–computer method except that mathematical models can be fitted to specific elemental composition intensities as well as nominal mass intensities.

MATERIALS AND METHODS

^2H- and ^{13}C,^2H-enriched ethanols were synthesised in the desired combinations by lithium aluminium hydride or lithium aluminium deuteride reduction of unenriched acetic acid, [1-^{13}C]-acetic acid, [2-^{13}C]-acetic acid or [2,2,2-$^2H_3$]-

acetic acid. The ethanol was analysed by ^{13}C NMR and GC–MS of the benzoate ester, resulting in the enrichments shown in Table 18.1. As in procedures previously described (Cronholm *et al.*, 1974), one, two or three Sprague–Dawley rats were provided with bile fistulae and thin catheters for intraperitoneal injections of labelled ethanol (in saline), and the concentration of ethanol in bile was monitored by gas chromatography. In one experiment, (–)-hydroxycitrate, which has been shown to be an effective inhibitor of citrate cleavage enzyme (Watson, Fang and Lowenstein, 1969), was administered in conjunction with [2,2,2-$^{2}H_3$]-ethanol. The volumes injected were adjusted to maintain a concentration of ethanol above 10 mM. Bile was collected for up to 96 h, and bile acids were extracted, saponified, methylated and separated on hydroxyalkoxypropyl Sephadex or silicic acid (Cronholm *et al.*, 1974; Cronholm, Makino and Sjövall, 1972).

Table 18.1 Isotopically labelled ethanols administered in separate experiments in this work

Ethanol administered	Enrichment
[2,2,2-$^{2}H_3$]	99.5% ^{2}H
[1,1-$^{2}H_2$]	99.5% ^{2}H
[1-^{13}C-1,1-$^{2}H_2$]	60% ^{13}C >95% ^{2}H
[2-^{13}C-1,1-$^{2}H_2$]	60% ^{13}C >95% ^{2}H

A Varian XL-100-15 Fourier transform NMR spectrometer in conjunction with an XDS Sigma 7 computer system was used to record ^{13}C spectra of the bile acid methyl esters in pyridine at 50°, using quadrature detection and sample microcells (Wilson, Olsen and Burlingame, 1974b). $^{13}C\{^{2}H,^{1}H\}$ difference spectra were recorded by decoupling ^{2}H only on alternate scans, and storing the two types of accumulated free-induction decays in separate computer memory locations (^{1}H was decoupled continuously). By pre-subtracting these accumulated responses prior to Fourier transformation, difference spectra result, in which ^{13}C resonances of carbons not coupled to deuterons are completely suppressed, decreasing the threshold of ^{2}H-incorporation detectability to less than 1 per cent (Wilson *et al.*, 1975).

HRMS data of cholate samples were obtained on an AEI MS 50 mass spectrometer, using the *m/e* 386 (M-36) ion for determination of isotope excess. A computer program was written in FORTRAN to generate elemental compositions and intensities from site-specific ^{13}C and ^{2}H abundance models or experimental information. Thus for each carbon C^i of 25 carbon atoms in the molecule, relative abundances $a^i[c,d]$ of each of 8 species ($c = 0$ or 1 for ^{12}C or ^{13}C; $d = 0, 1, 2$ or 3 for the number of ^{2}H atoms) are assigned, normalised to unity.

For example, a hypothetical model (*vide infra*) assuming uniform ^{2}H incorporation from [1,1-$^{2}H_2$]-ethanol of 5.23 per cent at 11 sites, and compatible with GC–MS–computer data, had for ^2H incorporation at C_{16}:

$$a^{16}[c,d] = \begin{bmatrix} (^{12}C)\cdot(^1H)^2 & 2(^{12}C)\cdot(^1H)\cdot(^2H) & (^{12}C)\cdot(^2H)^2 & 0 \\ (^{13}C)\cdot(^1H)^2 & 2(^{13}C)\cdot(^1H)\cdot(^2H) & (^{13}C)\cdot(^2H)^2 & 0 \end{bmatrix}$$

where $^{12}C = 0.988\,93$, $^{13}C = 0.011\,07$, $^1H = 0.9477$, $^2H = 0.0523$. The relative abundance of any particular elemental composition $A[m,n]$ is then given by

$$A[m,n] = \sum_{\substack{\Sigma c=m \\ \Sigma d=n}} \prod_{i=1}^{25} a^i[c,d]$$

where the products of all array elements whose summed values of c and d are equal to m and n, respectively, are summed. These terms are corrected for the presence of three oxygen atoms in the m/e 386 ion, yielding abundances $B[m,n,j,k]$ (j and k correspond to the presence of ^{17}O and ^{18}O atoms, respectively). These abundances are sorted according to increasing accurate mass within a nominal mass, to compare with HRMS data.

RESULTS AND DISCUSSION

The $^{13}C\{^2H,^1H\}$ NMR difference technique is exemplified in Figure 18.2, for the case of methyl cholate derived from [2,2,2-$^2H_3$]-ethanol administration. Spectra a and b are the $^{13}C\{^1H\}$ and $^{13}C\{^2H,^1H\}$ spectra, respectively, which result from separate Fourier transforms of the two accumulated responses generated during difference spectroscopy. Spectrum c is the $^{13}C\{^2H,^1H\}$ difference spectrum, wherein particular sites and relative extent of ^2H incorporation along the steroid skeleton are quite evident. The main point to note is that particular sites and relative ^2H incorporation levels are immediately obvious. Furthermore, because of increasing displacement of ^{13}C resonances to higher field with the number of attached deuterons, the relative extent of mono- and di-deuteron incorporation in methylene groups C_1, C_{15} and C_{22} is seen, as well as relative abundances of mono-, di- and tri-deuterium incorporation in methyl groups C_{18}, C_{19} and C_{21}. This result, showing the loss of ^2H from acetate, was surprising because it contradicts one of the premises of the mathematical model which appeared to fit the earlier GC–MS data, as shown in Figure 18.1.

To probe this point further, we have investigated the possibility that this loss of deuterium from labelled acetate might occur before mevalonate synthesis by reversible coupling of acetyl-CoA to form citrate. The [2,2,2-$^2H_3$]-ethanol experiment was repeated with simultaneous administration of (–)-hydroxy-citrate, which has been shown to be a powerful inhibitor of citrate cleavage enzyme (Watson *et al.*, 1969) and should prevent loss of deuterium from acetyl-CoA, which is used in steroid biosynthesis. The results are shown in Figure 18.3, where it is seen that the addition of this inhibitor had no effect on the apparent loss of ^2H from acetate. (–)-Hydroxycitrate solution was

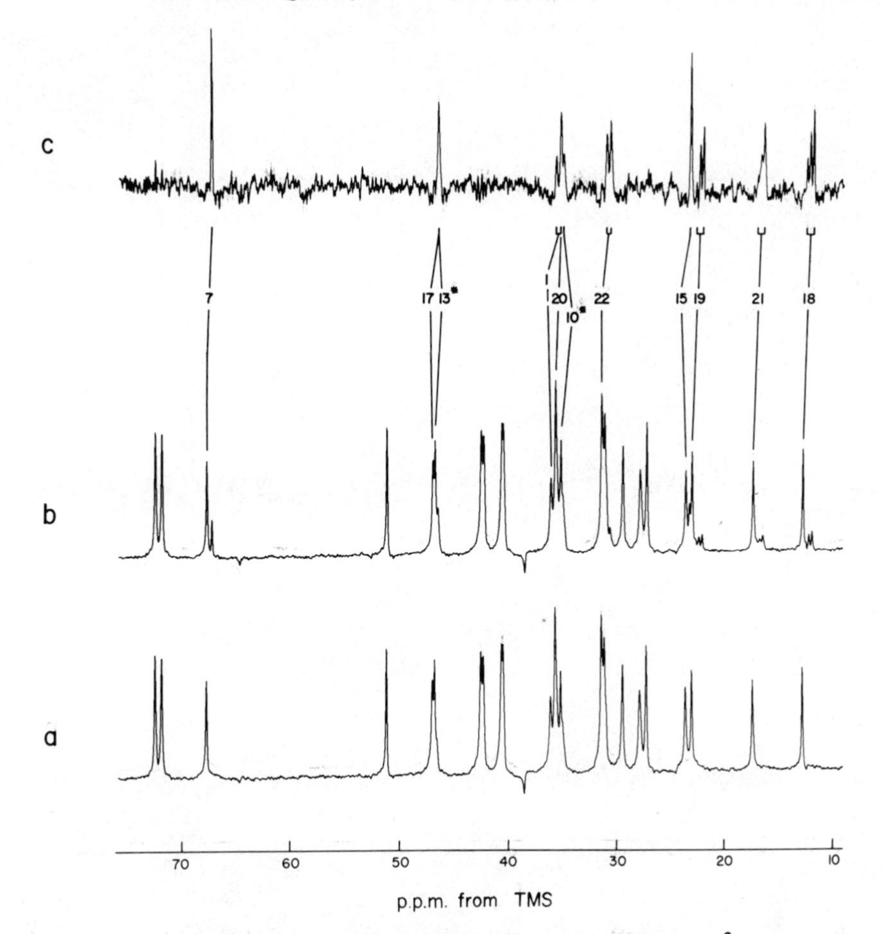

Figure 18.2 (a) $^{13}C\{^1H\}$ spectrum of methyl cholate resulting from $[2,2,2\text{-}^2H_3]$-ethanol administration (up field region); (b) $^{13}C\{^2H,^1H\}$ spectrum; (c) $^{13}C\{^2H,^1H\}$ difference spectrum, with connections to assigned protonated peaks in the b spectrum. Asterisks denote carbon sites which are not deuterated, but whose resonances appear because of geminal coupling to methyl deuterons

administered during the first 48 h, while $[2,2,2\text{-}^2H_3]$-ethanol was continuously infused for the first 72 h. Spectrum A is the $^{13}C\{^2H,^1H\}$ difference spectrum of cholate isolated from a 0–36 h collection, while spectrum B is of a 54–78 h collection taken after terminating the inhibitor infusion. Spectrum C is of a previous sample obtained without administration of (–)-hydroxycitrate. All three spectra show approximately the same loss of deuterium at C_{18}, C_{19}, C_{21} and C_{22}.

$^{13}C\{^2H,^1H\}$ NMR difference spectra were also taken of cholate derived from $[1,1\text{-}^2H_2]$-, $[1\text{-}^{13}C\text{-}1,1\text{-}^2H_2]$- and $[2\text{-}^{13}C\text{-}1,1\text{-}^2H_2]$-ethanol administration. The results of the $[1,1\text{-}^2H_2]$-ethanol experiment are shown in Figure 18.4, which suggests another use of the technique, namely that the percentage of deuterium incorporation per site should be calculable by integrating the appropriate

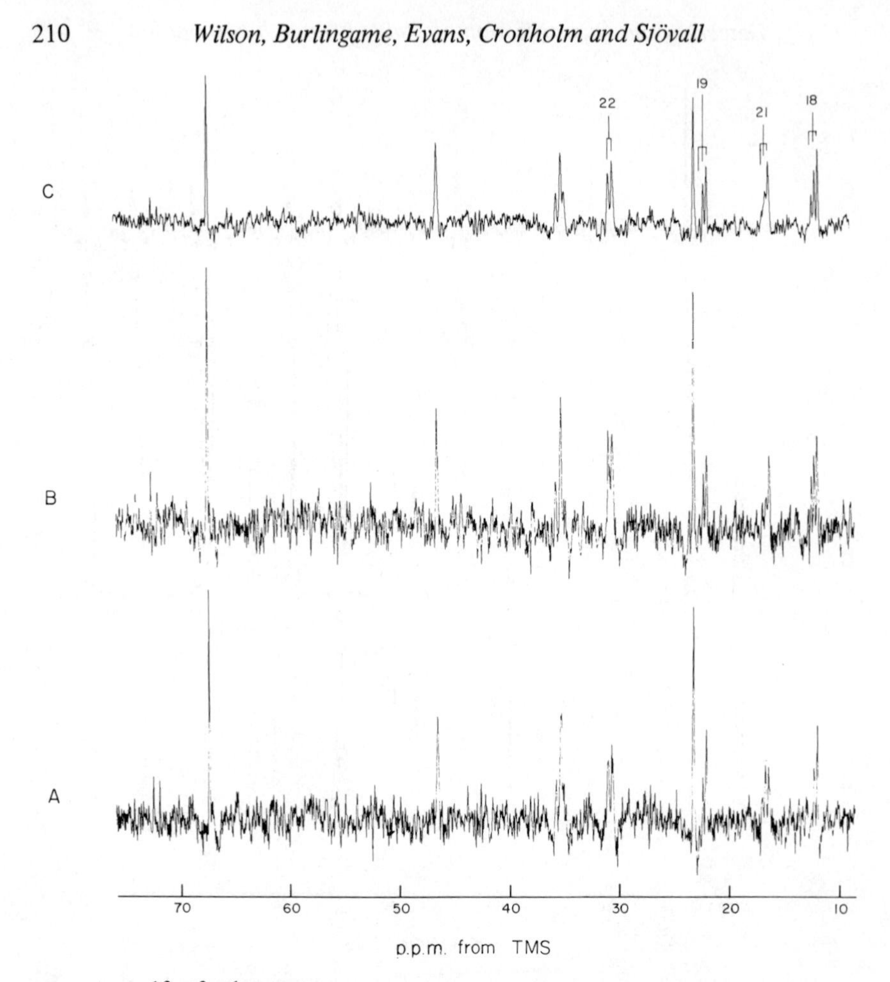

Figure 18.3 $^{13}C\{^2H,^1H\}$ difference spectra to test the hypothesis of loss of deuterium from acetyl-CoA by reversible citrate formation. (A) and (B) are spectra of cholate samples recovered from a 0–36 h and 54–78 h collection, respectively, in an experiment which included the administration of (–)-hydroxycitrate solution from 0–48 h and [2,2,2-$^2H_3$]-ethanol from 0–72 h. (C) Spectrum of cholate recovered during a similar experiment but without the infusion of (–)-hydroxycitrate. All spectra show approximately the same loss of deuterium at C-18, C-19, C-21 and C-22

resonances in the $^{13}C\{^1H\}$ and $^{13}C\{^2H,^1H\}$ difference spectra; the lower and upper spectra are the $^{13}C\{^1H\}$ spectrum and $^{13}C\{^2H,^1H\}$ difference spectrum, respectively. C_{10} is not deuterated, but an apparent resonance appears because of vicinal coupling to ^2H at C_2 and C_{11}. However, by this method the incorporation levels are not in agreement with the total incorporation as measured by the earlier GC–MS method (0.5–1.3 per cent by NMR, 5.23 per cent by GC–MS).

Although similar studies were made of cholate derived from [1-^{13}C-1,1-$^2H_2$]- and [2-^{13}C-1,1-$^2H_2$]-ethanol, such a comparison cannot be made because the GC–MS method cannot break down heavy isotope excess to ^2H and ^{13}C

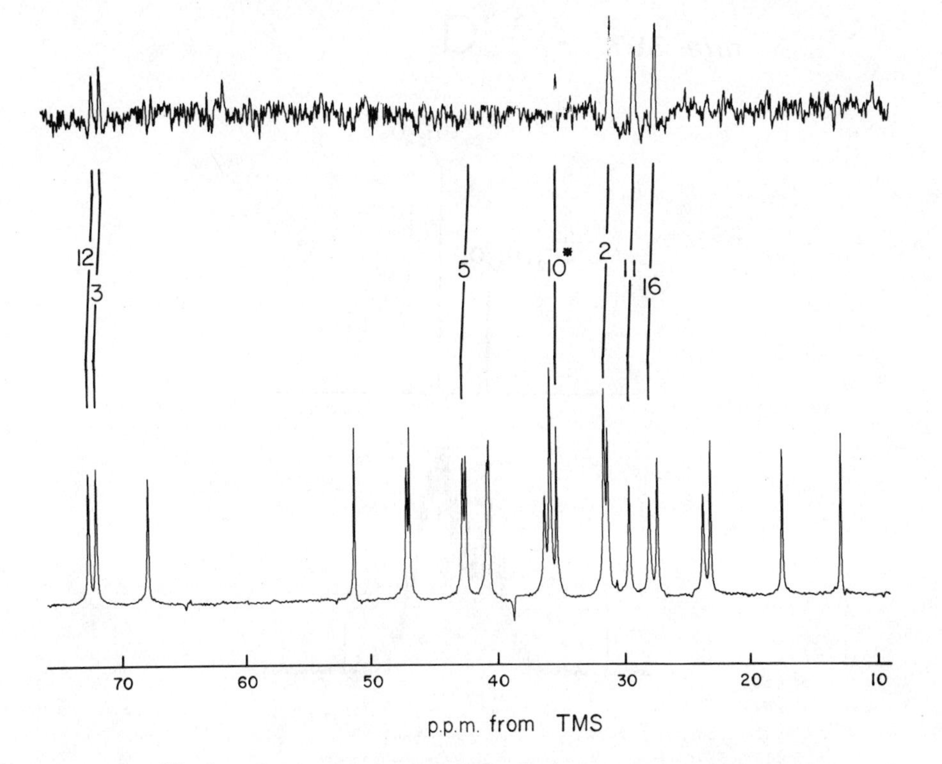

Figure 18.4 $^{13}C\{^1H\}$ spectrum (bottom) and $^{13}C\{^2H,^1H\}$ difference spectrum (top) of methyl cholate resulting from $[1,1\text{-}^2H_2]$-ethanol administration, with connections to assigned protonated peaks in the $^{13}C\{^1H\}$ spectrum. C-10 is not deuterated, but an apparent resonance appears because of vicinal coupling to ^2H at C-2 and C-11

contributions. Furthermore, in questions of this type an important distinction should be made in that mass spectrometry does not distinguish molecules according to their ^{13}C incorporation, whereas the $^{13}C\{^2H,^1H\}$ NMR difference method observes *only* ^{13}C–^2H bonding pairs. Therefore quantitative differences of the type observed could be caused by an isotope effect. To probe questions of this type, very-high-resolution mass spectra have been obtained with some of these samples. Figures 18.5(a–d) are results obtained for cholate derived from $[1,1\text{-}^2H_2]$-ethanol, along with elemental compositions and intensities computed on the basis of uniform incorporation of 5.23 per cent at 11 sites. Since the computer program was designed to accommodate isotope effects as well as site-specific heavy isotope abundances, we have tested the hypothesis of a very large isotope effect that would resolve the differences between the GC–MS and NMR data. The results for *m/e* 389 are given in Table 18.2. Comparing the actual data (Figure 18.5c), it is apparent that the low signal-to-noise ratio does not permit any conclusion to be drawn between the two possibilities.

Results have also been obtained for cholate derived from $[2\text{-}^{13}C\text{-}1,1\text{-}^2H_2]$-ethanol (Figures 18.6a–f). The accompanying computed results are based on a hybrid model of GC–MS and $^{13}C\{^1H\}$ results. GC–MS finds 0.94 heavy

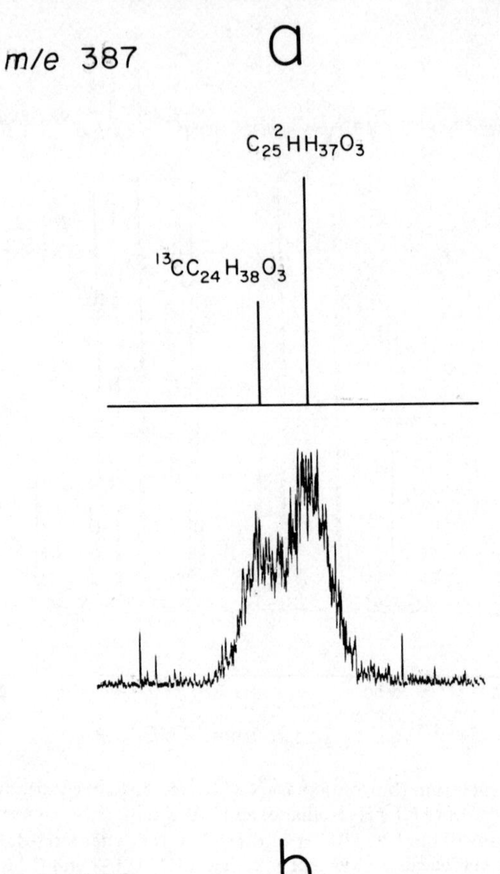

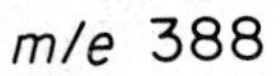

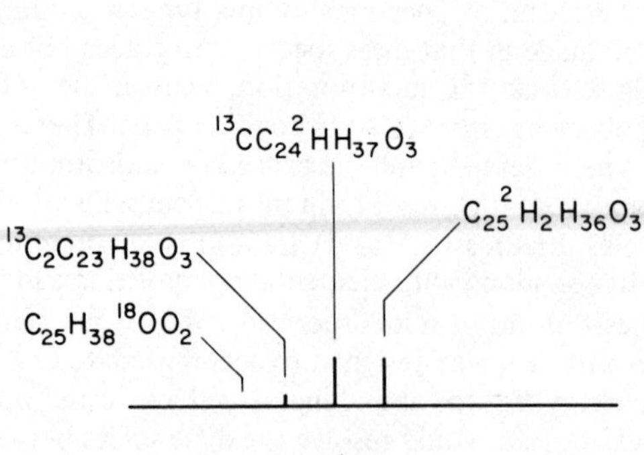

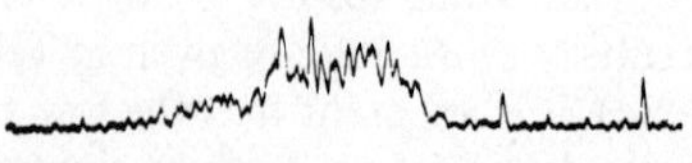

Figure 18.5 (a–d) High-resolution mass spectra data (bottom) and computed elemental compositions and intensities (top) of ions above *m/e* 386 (M-36) for methyl cholate derived from [1,1-$^2H_2$]-ethanol infusion. The computed model is based upon input data suggested

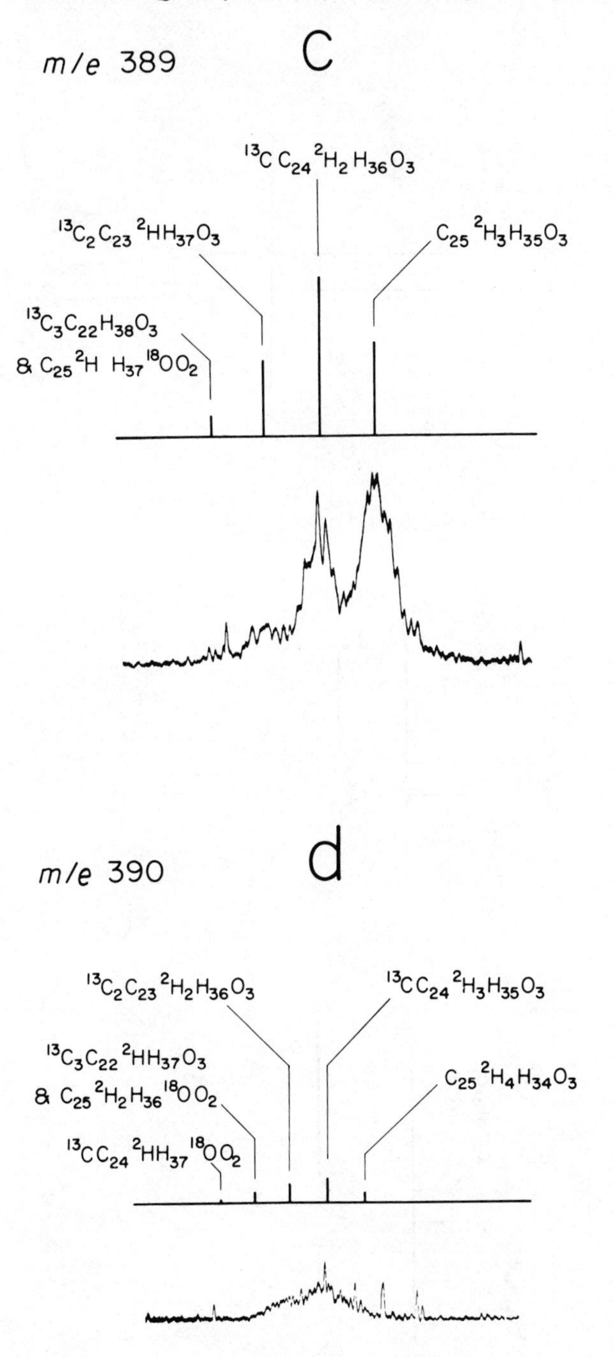

by GC–MS computer results, viz. 5.23 per cent uniform ^{2}H incorporation over the 11 deuterated sites C-2(2), C-3, C-5, C-6, C-11(2), C-12, C-14 and C-16(2). Natural abundance data for carbon, hydrogen and oxygen isotopes were from Biemann (1962)

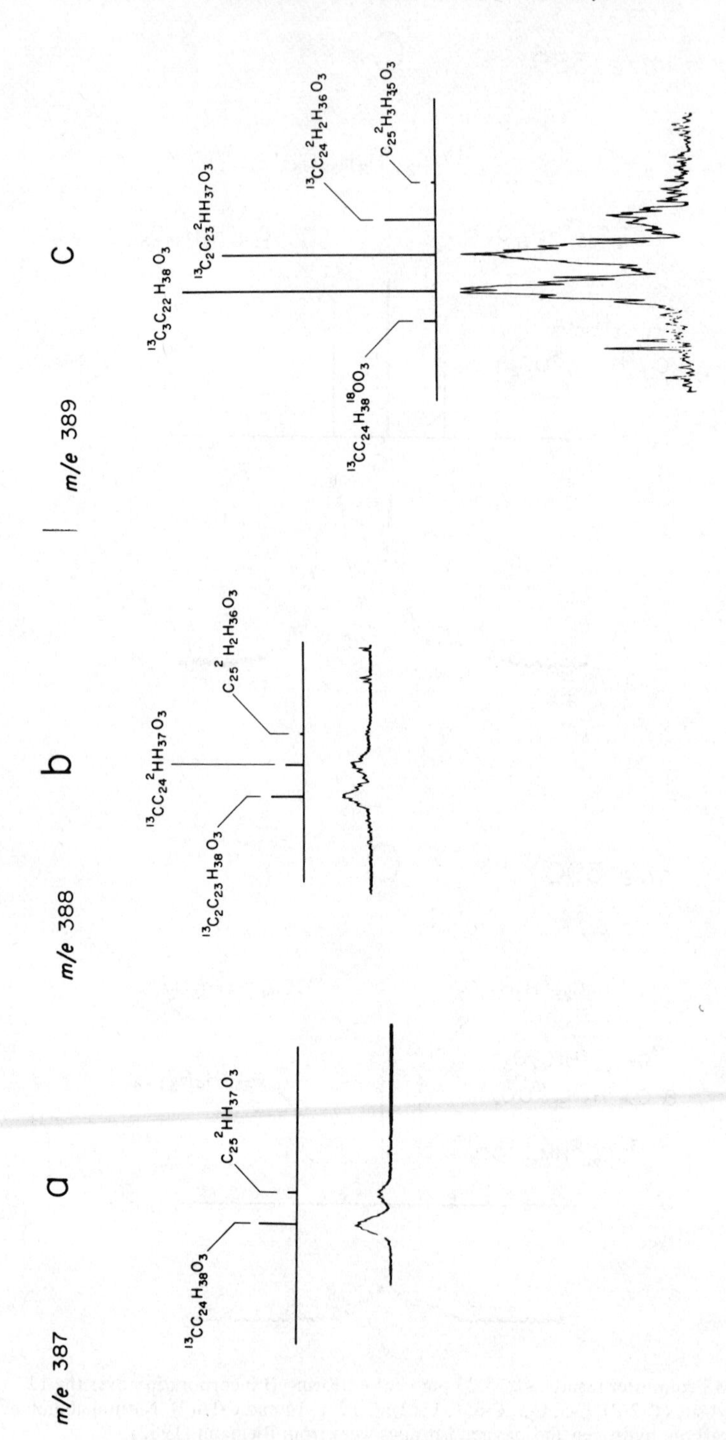
m/e 387
a
13CC24H38O3
C25 2HH37O3
m/e 388
b
13C2C23H38O3
13CC24 2HH37O3
C25 2H2H36O3
m/e 389
c
13C3C22H38O3
13C2C23 2HH37O3
13CC24H38 18OO3
13CC24 2H2H36O3
C25 2H3H35O3

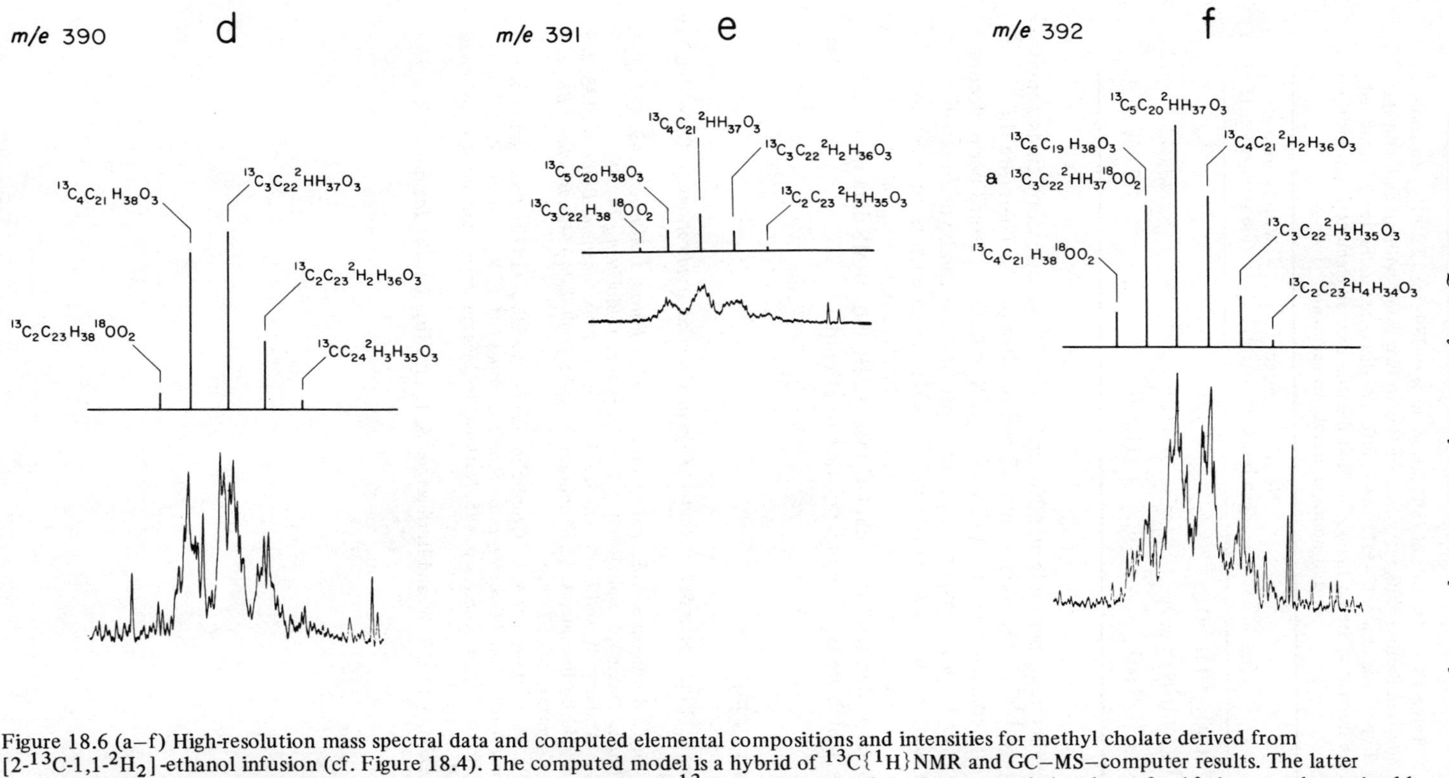

Figure 18.6 (a–f) High-resolution mass spectral data and computed elemental compositions and intensities for methyl cholate derived from [2-^{13}C-1,1-$^2H_2$]-ethanol infusion (cf. Figure 18.4). The computed model is a hybrid of $^{13}C\{^1H\}$NMR and GC–MS–computer results. The latter measured 0.94 heavy isotopes in excess per molecule, whereas an overall ^{13}C incorporation of six times natural abundance for 13 sites was determined by $^{13}C\{^1H\}$ NMR. Using the remaining heavy isotope excess per molecule and a hypothesis of uniform ^{2}H incorporation, over 11 deuterated sites gives an average incorporation per site of 2.09 per cent.

Table 18.2 Result of calculation for *m/e* 389, for [2,2-$^2H_2$]-ethanol-derived methyl cholate, corresponding to two cases: without and with the hypothesis of a $^{13}C-^2H$ isotope effect as discussed in the text. Calculated ion intensities are normalised so that the sum over all elemental compositions and nominal masses for each case = 1

	No isotope effect (ion intensity x 10^3)	With isotope effect (ion intensity x 10^3)
$^{13}C_2C_{23}{}^2H\,H_{37}O_3$	9.524	9.274
$^{13}C\,C_{24}{}^2H_2H_{36}O_3$	19.719	19.042
$C_{25}{}^2H_3H_{35}O_3$	11.780	12.068

isotopes in excess per molecule. Normal $^{13}C\{^1H\}$ spectroscopy finds a multiple of average ^{13}C incorporation over 13 carbon sites of 6X. Therefore the deuterium content is $0.94 - (13)(0.011)(5X) = 0.23$ ^2H atoms per molecule, spread over 11 positions, or ~2 per cent per site that can be deuterated.

Our work in progress involves more HRMS measurements on these and other similarly obtained samples. Of particular interest is the optimisation of the fit of computed models with the total GC–MS, NMR and HRMS data from each sample, which we are now in a position to attempt.

REFERENCES

Biemann, K. (1962). *Mass Spectrometry: Organic Chemical Applications*, McGraw-Hill, New York

Cronholm, T., Burlingame, A. L. and Sjövall, J. (1974). *Europ. J. Biochem.*, **49**, 497

Cronholm, T., Makino, I. and Sjövall, J. (1972). *Europ. J. Biochem.*, **24**, 507

Watson, J. A., Fang, M. and Lowenstein, J. M. (1969). *Arch. Biochem. Biophys.*, **135**, 209

Wilson, D. M., Burlingame, A. L., Cronholm, T. and Sjövall, J. (1974a). *Biochem. Biophys. Res. Commun.*, **56**, 828

Wilson, D. M., Burlingame, A. L., Cronholm, T. and Sjövall, J. (1975). *Proceedings of the Second International Conference on Stable Isotopes* (ed. E. R. Klein and P. D. Klein), Oak Brook, Illinois, U.S.A., p. 485. National Technical Information Service Document CONF-751027

Wilson, D. M., Olsen, R. W. and Burlingame, A. L. (1974b). *Rev. Sci. Instrum.*, **45**, 1095

19

Teratological studies on effects of carbon-13 incorporation into preimplantation mouse embryos on development after implantation *in vivo* and *in vitro*

H. Spielmann, H.-G. Eibs, U. Jacob and D. Nagel (Pharmakologisches Institut der Freien Universität Berlin, Abteilung Embryonalpharmakologie, West Germany) and C. T. Gregg (Los Alamos Scientific Laboratory, University of California, Los Alamos, New Mexico, USA)

INTRODUCTION

Since the use of radioactive isotopes is virtually prohibited in studies on drug metabolism in pregnant women and children, stable isotopes have increasingly been employed in investigations where the risk of radioactive damage is unacceptable (Klein and Peterson, 1973; Klein and Klein, 1975). However, the general knowledge of harmful effects of stable isotope enrichment in rapidly developing mammalian systems is rather limited. We have, therefore, studied the effects of high levels of ^{13}C substitution at different stages of embryogenesis on subsequent development of mouse embryos.

When pregnant mice received uniformly labelled ^{13}C-glucose (U-^{13}C-glucose) by stomach tube during organogenesis, no abnormalities of embryonic development due to stable isotope enrichment were found (Gregg *et al.*, 1973). ^{13}C enrichment in fetal tissues in these experiments was 2–5 atoms %. In other investigations, development and differentiation of mouse limb buds were unaffected by incubation with 83 mol % U-^{13}C-glucose as judged by morphological or biochemical criteria, even though some 60 per cent of DNA carbon was replaced by ^{13}C (Gregg *et al.*, 1975). Finally, we cultured preimplantation mouse embryos *in vitro* for 48 h from the 8-cell to the blastocyst stage in media containing U-^{13}C-glucose as the only energy source (Spielmann *et al.*, 1976). The ^{13}C content of such blastocysts was 20 atoms % according to U-^{14}C-glucose incorporation studies. When comparing 8-cell embryos that were cultured in media with U-^{13}C-glucose or ^{na}C-glucose (^{na}C indicates natural carbon isotope abundance, i.e. about 1.1 atoms % ^{13}C), the percentage of embryos reaching the blastocyst stage, as well as the cell number of the blastocysts, did not indicate any embryotoxic effects of ^{13}C incorporation. In the present study we investigated the developmental potential of such blastocysts both *in vivo* after transplantation to pseudopregnant foster mothers and *in vitro* by allowing the embryos to differentiate in recently developed culture systems beyond the blastocyst stage (Spindle and Pedersen, 1973; Pienkowski, Solter and Koprowski, 1974; Sherman, 1975).

MATERIALS AND METHODS

In vitro culture of 8-cell mouse embryos

Eight-cell embryos were flushed from the oviducts of female mice (strain NMRI, breeder: Schwencke and Co., Nauheim, West Germany) 60 h after copulation. The standard culture medium was prepared according to Whitten (Whitten, 1971) with 0.4 per cent bovine serum albumin (BSA) (W-BSA). Besides glucose, W-BSA contains the following substrates supporting *in vitro* development of 8-cell mouse embryos: lactate, pyruvate and BSA. To enrich the embryos as highly as possible with carbon-13 from U-^{13}C-glucose, we used a medium described by Donahue (1968) in which glucose (concentration 5.5 mM) was the only energy source and in which BSA was replaced by 0.5 per cent Ficoll (DON-FIC). The ^{13}C content of the U-^{13}C-glucose used in these culture systems was 86 atoms %. When 8-cell mouse embryos were cultured in 5 per cent CO_2 in air under paraffin oil (according to Brinster, 1963), they developed to the morula stage after 24 h and reached the blastocyst stage after 48 h.

Transplantation of mouse blastocysts after *in vitro* development from the 8-cell stage

Blastocysts which had developed *in vitro* from 8-cell embryos during a 48 h culture period, together with normal blastocysts which were obtained from the uteri of female mice 84 h after mating, were surgically transferred in groups of five to each uterine horn of pseudopregnant females on day 2 of pseudopregnancy. Pseudopregnancy was induced by mating normal females with vasectomised males. The foster mothers were sacrificed on day 17 of gestation and the success rate of the transplantations was determined by the percentage of resorbed and live embryos. The embryos were weighed, carefully inspected for growth retardations and malformations, and stained for skeletal abnormalities with Alizarin Red S (according to Lorke, 1965).

In vitro growth and differentiation of mouse blastocysts

Blastocysts that had either developed *in vivo* or *in vitro* were transferred to different media at the same developmental stage as in the transplantation experiments and cultured in Falcon plastics dishes at 37° in 5 per cent CO_2 in air for up to 120 h. Development and differentiation of the blastocysts were checked every 24 h and the following criteria were determined: hatching from the zona pellucida, attachment to the surface of the culture dish, outgrowth and differentiation into trophoblast and inner cell mass (ICM); the ICM either consists of one or two germ layers. Blastocyst culture was carried out in the following media: Whitten's medium with 0.3 per cent BSA (W-BSA), Eagle's minimal essential medium (MEM, Flow Laboratories) supplemented with 0.3 per cent BSA, as described by Pienkowski *et al.* (1974) (MEM-BSA), MEM with BSA being replaced by 10 per cent fetal calf serum (FCS, Gibco) (MEM-FCS), and NCTC-109 (Difco) supplemented with 10 per cent FCS according to Sherman (1975) (NCTC-109-FCS).

Photomicrographs of the embryos were taken at the end of the culture period on a Biovert-Photomicroscope (Reichert AG, Austria) at a magnification of x160.

RESULTS

In vitro culture of 8-cell mouse embryos

The influence of U-^{13}C-glucose on the differentiation of 8-cell mouse embryos into blastocysts was studied in media containing energy sources other than glucose and in media with glucose as the only source. Table 19.1 indicates that blastulation of 8-cell mouse embryos was not inhibited when Whitten's medium (W-BSA) was replaced by DON-FIC. It is of particular importance that the 48 h

Table 19.1 Development of 8-cell mouse embryos for 48 h in media containing ^{na}C- or U-^{13}C-glucose

Medium	Glucose	B	M	Deg.	Number of determinations (experiments)
W-BSA	^{na}C	94%	5%	1%	340 (15)
W-BSA	^{13}C	97%	3%	0	286 (12)
DON-FIC	^{na}C	97%	2%	1%	160 (5)
DON-FIC	^{13}C	94%	2%	4%	163 (5)

Abbreviations used for the media are given in the Materials and Methods section. Developmental stage of the embryos: B = blastocysts, M = morulae (retarded for 24 h). Deg. = degenerated.

in vitro development of the embryos was not any different in the presence of U-^{13}C-glucose, even when this substrate was the only energy source (DON-FIC).

For better evaluation of the *in vitro* development of the embryos, the cell number of the blastocysts was determined at the end of the culture period. As outlined in detail elsewhere (Spielmann *et al.*, 1976), no inhibitory effects of the stable isotope on the cell number of the embryos was detected. However, the cell number of blastocysts was significantly higher after incubation in the optimal medium (W-BSA) than after culture in the medium with glucose as the only energy source (DON-FIC). The ^{13}C enrichment during the *in vitro* culture was calculated from the incorporation of ^{14}C from U-^{14}C-glucose (Spielmann *et al.*, 1976). According to these data, the ^{13}C content of the embryos was 20 atoms % at the end of the 48 h culture period. The incorporation of ^{14}C from U-^{14}C-glucose was identical in the optimal medium (W-BSA) and in DON-FIC.

Development of cultured blastocysts after transplantation to foster mothers
In the presence of metabolic inhibitors, preimplantation mouse embryos may cleave *in vitro* without visible or measurable retardation, but they may be unable to develop normally after transfer to pseudopregnant foster mothers (Fisher and Smithberg, 1972; Bell and Glass, 1975; Spielmann, 1976). Blastocysts that had been cultured *in vitro* from the 8-cell stage were, therefore, transplanted to pseudopregnant hosts. Transfers of normal blastocysts served as controls. Experiments in which the foster mothers were not found pregnant at term were excluded from the calculations of the success rate of the transplantations shown in Table 19.2. On day 17 the percentage of implantations (living and resorbed

Table 19.2 Success rate of blastocyst transfer experiments

A	B	C	D	E	F
Development before transfer	Culture medium (glucose)	Total number of embryos	Implantation sites (% of C)	Living embryos (% of C)	Resorbed embryos (% of C)
in vivo	–	400	52	38	14
in vitro	W-BSA (^{na}C)	75	53	27	26
in vitro	W-BSA (U-^{13}C)	75	52	25	27
in vitro	DON-FIC (^{na}C)	80	47	21	26
in vitro	DON-FIC (U-^{13}C)	80	45	19	26

Blastocysts were transferred in groups of five to each uterine horn of a pseudopregnant recipient. Evaluation of success rate on day 17, 24 h before term.

embryos) was identical after transplantation of normal blastocysts and of blastocysts that had been cultured in medium W-BSA. However, a lower percentage of living embryos resulted from the *in vitro* developed blastocysts. Compared with these two groups of embryos, the percentage of both implantation sites and living fetuses was decreased for blastocysts cultured in medium DON-FIC. Table 19.2 furthermore demonstrates that there were no differences in the success rates of transfers of blastocysts cultured in media with either ^{na}C-glucose or U-^{13}C-glucose. Finally, no signs of malformations could be detected in any organ or in the skeleton of the 44 living fetuses which had developed to the blastocyst stage in media containing U-^{13}C-glucose and either W-BSA or DON-FIC.

Growth and differentiation of blastocysts *in vitro*
Before testing the viability of *in vitro* developed blastocysts in a culture system which allows outgrowth and differentiation of mouse blastocysts, the development of normal blastocysts was studied in media which had already been used by previous investigators (Spindle and Pedersen, 1973; Pienkowski *et al.*,

1974; Sherman, 1975). Figure 19.1 illustrates the extent to which blastocysts of the strain NMRI were able to differentiate during a culture period of 120 h in the various media. W-BSA, the standard culture medium for preimplantation embryos, allowed the blastocysts only to hatch and attach to the surface of the culture dish. A low degree of trophoblast outgrowth and inner cell mass (ICM) development occurred in MEM-BSA. When BSA was replaced in MEM by FCS, differentiation and growth of the trophoblast were significantly better. Medium NCTC-109-FCS, however, supported growth and differentiation of the blastocysts significantly better than any of the other media. More than 90 per cent of the blastocysts had an ICM at the end of the culture period in NCTC-109-FCS.

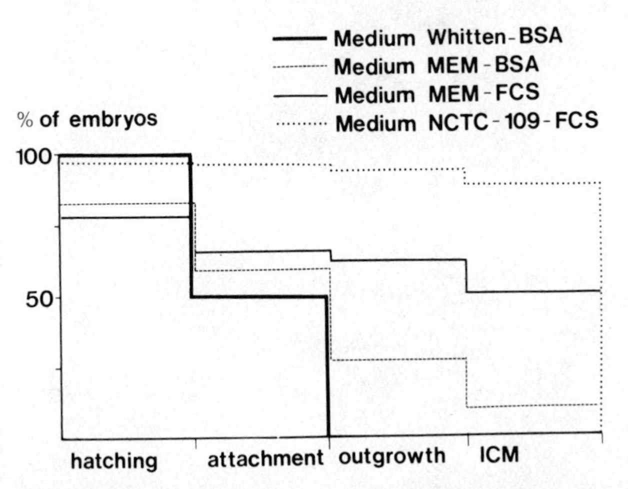

Figure 19.1 Development of mouse blastocysts *in vitro* in different media. The percentage of mouse blastocysts reaching progressive stages of differentiation after 120 h in culture is given for the following media: W-BSA, MEM-BSA, MEM-FCS, and NCTC-109-FCS; see Materials and Methods for abbreviations

Development in medium NCTC-109-FCS, which was used for routine culture experiments, is characterised by the following time-dependent steps of differentiation: After 24 h 10 per cent of the blastocysts had hatched; after 48 h 85 per cent had attached to the surface of the culture dish and 32 per cent showed trophoblast outgrowth; after 72 h 82 per cent had an ICM and 52 per cent an ICM consisting of two germ layers (entoderm and ectoderm = extensive ICM); after 96 h 87 per cent had an extensive ICM; and after 120 h the ICM of 91 per cent of the blastocysts consisted of two germ layers. Examples of the three characteristic stages of trophoblast and ICM differentiation are shown in Figure 19.2.

Subsequently blastocysts which had cleaved from the 8-cell to the blastocyst stage *in vitro* in W-BSA were transferred to NCTC-109-FCS and the degree of differentiation was scored. As shown in Figure 19.3, hatching, attachment, trophoblast outgrowth and development of an ICM occurred in the same percentage for *in vivo* and *in vitro* developed blastocysts. Differentiation of the ICM into two germ layers, however, only occurred in 50 per cent of the *in vitro*

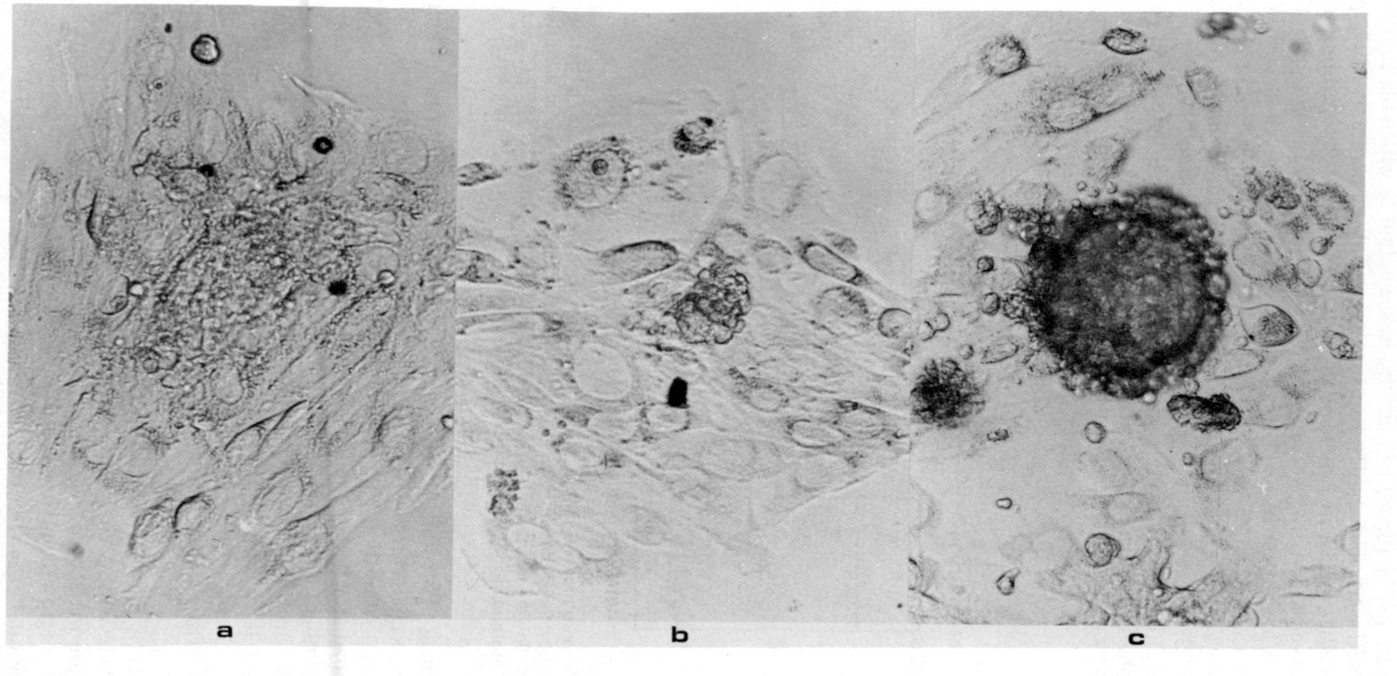

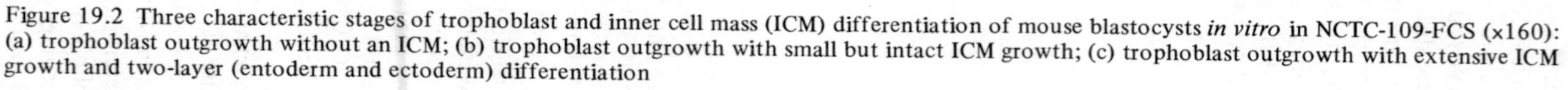
Figure 19.2 Three characteristic stages of trophoblast and inner cell mass (ICM) differentiation of mouse blastocysts *in vitro* in NCTC-109-FCS (×160): (a) trophoblast outgrowth without an ICM; (b) trophoblast outgrowth with small but intact ICM growth; (c) trophoblast outgrowth with extensive ICM growth and two-layer (entoderm and ectoderm) differentiation

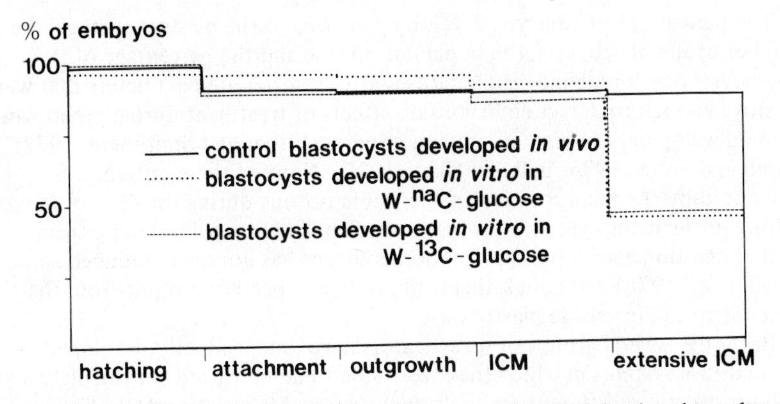

Figure 19.3 *In vitro* differentiation of mouse blastocysts which had developed in culture from 8-cell embryos. The percentage of mouse blastocysts reaching progressive stages of differentiation in medium NCTC-109-FCS after 120 h is given for blastocysts which had developed either *in vivo* or *in vitro* from 8-cell embryos in medium W-BSA containing ^{na}C-glucose or U-^{13}C-glucose. The differentiation of normal blastocysts in the same medium serves as control

developed blastocysts compared with more than 90 per cent for normal blastocysts. Figure 19.3 furthermore demonstrates that in this *in vitro* culture system there was no difference in the viability of blastocysts previously cultured in W-BSA in the presence of either ^{na}C-glucose or U-^{13}C-glucose. This result again confirms that the enrichment of the stable isotope in the embryos up to 20 atoms % ^{13}C has no detectable effects on subsequent differentiation.

DISCUSSION

In early studies with deuterium on the toxicity of stable isotopes in mammals, Thomson (1963) reported that after treatment during pregnancy all fetuses died at levels of deuterium enrichment which did not affect the longevity of adult animals. This strong evidence of the remarkable sensitivity of fetal systems to stable isotopes and the increasing clinical use of ^{13}C led to the present investigation. Whole animal feeding experiments are limited by the large quantities of isotopic compounds required and the correspondingly small numbers of animals which can be used. The ^{13}C enrichment which was reached in mouse embryos after application of U-^{13}C-glucose by stomach tube to the mother on days 9, 12 or 16 of pregnancy was 2–5 atoms % (Gregg *et al.*, 1975). A considerably higher degree of ^{13}C enrichment in mammalian embryos was reached by incorporating U-^{13}C-glucose into preimplantation mouse embryos during *in vitro* culture for 48 h, as described in this and a previous report (Spielmann *et al.*, 1976). According to our calculations the embryos contained approximately 20 atoms % ^{13}C at the end of the culture period. The recently developed methods for *in vitro* culture of blastocysts and, more importantly, the blastocyst transplantation technique permitted investigations to be carried out on the development of preimplantation mouse embryos containing a high amount of the stable isotope.

The percentage of embryos developing *in vitro* to the blastocyst stage, the cell number of the blastocysts, the implantation rate and the percentage of blastocysts developing into living fetuses after transfer are parameters that were sensitive enough to detect embryotoxic effects of treatment during preimplantation development *in vitro* in previous studies (Fisher and Smithberg, 1972; Ansell and Snow, 1975; Bell and Glass, 1975). None of these criteria was influenced by the incorporation of the stable isotope during the 48 h of *in vitro* culture. Convincing evidence for malformations in fetuses developing from treated and transferred preimplantation embryos has not been obtained so far (Spielmann, 1976) and could, therefore, not be expected to result from the ^{13}C enrichment in our mouse blastocysts.

Recently, several groups of investigators have independently developed *in vitro* culture systems in which they were able to demonstrate convincingly a very high sensitivity of differentiating blastocysts towards various metabolic inhibitors (Spindle and Pedersen, 1973; Ansell and Snow, 1975; Sherman and Atienza, 1975; Rowinski, Solter and Koprowski, 1975). In our routine cultures of mouse blastocysts in medium NCTC-109-FCS we achieved growth and differentiation to a degree which was as successful as the best of the previous reports. Even in this sensitive system there was, in accordance with all earlier studies, no indication of toxic isotope effects of ^{13}C enrichment in mammalian embryos. It is of particular importance that the ^{13}C level of the embryos studied in our system (20 atoms % ^{13}C) is tenfold higher than that usually achieved in clinical use (Gregg, 1974).

ACKNOWLEDGEMENTS

This work was supported by grants of the Deutsche Forschungsgemeinschaft awarded to the Sonderforschungsbereich 29, by the US Energy Research and Development Administration and by the Scientific Affairs Division, NATO.

REFERENCES

Ansell, J. D. and Snow, M. H. L. (1975). *J. Embryol. Exp. Morphol.*, **33**, 177

Bell, P. S and Glass, R. H. (1975). *Fertil. Steril.*, **26**, 449

Brinster, R. L. (1963). *Exp. Cell Res.*, **32**, 205

Donahue, R. P. (1968). *J. Exp. Zool.*, **169**, 237

Fisher, D. L. and Smithberg, M. (1972). *Teratology*, **6**, 159

Gregg, C. T. (1974). *Europ. J. Clin. Pharmacol.*, **7**, 315

Gregg, C. T., Hutson, J. Y., Prine, J. R., Ott, D. G. and Furchner, J. E. (1973). *Life Sci.*, **13**, 775

Gregg, C. T., Ott, D. G., Deaven, L., Spielmann, H., Krowke, R. and Neubert, D. (1975): *Proceedings of the Second International Conference on Stable Isotopes*, (ed. E. R. Klein and P. D. Klein) Oak Brook, Illinois, U.S.A., p. 61. National Technical Information Service Document CONF-751027

Klein, E. R. and Klein, P. D. (1975). *ibid.*

Klein, P. D. and Peterson, S. V. (1973): *Proceedings of the First International Conference on Stable Isotopes*, Argonne, Illinois, U.S.A. National Technical Information Service Document CONF-730525

Lorke, D. (1965). *Naunyn-Schmiedebergs Arch. Exp. Path. Pharmakol.*, **250**, 360

Pienkowski, M., Solter, D. and Koprowski, H. (1974). *Exp. Cell Res.*, **85**, 424
Rowinski, J., Solter, D. and Koprowski, H. (1975). *J. Exp. Zool.*, **192**, 133
Sherman M. I. (1975). *Differentiation*, **3**, 51
Sherman, M. I. and Atienza, S. B. (1975). *J. Embryol. Exp. Morphol.*, **34**, 467
Spielmann, H. (1976). *Current Topics in Pathology*, Vol. 62 (ed. A. Gropp and K. Benirschke). Springer-Verlag, Berlin–Heidelberg, p. 87
Spielmann, H., Eibs, H.-G., Nagel, D. and Gregg, C. T. (1976). *Life Sci.*, **19**, 633
Spindle, A. I. and Pedersen, R. A. (1973). *J. Exp. Zool.*, **186**, 305
Thomson, J. F. (1963). *Biological Effects of Deuterium*, Macmillan, New York
Whitten, W. K. (1971). *Advances in the Biosciences*, Vol. 6 (ed. G. Raspé), Pergamon Press–Vieweg Verlag, New York–Braunschweig, p. 129

20

$^{13}CO_2$ breath tests in normal and diabetic children following ingestion of ^{13}C-glucose

H. Helge, B. Gregg, C. Gregg*, S. Knies, I. Nötges-Borgwardt, B. Weber and D. Neubert (Department of Paediatrics and Department of Toxicology, Free University, Berlin, West Germany)

Recent developments in mass spectrometric technology and the availability of ^{13}C-enriched substrates for clinical research have enabled the clinician to perform $^{13}CO_2$ breath tests in patients with various diseases. Several research groups have compared $^{14}CO_2$ and $^{13}CO_2$ experiments in animals and man, and similar findings have been reported with both methods (Schoeller *et al.*, 1975; Shreeve, 1973).

Since radioisotopes cannot be used in pregnant women and in children, these groups of patients will benefit most from the substitution of stable isotopes in tests of diagnostic and investigative value (Bier *et al.*, 1975; Kalhan, Savin and Adam, 1975; Kerr, Stevens and Picou, 1975). When glucose-U-^{13}C was made available to us through the courtesy of the Los Alamos Scientific Laboratory (Matwiyoff and Ott, 1973), we planned to evaluate the applicability of the non-invasive $^{13}CO_2$ breath test for glucose tolerance studies and for the diagnosis of disorders of carbohydrate metabolism in childhood. Differences in the rate of $^{13}CO_2$ exhalation after an oral load of natural glucose between healthy, obese and diabetic adults have been described previously (Duchesne *et al.*, 1975; Lacroix *et al.*, 1973; Lefebvre *et al.*, 1975), and the first results of studies in normal and diabetic children with a ^{13}C-enriched glucose are the subject of this presentation.

METHODS

Breath samples of at least 250 ml were collected in plastics bags, and aliquots were transferred to leakproof 50 ml glass vessels. In newborns and small infants a

* Guest professor on leave of absence from the Los Alamos Scientific Laboratory.

mask and valve were used for the collection, which originally had been designed for respiratory studies in cats. The flasks were vacutainers with an outlet fitting to the vacuum pump, and CO_2 was separated cryogenically. The cool trap consisted of liquid nitrogen for the first and a mixture of dry ice in acetone for the second step. A mass spectrometer (Varian MAT 230), equipped with a double inlet system and a double ion collector, was used to determine the $^{13}CO_2/^{12}CO_2$ ratio of a breath sample. Simultaneous measurement of a standard mixture permitted us to calculate the $^{13}CO_2$ abundance as percentage of the total breath CO_2 with an accuracy of ± 0.0005 per cent.

Breath tests were performed in 10 healthy boys and girls, 1 to 10 years old, and in 9 freshly diagnosed diabetic children, 3 to 13 years old. Many of them were moderately hyperglycaemic and ketotic. Urinary losses of glucose were small and have been neglected. All were at rest, most had been fasted overnight, and only a few insulin-treated diabetics and infants had their last meal 4 h before the application of ^{13}C-glucose. The glucose tolerance tests (GTT) – in order to be comparable with those in adults – were done by oral administration of 40 g glucose per m^2 of body surface area (BS) in the form of a glucose mixture containing 1.6 atoms % ^{13}C, i.e. an excess of 0.5 atoms % ^{13}C over that of natural glucose. Since the ^{13}C content of our uniformly labelled ^{13}C-enriched glucose preparation was 65.2 atoms %, a dose of 40 g had to be prepared from 39.688 g of natural glucose and 0.312 g of enriched glucose.

For other investigations in the same group of children, a dose of 0.312 g of ^{13}C-enriched glucose (65.2 atoms % ^{13}C) per m^2 BS was used orally or intravenously without the concomitant administration of natural glucose ('tracer' studies). Although this amount of glucose is greater than that commonly taken for true tracer studies, in this communication we wish to use the term to distinguish these low doses of glucose from those applied for the loading tests.

In attempting to establish a dose–effect relationship and in order to find an optimal dose for diagnostic purposes in small children, we have subsequently further decreased the dose and based the calculation on body weight (BW) instead of surface area. More recently, the dose was 3 mg (*ca* 10 μmol) glucose-U-^{13}C (65.2 atoms % ^{13}C) per kg BW.

Breath samples were regularly collected for a time period of 10–12 h at intervals of from 15 min to 2 h. For an exact calculation of the cumulative $^{13}CO_2$ exhalation, the total respiratory CO_2 excretion of the patient has to be known. We have not measured total CO_2, and therefore we have based our estimates on data from the literature (Lister, Hofmann and Rudolph, 1974; Shreeve, Cerasi and Luft, 1970; Slanger, Kusubov and Winchell, 1970; Winchell *et al.*, 1970).

For blood sugar determinations, the glucose oxidase method was used, while insulin was measured radioimmunologically.

RESULTS

In several instances, $^{13}CO_2$ exhalation was measured after intravenous injection or oral administration of a 'tracer' dose of glucose-U-^{13}C (312 mg per m^2; 65.2 atoms % ^{13}C) and the time course of this process was compared with that after a loading test in the same children (Figure 20.1). With the small amount of glucose, the elimination of $^{13}CO_2$ reaches a maximum 1 h after ingestion. Surprisingly, the peak height and time course of the $^{13}CO_2$ exhalation curve are not significantly different after intravenous injection of the same dose. Blood glucose and insulin determinations in these 'tracer' experiments did not show any consistent fluctuations attributable to the oral or intravenous administration of the glucose.

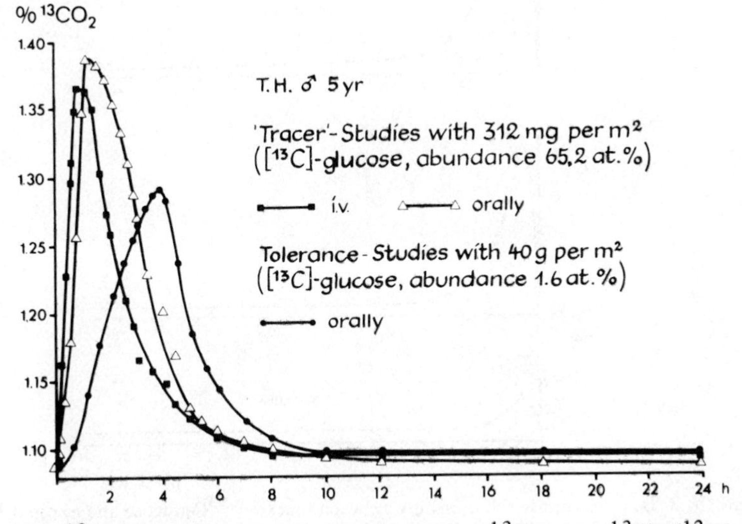

Figure 20.1 $^{13}CO_2$ exhalation following administration of ^{13}C-glucose. $^{13}CO_2/^{12}CO_2$ ratios are compared after intravenous (i.v.) and oral administration of glucose-U-^{13}C to a healthy 5 year old boy. ■———■ 'tracer' study i.v., △———△ 'tracer' study orally; 312 mg per m^2; ^{13}C-abundance of glucose 65.2 atoms %. ●———● tolerance study, 40 g per m^2 orally, ^{13}C-abundance 1.6 atoms %

Following the ingestion of a glucose load (40 g ^{13}C-enriched glucose per m^2; 1.6 atoms % ^{13}C), breath $^{13}CO_2$ rises more gradually up to a peak at 2½ h, 90 min later than the maximum of blood glucose or insulin concentrations. By repeating these studies at different dose levels, a dose–effect relationship can be demonstrated between the amount of glucose ingested and total $^{13}CO_2$ expired (Figure 20.2). This relationship is seen in the tolerance studies as well as in the tracer studies. However, while the maximum values of the $^{13}CO_2/^{12}CO_2$ ratio seem to be dependent on the ^{13}C-excess of the glucose, the shape of the curve and its time course are determined by the total amount of glucose.

The differences between 'tracer' and loading tests for the group of control children are shown in Figure 20.3 (left side). In the glucose tolerance tests the average $^{13}CO_2/^{12}CO_2$ ratio in breath increased more slowly than in the 'tracer' studies and attained a lower maximum. The respective mean values were 1.274 ± 0.018 per cent at 210 min and 1.390 ± 0.002 per cent at 75 min. $^{13}CO_2$ peaks in the individual tests were reached at 75 or 90 min after a 'tracer' dose and between 150 and 225 min following a glucose load.

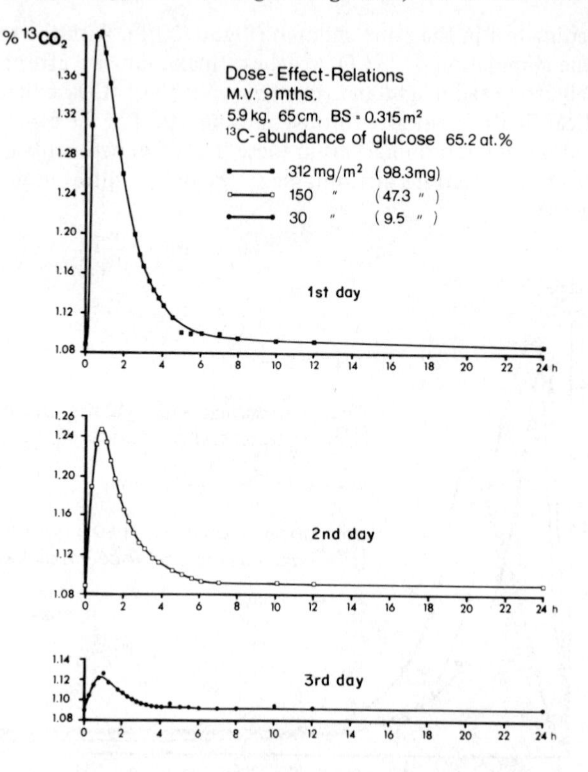

Figure 20.2 Dose–response relationships between ingested ^{13}C-glucose and expired $^{13}CO_2$ in a 9 month old infant without a disorder of carbohydrate metabolism. The tests were performed on consecutive days with different, metabolically inert 'tracer' doses of ^{13}C-glucose (^{13}C-abundance 65.2 atoms %): ▪———▪ 312 mg/m^2; □———□ 150 mg/m^2; •———• 30 mg/m^2

The results in diabetic children (Figure 20.3, right side) differ from those in the control group in at least two respects. All $^{13}CO_2$ maxima are significantly lower and occur later than in the control group, and the individual values are scattered more widely. Differences are obvious in the loading tests (maximum 1.179 ± 0.018 per cent at 240 min) as well as in the 'tracer' studies (maximum 1.239 ± 0.062 per cent at 150 min). Peaks were registered between 195 and 240 min or between 105 and 165 min, respectively. In some diabetic patients a plateau of the $^{13}CO_2$ exhalation curve was maintained over several hours, and there was no distinctive peak.

With respect to the time course of the $^{13}CO_2$ exhalation, the descending slope of the curve is also of interest. On a semilogarithmic plot two straight lines have been fitted to connect the single values, and the half-life times were estimated graphically, analogous to the determination of the blood glucose disappearance rate. In diabetic children the $^{13}CO_2$ elimination apparently was retarded. During the third and fourth hour a $t_{1/2}$ of 460 against 100 min was found, followed by a value of 65 against 40 min.

In order to calculate the cumulative $^{13}CO_2$ exhalation, we assumed that the

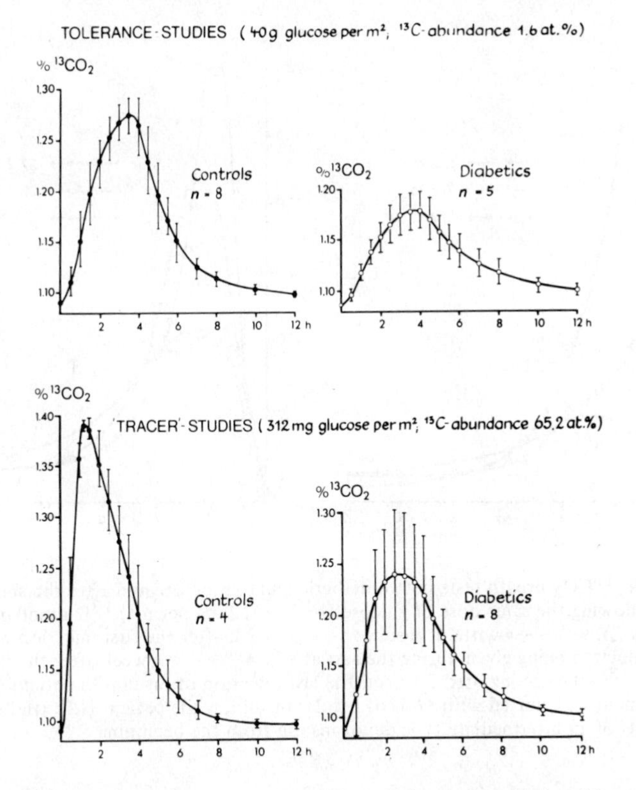

Figure 20.3 $^{13}CO_2$ exhalation after ingestion of small ('tracer' studies) and large (tolerance studies) amounts of glucose (^{13}C abundance 65.2 or 1.6 atoms %) in groups of healthy and diabetic children. The dose of glucose-U-^{13}C was the same in 'tracer' and tolerance studies. •——• Controls, ○——○ diabetics. The points represent means ±1 SD

minimal CO_2 production was 9 mmol kg^{-1} h^{-1} for the controls as well as for the diabetics. The calculations suggest that much less ^{13}C-glucose is oxidised to $^{13}CO_2$ in diabetic children. After 4 h, 19 per cent against 33 per cent of the ^{13}C administered with the 'tracer' dose of glucose had been eliminated as $^{13}CO_2$, and after a glucose load the comparable figures were 9 per cent in diabetic and 20 per cent in control children. Six hour estimates were 26 per cent against 38 per cent for 'tracer' studies and 15 per cent against 29 per cent for loading tests, respectively, and only minor changes occurred thereafter.

In diabetic children the rate of $^{13}CO_2$ exhalation following a small dose of ^{13}C-glucose is changed considerably by a single injection of soluble insulin (Figure 20.4). Even after a few days of effective insulin treatment the elimination curve of $^{13}CO_2$ may be found to resemble that of control children.

When a ^{13}C-glucose dose of 3 mg per kg BW was used instead of 312 mg per m^2 BS, analogous results were achieved. Accordingly, the respiratory $^{13}CO_2$ concentrations were much lower, but the relative differences between alternate tests in the same individual or between controls and diabetic children were similar.

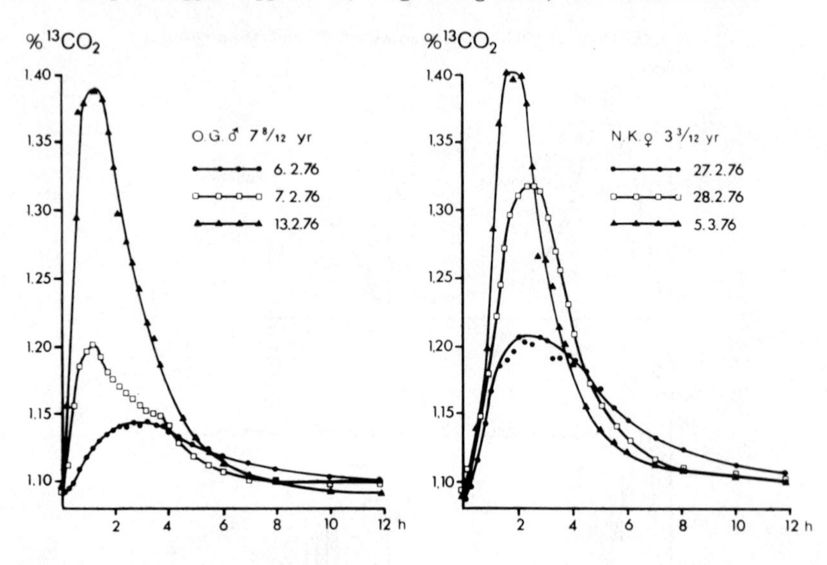

Figure 20.4 $^{13}CO_2$ breath tests in two diabetic children before and after the start of insulin therapy following the same dose of glucose-U-^{13}C (312 mg per m^2; ^{13}C-abundance 65.2 atoms %), •——• without insulin; □——□ 2 h after the first injection of insulin, the next injection being given during the test at 4 h; ▲——▲ 1 week after the initiation of treatment, ^{13}C-glucose ingested 3 h after the last injection of insulin. In patient O.G. (left side) treatment was started with 16 U of regular insulin, while patient N.K. (right side) received 8 U of an intermediate type depot insulin from the beginning

DISCUSSION

The results of the $^{13}CO_2$ breath tests following the ingestion of uniformly labelled ^{13}C-glucose in normal and diabetic children seem to indicate that untreated, insulin-deficient diabetics oxidise glucose less efficiently than healthy controls. This metabolic disturbance can be observed in 'tracer' studies under basal conditions, and is evident also in the GTT, irrespective of the amount of glucose administered.

These findings are in accordance with those reported for another group of insulin-dependent diabetics (Lefebvre *et al.*, 1975). In these patients the respiratory $^{13}CO_2$ was measured during a GTT performed with maize glucose, a natural product known for its relatively high ^{13}C-abundance. However, contradictory results were obtained in several investigations regarding the glucose production and oxidation in obese and non-obese adult diabetics (Bowen and Moorhouse, 1973; Magnenat *et al.*, 1975; Shreeve, 1973). The discrepancies are striking, but may well be explained by the insufficiently controlled or even disparate disease states of the patients, the inconsistent experimental conditions and the different methods used by the various authors.

The shape of the $^{13}CO_2$ exhalation curves found in adults receiving

^{13}C-glucose was very similar to those obtained from our patients (Duchesne *et al.*, 1975). It was apparent that the measurement of the $^{13}CO_2$ production rate and the comparison of the cumulative $^{13}CO_2$ excretion could have offered additional evidence for a disturbance of glucose oxidation. Because total CO_2 excretion had not been determined in our studies, we have tried to estimate the cumulative $^{13}CO_2$ exhalation on the basis of reference data published for normal adults (Shreeve *et al.*, 1970; Slanger *et al.*, 1970; Winchell *et al.*, 1970), children (Lister *et al.*, 1974) and diabetics (Bowen and Moorhouse, 1973; Duchesne *et al.*, 1975; Magnenat *et al.*, 1975). There is good agreement between our values estimated in this manner and the calculations of other authors with respect to the ^{13}C fraction of glucose oxidised to $^{13}CO_2$. As long as the total amount of glucose ingested is a tracer dose or a constant load, a dose–response relationship between the ^{13}C-glucose and respiratory $^{13}CO_2$ produced from it can be postulated. The close resemblance between the $^{13}CO_2$ exhalation curves of our control and diabetic children and those of adults (Lefebvre *et al.*, 1975), who ingested glucose much less enriched in ^{13}C, is taken to support this view.

In loading tests additional metabolic effects have to be considered. Important changes can be induced in the pool size and tissue distribution of glucose, the fractional glucose turnover and the oxidation rate (Bowen and Moorhouse, 1973; Magnenat *et al.*, 1975; Shreeve, 1973). It is not surprising, therefore, that the time course of the ^{13}C exhalation, the maximum value for the $^{13}CO_2/^{12}CO_2$ ratio, and the cumulative $^{13}CO_2$ production from an equal amount of ^{13}C-glucose are altered according to the amount of unlabelled glucose ingested at the same time.

For this type of $^{13}CO_2$ breath test following the administration of ^{13}C-glucose, and used instead of the established GTT, the intravenous injection of ^{13}C-glucose does not offer any advantage over the oral route. There are a few occasions, however, in which intravenous injection of labelled glucose and successive breath analysis for $^{13}CO_2$ might be desirable. One instance would be in the rare patient with glucose malabsorption, when the $^{13}CO_2$ exhalation following intravenous injection and oral ingestion would have to be compared for diagnostic purposes.

Another example could be a planned clinical research project, in which the intravenous injection or infusion of ^{13}C-glucose and subsequent breath analysis of $^{13}CO_2$ are combined for metabolic studies. The determination of systemic glucose production (Kalhan *et al.*, 1975) and glucose fluxes (Bier *et al.*, 1975) may help to define changes and differences in the descending slope of the $^{13}CO_2$ elimination curve that can be correlated with altered rates of glucose production and oxidation in diabetes and other states of carbohydrate intolerance or hypoglycaemia (Kerr *et al.*, 1975).

Our experience with pregnant women, infants of diabetic mothers and children with hypoglycaemia is limited. Further studies are required to show whether the $^{13}CO_2$ breath test will be sensitive enough to detect minor disturbances of glucose metabolism. For diabetic children our results have confirmed other data collected with more laborious methods. In this respect the non-invasive $^{13}CO_2$ breath test is considered to be a valuable and straightforward procedure for the study of disorders of carbohydrate metabolism.

REFERENCES

Bier, D., Leake, R., Gruenke, L. and Sperling. M. (1975). *Proceedings of the Second International Conference on Stable Isotopes* (ed. E. R. Klein and P. D. Klein), Oak Brook, Illinois. National Technical Information Service Document CONF 751027, p.344

Bowen, H. F. and Moorhouse, J. A. (1973). *J. Clin. Invest.*, **52**, 3033

Duchesne, J., Lacroix, M., Lefebvre, P., Mosora, F., Luyckx, A. and Pirnay, F. (1975). *Proceedings of the Second International Conference on Stable Isotopes* (ed. E. R. Klein and P. D. Klein), Oak Brook, Illinois. National Technical Information Service Document CONF 751027, p. 282

Kalhan, S. C., Savin, S. M. and Adam, P. A. (1975). *Proceedings of the Second International Conference on Stable Isotopes* (ed. E. R. Klein and P. D. Klein), Oak Brook, Illinois. National Technical Information Service Document CONF 751027, p. 330

Kerr, D., Stevens, M. and Picou, D. (1975). *Proceedings of the Second International Conference on Stable Isotopes* (ed. E. R. Klein and P. D. Klein), Oak Brook, Illinois. National Technical Information Service Document CONF 751027, p. 336

Lacroix, M., Mosora, F., Pontus, M., Lefebvre, P., Luyckx, A. and Lopez-Habib, G. (1973). *Science, N.Y.*, **181**, 445

Lefebvre, P., Mosora, F., Lacroix, M., Luyckx, A., Lopez-Habib, G. and Duchesne, J. (1975). *Diabetes*, **24**, 185

Lister, G., Hoffman, J. I. E. and Rudolph, A. M. (1974). *Pediatrics*, **53**, 656

Magnenat, G., Felber, J. P., Curchod, B., Gomez, F., Pittet, P., Bonjour, J. P., Weissbrodt, P., Geser, C. A., Müller-Hess, R. and Jequier, E. (1975). *Schweiz. Med. Wschr.*, **105**, 69

Matwiyoff, N. A. and Ott, D. G. (1973). *Science, N.Y.*, **181**, 1125

Schoeller, D. A., Klein, P. D., Schneider, J. F., Solomons, N. W. and Watkins, J. B. (1975). *Proceedings of the Second International Conference on Stable Isotopes* (ed. E. R. Klein and P. D. Klein), Oak Brook, Illinois. National Technical Information Service Document CONF 751027, p. 246

Shreeve, W. W. (1973). *Proceedings of the First International Conference on Stable Isotopes in Chemistry, Biology and Medicine* (ed. P. D. Klein and S. V. Peterson), Argonne, Illinois. National Technical Information Service Document CONF. 730525, p. 390

Shreeve, W. W., Cerasi, E. and Luft, R. (1970). *Acta Endocrinol.*, **65**, 155

Slanger, B. H., Kusubov, N. and Winchell, H. S. (1970). *J. Nucl. Med.*, **11**, 716

Winchell, H. S., Stahelin, H., Kusubov, N., Slanger, B., Fish, M., Pollycove, M. and Lawrence, J. H. (1970). *J. Nucl. Med.*, **11**, 711

21
Studies on catecholamine metabolism in man using *in vivo* deuterium labelling

E. Änggård, B. Sjöqvist and T. Lewander (Department of Pharmacology, Karolinska Institutet, 104 01 Stockholm, and Psychiatric Research Centre, Ulleråker Hospital, 750 17 Uppsala, Sweden)

INTRODUCTION

Studies in animals have shown that the rates of synthesis and turnover of catecholamines (CAs), rather than their levels, reflect changes in the functional activity of adrenergic nerves both in the central nervous system (CNS) and periphery (Costa and Neff, 1966; Costa, 1970; Sedvall, Weise and Kopin, 1968). Since a role for CAs has been implied in many physiological and pathological processes in man such as stress, hormone release, temperature regulation, affective disorders, schizophrenia and hypertension, there has been an interest in developing methods to study CA turnover in humans. The techniques which have been successfully utilised in animals can be divided into 'steady state' and 'non-steady state' methods. The former methods have advantages over the latter in as much as the analysis does not necessarily involve a perturbation of the system. The most common 'steady state' methods are: (1) labelling of the CA stores with radioactive amine and following decay; (2) measuring rate of formation of labelled amine after injection or infusion of the labelled precursor amino acid. 'Non-steady state' methods usually employ inhibitors of catecholamine synthesis, metabolism or metabolite transport and measure rates of accumulation or disappearance of the CA or metabolite.

In the 'steady state' methods, radioactively labelled compounds have been employed. The use of radioactive chemicals in humans is not readily feasible owing to the possible radiation hazards involved, particularly with the precursor amino acids. These are in part incorporated into proteins and other molecules with a low turnover. Also, very high amounts of radioactive precursor amino acid would have to be given due to low conversion to catecholamines and to the low levels of the metabolites present in blood and cerebrospinal fluid. By the use of labelling with stable isotopes together with the sensitive ion-specific detection methods inherent in gas chromatography–mass spectrometry (GC–MS), these problems can now be tackled in humans.

In this communication we shall describe the use of deuterium labelled CAs, precursors and metabolites in the study of pathways and kinetics of CA metabolism in man.

DEUTERATED CAs, PRECURSORS AND METABOLITES

A list of available CAs, precursors and metabolites is given in Table 21.1. These are either commercially available or can be prepared by simple chemical procedures. L-Tyrosine is the major precursor of the catecholamines. A suitably labelled tyrosine should have the following properties:

(1) It should be the L-isomer rather than the racemate.
(2) The label should not be in a position affected by rate limiting steps in the synthesis and/or metabolism of the CAs.
(3) The label should be in a chemically stable position, to prevent loss of label *in vivo* or during work-up of samples.
(4) The labelled tyrosine should be relatively inexpensive. Large amounts ($>$10 mg/kg) are needed for metabolic studies.

L-Tyrosine-3,5-d_2 has been prepared and given to animals and humans (Lindström, Sjöqvist and Änggård, 1974; Curtius, Baerlocher and Völlmin, 1972a,b; Curtius *et al.*, 1975; Faull *et al.*, 1975). Although suitable for studies on tyrosine metabolism, it is not a good choice for the CA pathway since a deuterium in the 3-position would give an isotope effect in the rate-limiting step in the CA biosynthesis. Furthermore the CAs and their metabolites would only have one deuterium.

D,L-Tyrosine-β-d_2 has the disadvantage of being a racemic mixture and, furthermore, one deuterium will be lost in the conversion of dopamine to noradrenaline. The L-tyrosine-2,6,α-d_3 seems to be the best choice since the deuterium in the 3 position would give an isotope effect in the rate-limiting step metabolism. One deuterium will be removed during deamination. It is not expected that the deuterium present at the α position will introduce any noticeable isotope effect on the rate of deamination or introduce a metabolic shift, e.g. in the ratio of acidic and alcoholic metabolites.

CATECHOLAMINES

Dopamine can be obtained labelled with deuterium both in the ring and side chain (Table 21.1). The α-methylated analogues of dopamine and noradrenaline are also commercially available and have been used as internal standards in the mass fragmentographic determination of catecholamines.

Table 21.1 Deuterium-labelled CAs, precursors and metabolites

Amino acids	Source
3,4-Dihydroxyphenylalanine = DOPA	
D,L-DOPA-α-d_1·HBr	MSD
D,L-DOPA-β-d_2·HBr	MSD
D,L-DOPA-(α-d_1,β-d_2)·HBr	MSD
L-DOPA-(ring d_3)	MSD
DL-Phenylalanine-(ring d_5, α-d_2)	MSD
DL-Phenylalanine-(ring d_5)	MSD
L-Phenylalanine-d_8	MSD, SIC
DL-Tyrosine-α-d_1	MSD
DL-Tyrosine-β-d_2	MSD
DL-Tyrosine-(α-d_1, β-d_2)	MSD
DL-Tyrosine-2,6-d_2	MSD
DL-Tyrosine-3,5-d_2	MSD
L-Tyrosine-d_7	MSD, SIC
L-Tyrosine-2,6,α-d_3	MSD
Amines	
Dopamine-α-d_2·HCl	MSD
Dopamine-β-d_2·HCl	MSD
Dopamine-(α-d_2, β-d_2)·HCl	MSD, SIC
3-Methoxytyramine-(α-d_2, β-d_1)·HCl	SIC
3-Methoxytyramine-(ring d_3)·HCl	PK
DL-Norepinephrine-(α-d_2, β-d_1)	MSD, SIC
Acids	
Homovanillic acid-2-d_2	MSD, PK
Homovanillic acid-(ring d_3,2-d_2)	MSD, PK
Homovanillic acid-(methyl d_3)	PK
Homovanillic acid-d_8	PK
4-Hydroxy-3-methoxymandelic acid-(methyl d_3)	PK
4-Hydroxy-3-methoxymandelic acid-(ring d_3,2-d_1)	SIC
4-Hydroxy-3-methoxymandelic acid-2-d_1	MSD
3,4-Dihydroxyphenylacetic acid-2-d_2	MSD
3,4-Dihydroxyphenylacetic acid (ring d_3,2-d_2)	MSD, SIC
Alcohols	
4-Hydroxy-3-methoxyphenylglycol piperazine salt (methyl d_3)	PK
4-Hydroxy-3-methoxyphenylglycol(1-d_2,2-d_1)	Lindström *et al.*, 1974

MSD = Merck-Sharp and Dohme GmbH, 8000 München 80, Leuchtenbergring 20, Germany.
PK = Produktkontroll AB, Banvaktsvägen 22, 17148 Solna, Sweden.
SIC = Stohler Isotope Chemicals, 49 Jones Road, Waltham, Mass. 02154, USA.

CATECHOLAMINE METABOLITES

The O-methylated metabolites of 3-methoxy-tyramine are available labelled in the ring or in the β position and normetanephrine can be obtained labelled with three deuterium atoms in the α and β positions. The corresponding α-methyl analogues are also commercially available.

Homovanillic acid and 3,4-dihydroxyphenyl acetic acid (DOPAC) are important dopamine metabolites. These compounds are now available labelled in a variety of forms as shown in Table 21.1. The corresponding metabolites of noradrenaline, 4-hydroxy-3-methoxy mandelic acid (vanillylmandelic acid, VMA) and 3,4-dihydroxymandelic acid, are also available labelled with three deuterium atoms. The glycolic metabolite of noradrenaline, 4-hydroxy-3-methoxyphenylethylene glycol (HMPG), is of importance since it is the major noradrenaline metabolite formed in the central nervous system.

MEASUREMENT OF TURNOVER OF HOMOVANILLIC ACID IN MAN

In the present study, HVA-d_5 was used to label the body pool of endogenous HVA and to calculate its turnover rate. Details of this study have been reported previously (Änggård, Lewander and Sjöqvist, 1975).

Five healthy men volunteered to participate in the study. The fasted resting subjects received an injection of 5.5 μmol HVA-d_5 and blood samples were collected at 5, 10, 20, 30, 60, 120, 180, 240, 360 and 420 min following the injection. Urine was collected every hour. The levels of HVA-d_5 and HVA-d_0 in urine and plasma were determined by previously described mass fragmentographic techniques using HVA-d_2 as internal standard (Sjöqvist, Lindström and Änggård, 1973). The coefficient of variation for determination of HVA-d_0 was determined to be ±4.5 per cent at a level of 48 pmol/ml (n = 15) in plasma and ±5.4 per cent in urine (n = 10) at a level of 20 nmol/ml.

Following the i.v. injection of HVA-d_5, the elimination of HVA-d_5 was exponential after 10–20 min. This is shown for one subject in Figure 21.1. The levels of HVA-d_0 were relatively constant during the experiment with the

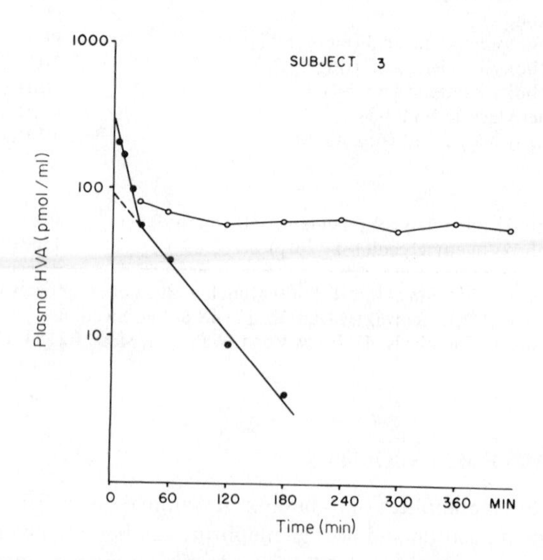

Figure 21.1 Blood levels of HVA-d_5 (●——●) and unlabelled HVA (○——○) following i.v. administration of 5.5 μmol of HVA-d_5 to a human volunteer

exception of the first 20 min period, when in subject 5 an elevation in the HVA-d_0 levels was observed. The slope of the line obtained by plotting log (plasma level of HVA-d_5) against time yields the rate constant (K_{el}), which is identical with the HVA turnover rate constant since the HVA-d_0 level was unchanged during this time period. The body turnover and body clearance can be calculated from K_{el}, the extrapolated concentration of HVA-d_5 at zero time (C_0) and the apparent volume of distribution (V_D). The results are given in Table 21.2.

The V_D was found to be relatively small. The mean of 0.59 l/kg of body weight indicates a distribution in the body water, with the exception of CNS water, since HVA is supposed not to pass the blood–brain barrier. The size of the peripheral body pool of HVA was 3.4 ± 1.1 μmol. This figure should be compared with the figure for body turnover, 3.8 ± 1.9 μmol/h. However, the latter figure also includes the CNS production of HVA since HVA produced in the CNS pool empties and presumably mixes with the peripheral pool before elimination.

If HVA is a metabolic end-product, its rate of excretion should equal its turnover. However the urinary excretion rate was only 1.70 ± 0.5 μmol/h, i.e. about half of the turnover. Similarly, the urinary recovery of HVA-d_5 was about 60 per cent and the body clearance (50 ± 22 l/h), i.e. about double the renal clearance (24 ± 7.2 l/h). Together these results indicate that renal excretion is not the only mechanism by which HVA is eliminated.

In vivo LABELLING OF CATECHOLAMINES USING ORAL LOADING WITH DEUTERATED TYROSINE

These experiments were undertaken to test the methods for determination of labelled and unlabelled tyrosine, HVA and HMPG in blood and urine and to assess the dose of deuterium needed for a useful degree of incorporation. In these experiments D,L-tyrosine-β-d_2 was used for reasons of economy.

Previous authors (Curtius *et al.*, 1975; Faull *et al.*, 1975) used very high doses, 150–300 mg/kg, of deuterated tyrosine. We thought this would represent an unphysiological and possibly hazardous expansion of the tyrosine pool. D,L-Tyrosine-β-d_2 (40 mg/kg) was mixed with two teaspoonfulls of chocolate-flavoured Meritene and ingested together with 500 ml of water. Blood and urine samples were collected every hour for 8–10 h following the intake of tyrosine. The levels of tyrosine-d_0 and tyrosine-d_2 in plasma were analysed using α-methyl-*p*-tyrosine (α-MT) as internal standard. To 500 μl of plasma was added 200 nmol of α-MT. After precipitation of proteins with 500 μl 10 per cent trichloroacetic acid and 2.5 ml 0.1M HCl, the mixture was centrifuged and the clear supernatant was applied to a column of Amberlite-IR-120 (5 x 50 mm). The column was washed with H_2O until the effluent became neutral and tyrosine was then eluted with 20 ml 0.5M HCl in methanol. The eluate was evaporated to dryness, the residue was dissolved in 2 ml of 5 per cent HCl in *n*-butanol and heated to 110° for 3 h. The solvent was evaporated and the dried residue reconstituted in 50 μl of ethyl acetate and 50 μl of pentafluoropropionyl anhydride. After 30 min, the reagent and solvent were removed by a stream of nitrogen and the sample was dissolved in 50 μl of ethyl acetate of which 1–5 μl

Table 21.2

Subject	Weight (kg)	HVA-d_5[a] C_0 (μmol/l)	V_D (*l*)	V_D (l/kg)	Plasma[b] conc. HVA-d_0 (nmol/l)	HVA[c] pool size (μmol)	HVA-d_5 $t_{1/2}$ (h)	HVA-d_5 K_{el} (h^{-1})	HVA[d] turnover rate ($nmol\ l^{-1}\ h^{-1}$)	HVA[e] body turn-over (μmol/h)	Body[f] clearance (1/h)
1	75	0.100	55	0.73	89 ± 5	4.9	0.50	1.38	123	6.8	76
2	86	0.119	46	0.53	59 ± 15	2.7	0.57	1.21	71	3.3	56
3	74	0.093–	59	0.80	72 ± 18	4.2	0.64	1.08	78	4.6	64
4	67	0.158	35	0.52	87 ± 15	3.0	0.87	0.79	69	2.4	28
5	64	0.218	25	0.39	82 ± 20	2.1	0.70	0.99	81	2.0	25
Mean	73	0.14	44	0.59	78	3.4	0.66	1.1	84	3.8	50
± SD	9	0.05	14	0.17	12	1.1	0.14	0.22	22	1.9	22

[a] Concentration of HVA-d_5 at zero time, read from the intercept of the exponential elimination curve with the *y* axis.
[b] Mean ± SD of analyses on 6–9 samples collected during the experiment.
[c] Obtained by multiplication of V_D by plasma HVA-d_0 conc.
[d] Obtained by multiplication of the rate constant of elimination for HVA-d_5 with plasma HVA-d_0 conc.
[e] Obtained by multiplication of turnover rate with V_D.
[f] By definition $K_{el} \times V_D$.

was injected into a LKB 2091 GC–MS instrument, focused on *m/e* 442 (α-MT), 430 (tyrosine-d_2) and 428 (tyrosine-d_0). These ions represent loss of $COOC_4H_9$ from the butyl ester pentafluoropropionyl derivatives of these tyrosines. Replicate analyses of tyrosine-d_0 from a plasma pool gave a value of 80 ± 4 nmol/ml (14.5 ± 0.7 μg/ml).

Levels of HVA-d_0, HVA-d_2, HMPG-d_0 and HMPG-d_1 in urine were determined using previously published procedures (Sjöqvist *et al.*, 1973, 1975). The results are shown in Figure 21.2 and Table 21.3. After oral administration of D,L-tyrosine-β-d_2, the concentration of this deuterium-labelled tyrosine in

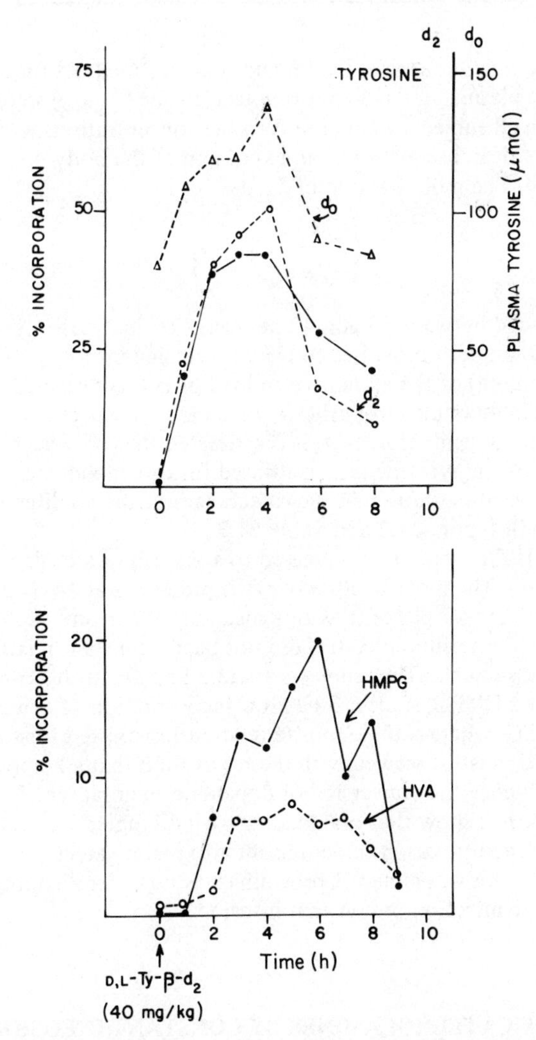

Figure 21.2 Incorporation of deuterium into urinary catecholamine metabolites following oral administration of D,L-tyrosine-β-d_2 (40 mg/kg) to a human volunteer. The mole fraction of tyrosine-d_2 in plasma tyrosine is shown in the upper frame (●———●) and the mole fraction of urinary HMPG (●———●) and HVA (○———○) in the lower frame.

Table 21.3 Oral administration of D,L-tyrosine-β-d_2 to healthy human subjects

Subject	Dose (mg/kg)	% Incorporation		
		Tyrosine-d_2	HVA-d_2	HMPG-d_1
1	20	–	5.5	–
1	40	–	10	8
2	40	35	22	17
3	40	43	8	20
4	40	83	37	3.6

plasma rose to a maximum after 3–4 h and then declined. At the same time an elevation of the plasma levels of endogenous tyrosine (T_y-d_0) were noted. This was seen in all subjects where the tyrosine concentrations were measured and is probably the consequence of an expansion of the body pool by the tyrosine loading. The mole fraction of T_y-d_2

$$\left(\frac{d_2}{d_0 \times d_2} \times 100\right)$$

incorporated varied between 35 and 43 per cent (Table 21.3).

Unless transamination processes convert D-tyrosine into L-tyrosine, only 50 per cent (= 20 mg/kg) of the administered load of D,L-tyrosine-d_2 should be available for catecholamine biosynthesis, which is stereospecific for the natural L-form of the amino acid. Moreover, it is estimated that only a small amount (<0.5 per cent) of the tyrosine pool is utilised for catecholamine biosynthesis. The incorporation of deuterium in the catecholamine metabolites HVA and HMPG is shown in Figure 21.2 and Table 21.3.

The level of HVA-d_2 in urine increased to a plateau or a peak between 4–6 h and then declined. The mole fraction of HVA present as HVA-d_2 at the peak varied between 8 and 37 per cent, with a mean of 19 per cent (Table 21.3). The incorporation of deuterium in HMPG did not bear a constant relationship to HVA in the four subjects. Thus subjects 1 and 2 had about the same incorporation in HVA and HMPG, subject 3 had low incorporation of deuterium in HVA but high in HMPG, whereas the opposite relationship existed for subject 4. These results on HMPG must be viewed with the reservation that a kinetic isotope effect could influence the conversion of dopamine to noradrenaline.

These results thus show that oral loading with 40 mg/kg D,L-tyrosine-β-d_2 gives reasonable incorporation of deuterium into major catecholamine metabolites. Further work was aimed at providing a more controlled mode of administration by injection or constant infusion.

LABELLING OF CATECHOLAMINES BY CONSTANT INFUSION OF DEUTERATED TYROSINE

The underlying theory behind these experiments was that a constant infusion of the labelled precursor amino acid would lead to a rate of incorporation of deuterium into catecholamines and their metabolites that was proportional to their rate of synthesis and release. This technique has been successfully used in animal experiments using radioactive tyrosine (Sedvall *et al.*, 1968; Neff *et al.*, 1969). The approach was first evaluated by infusion of deuterated tyrosine ethyl ester in dogs and rats.

Infusion of deuterated tyrosine ethyl ester in dogs and rats

The free amino acid has too low a solubility in physiological salt solutions (<0.5 mg/ml) to permit infusion in a reasonable volume. We therefore decided to evaluate the use of the ethyl ester of tyrosine for the infusion experiments. The ethyl ester was chosen in preference to the methyl ester since the methyl group is converted *in vivo* to methanol which is more toxic than ethanol. The total amount of ethanol formed from infusion of 40 mg/kg of tyrosine ethyl ester·HCl would be about 11 mmoles (506 mg) in a 70 kg person or 3.2 mmoles in a 20 kg dog.

Five male dogs, fasted overnight and weighing 8.5–19 kg, were anaesthetised with pentabarbital and prepared for sample collection by insertion of catheters in the cubital vein and urethers. The cerebroventricular system was perfused at a constant rate with a balanced salt solution ('artificial CSF') at 8.4 ml/min from cannulae inserted into the lateral ventricle and cisterna magna. The arterial blood pressure was recorded from the left carotid artery. Samples of blood, cerebrospinal fluid (CSF) and urine were collected every 30 min.

After a base-line period of 30 min, 20 mg/kg of D,L-tyrosine-β-d_2 was infused during 90 min. The levels of tyrosine-d_2 in plasma and CSF are shown in Figure 21.3. It can be seen that the concentration of tyrosine-d_2 in the plasma increases during the infusion period to 35 per cent of the total tyrosine level and thereafter declines in what appears to be an exponential fashion. During this infusion period no significant increase in tyrosine-d_0 was observed (not shown). In contrast, the incorporation of tyrosine-d_2 in the CSF was much lower.

The urinary HVA-d_2 appeared with a maximum at about 90 min after the plasma peak of tyrosine-d_2. The mole fraction incorporated in the peak was 11 per cent. The appearance of HVA-d_2 in cerebrospinal fluid was irregular. Only in one dog was a significant incorporation of about 20 per cent obtained. In two other animals an elevation of HVA-d_2 was noted in the last two samples collected.

The poor incorporation of deuterium into HVA of the CNS could be due to inefficient distribution of the exogenous tyrosine into the brain. This possibility was investigated in the rat. Groups of animals were given intravenous infusions of L-tyrosine-2,6,α-d_3 (L-ty-d_3) and then killed after various time intervals. The accumulation of L-tyrosine-d_3 was highest in the heart, intermediate in plasma and lowest in the brain. These results indicate that some organs, like the heart, show a preferential accumulation of exogenously administered tyrosine. If the infused L-tyrosine-d_3 is accumulated by peripheral organs then correspondingly less would be available for transport into the brain.

These studies are now being extended by studying the incorporation of deuterium into catecholamines and their metabolites during a constant intravenous infusion of L-tyrosine-2,6,α-d_3 in human subjects.

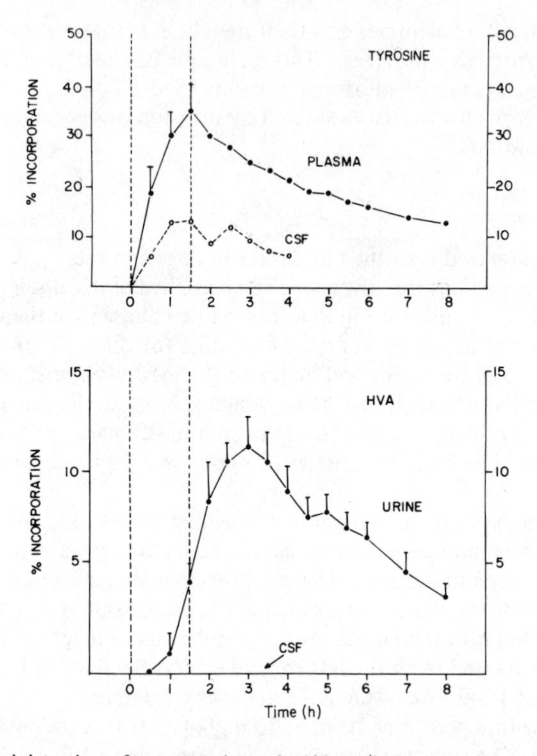

Figure 21.3 Administration of D,L-tyrosine-β-d_2 (40 mg/kg) by constant i.v. infusion to anaesthetised dogs (n = 5). The mole fraction of tyrosine-d_2 in plasma (●———●) and cerebrospinal fluid (○———○) is shown in the upper frame and the mole fraction of HVA-d_2 in urine is shown in the lower frame. No significant incorporation of deuterium could be demonstrated in cerebrospinal fluid HVA

ACKNOWLEDGEMENTS

These studies were supported by grants from the Swedish Medical Research Council (3872) and from Cilag Chemie Foundations. We are indebted to Berit Holmberg, Erik Magnusson and Hélène Leksell for technical assistance.

REFERENCES

Änggård, E., Lewander, T. and Sjöqvist, B. (1975). *Life Sci.*, **15**, 111

Costa, E. (1970). *Adv. Biochem. Psychopharmacol.*, **2**, 169

Costa, E. and Neff, N. E. (1966). *Biochemistry and Pharmacology of the Basal Ganglia* (ed. E. Costa, L. Côté and M. D. Yahr), Raven Press, New York, p. 141

Curtius, H. C., Baerlocher, K. and Völlmin, J. A. (1972b). *Clin. Chim. Acta*, **42**, 235

Curtius, H. C., Redeweik, U., Steinmann, B., Leimbacher, W. and Wegmann, H. (1975). *Proceedings of the Second International Conference on Stable Isotopes* (ed. E. R. Klein and P. D. Klein), Oak Brook, Illinois, USA, p. 385. National Technical Information Services Document CONF-751027

Curtius, H. C., Völlmin, J. A. and Baerlocher, K. (1972a). *Clin. Chim. Acta*, **37**, 277
Faull, K. F., Gan, I., Halpern, B. and Danks, D. M. (1975). Proceedings of the Second International Conference on Stable Isotopes (ed. E. R. Klein and P. D. Klein), Oak Brook, Illinois, USA, p. 392. National Technical Information Services Document CONF-751027
Lindström, B., Sjöqvist, B. and Änggård, E. (1974). *J. Labelled Comp.*, **10**, 187
Neff, N. H., Ngai, S. H., Wang, C. T. and Costa, E. (1969). *Mol. Pharmacol.*, **5**, 90
Sedvall, G. C., Weise, V. K. and Kopin, I. J. (1968). *J. Pharmacol. Exp. Ther.*, **159**, 274
Sjöqvist, B., Lindström, B. and Änggård, E. (1973). *Life Sci.*, **13**, 1655
Sjöqvist, B., Lindström, B. and Änggård, E. (1975). *J. Chromatogr.*, **105**, 309

22
The metabolism of deuterium-labelled *p*-tyramine in man

R. A. D. Jones, C. R. Lee and R. J. Pollitt (MRC Unit for Metabolic Studies in Psychiatry, University Department of Psychiatry, Middlewood Hospital, P.O. Box 134, Sheffield, S6 1TP, UK)

INTRODUCTION

In recent years interest in the metabolism of *p*-tyramine in man has been stimulated by its possible role in migraine (Hanington, 1967), in hepatic encephalopathy (Faraj *et al.*, 1976) and in the hypertensive crises experienced by some subjects taking monoamine oxidase inhibitors (Blackwell, 1963). It has also been reported that pressor response to infused tyramine is higher in endogenous depression (Ghose, Turner and Koppen, 1975). A number of studies using unlabelled tyramine (Horowitz *et al.*, 1964; Youdim *et al.*, 1971; Smith and Mitchell, 1974) and ^{14}C-labelled tyramine (Smith, March and Gordon, 1972; Boulton and Marjerrison, 1972; Tacker, Creaven and McIsaac, 1972) have been performed in man. They have revealed that the major pathways of metabolism are oxidation and conjugation and that a small proportion of the administered tyramine is excreted unchanged in the urine.

In animals, injected tyramine is taken up into adrenergic nerve endings where it acts as a substrate for dopamine-β-hydroxylase to form octopamine, the major end metabolite of which is *p*-hydroxymandelic acid (Figure 22.1). The conversion to octopamine has been shown many times *in vitro* and *in vivo*, examining tissue specimens. The overall pathway to *p*-hydroxymandelic acid has been demonstrated in rabbits using ^{14}C-labelled tyramine (Lemberger, Klutch and Kuntzman, 1966). A further possible minor pathway of tyramine metabolism involves ring hydroxylation to produce dopamine. This has been demonstrated *in vitro* using microsomal enzyme preparations from liver (Lemberger *et al.*, 1965; Axelrod, Inscoe and Daly, 1965), and using beef adrenal (Imaizumi *et al.*, 1958). *In vivo*, very small amounts of homovanillic and vanillylmandelic acid are formed from injected tyramine in rabbits (Lemberger *et al.*, 1966) and norepinephrine and normetanephrine were labelled in the urine of rats given injections of ^{14}C-tyramine (Creveling, Levitt and Udenfriend, 1962). None of these transformations has hitherto been demonstrated in man.

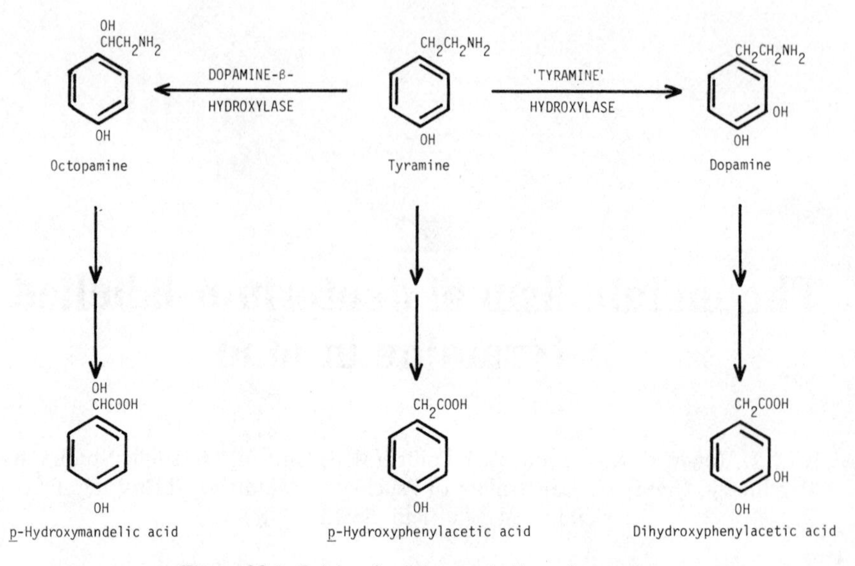

Figure 22.1 Origin of acidic metabolites of tyramine

As part of a projected study of the effects of monoamine oxidase status on the metabolism of tyramine in man we have been administering deuterium-labelled *p*-tyramine orally and intravenously to normal subjects. Although these studies are still at an early stage, the results are of considerable interest and demonstrate the superiority of stable isotopes over radioactive tracers for this type of work.

EXPERIMENTAL

Tyramine labelled with two or four deuterium atoms was prepared from unlabelled tyramine by heating in 5M ^{2}HCl in ^2H_2O at 100° for 6 h or at 170° for 7 days, respectively. After purification, the deuterated tyramine was administered, as the hydrochloride, to normal volunteers. Blood pressure was monitored at approximately 1 min intervals during intravenous infusion and also for the higher oral doses. On the day of the experiment the subjects were maintained on milk and water only until at least 6 h after the load, urine being collected hourly.

The unconjugated urinary acids were extracted using diethyl ether and converted to trimethylsilyl derivatives using standard methods. Gas chromatography–mass spectrometry (GC–MS) was carried out on a Perkin-Elmer 270 instrument which had been modified by the incorporation of a home-made multiple ion monitor (Lee, 1976). This recorded the peak heights of selected fragments obtained by repetitive scanning of a small section of the mass range using a commercial ramp generator to modulate the magnet control circuit. Packed columns of 3% OV-3, OV-17 or OV-225 on Gas Chrom Q, or OV-101 coated glass capillary columns (50 m long) were used for separations. The fragment at *m/e* 296 (natural compound) was used to monitor *p*-hydroxyphenylacetic acid, at *m/e* 267 for *p*-hydroxymandelic acid, at *m/e* 384 for 3,4-dihydroxyphenylacetic acid and at *m/e* 326 for homovanillic acid. For quantitation, deuterated internal standards ([$^2H_4$]-*p*-hydroxyphenylacetic acid, [$^2H_2$]-*p*-hydroxymandelic acid, [$^2H_5$]-3,4-dihydroxyphenylacetic acid and [^2H]-homovanillic acid) were added to the urine before extraction.

RESULTS

Using [$^2H_2$]-tyramine we have shown that in addition to the expected labelling of *p*-hydroxyphenylacetic acid there was also significant labelling of *p*-hydroxymandelic acid and 3,4-dihydroxyphenylacetic acid (Jones and Pollitt, 1976). This experiment has now been extended using [$^2H_4$]-tyramine and the greater separation of the fragments of the labelled product from the natural metabolite has enabled us to detect labelling in homovanillic acid in favourable cases (Figure 22.2). The proportion of these metabolites produced is dependent

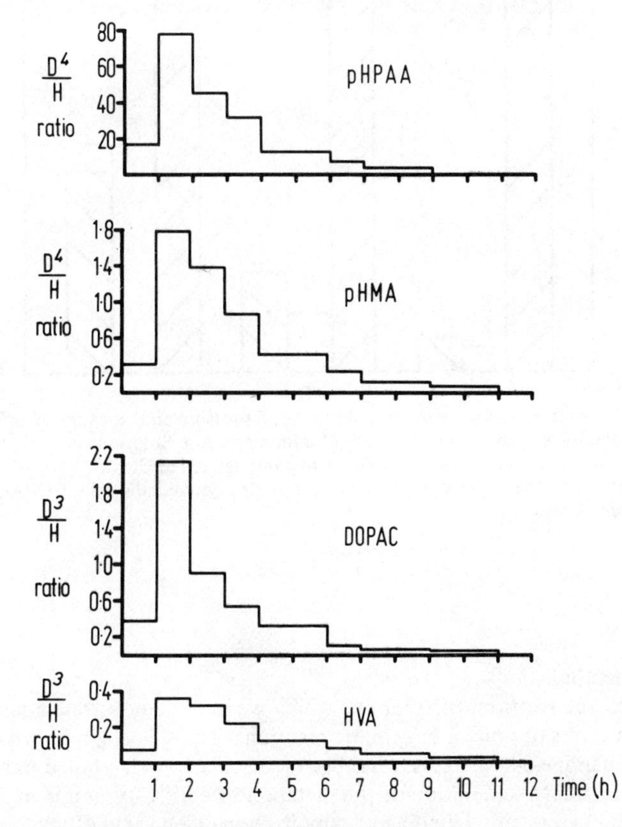

Figure 22.2 Excretion of labelled metabolites (molar ratios with natural metabolites) following an oral load of 277 mg of [$^2H_4$]-tyramine hydrochloride. pHPAA = *p*-hydroxyphenylacetic acid; pHMA = *p*-hydroxymandellic acid; DOPAC = 3,4-dihydroxyphenylacetic acid; and HVA = homovanillic (4-hydroxy-3-methoxyphenylacetic) acid

on the rate of administration, particularly in the case of 3,4-dihydroxyphenylacetic acid (Figure 22.3). This dose dependence also appears to apply to the oral route although the rate of absorption from an oral load, as assessed by the pressor response, is rather variable.

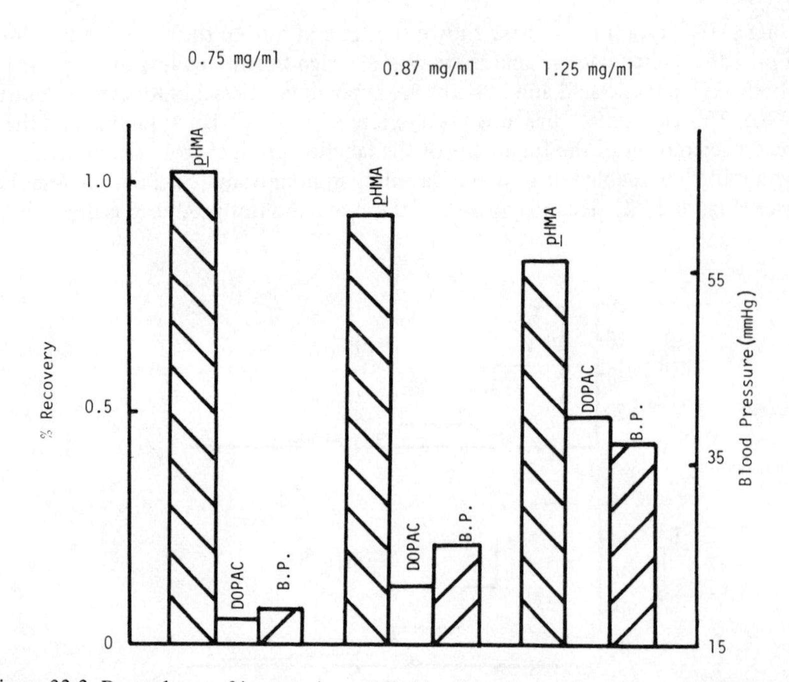

Figure 22.3 Dependence of increase in systolic blood pressure and recovery of infused tyramine as labelled metabolites on the rate of administration. Solutions of $[^2H_4]$-tyramine hydrochloride of the concentrations shown were administered at the rate of 1 ml/min on separate occasions to the same subject, pHMA = *p*-hydroxymandellic acid; DOPAC = 3,4-dihydroxyphenylacetic acid

DISCUSSION

Tyramine metabolism

The metabolic conversions of tyramine which we have demonstrated can be explained in terms of known enzymatic reactions. The side-chain hydroxylation involving dopamine-β-hydroxylase has been particularly well studied in animals, although a detailed examination of the metabolites of ^{14}C-tyramine in man failed to demonstrate any labelling of *p*-hydroxymandelic acid (Tacker *et al.*, 1972). The discrepancy between those results and ours is not easy to explain and it seems unlikely to be related to the smaller dose used in the ^{14}C-studies. However, the failure of Tacker and co-workers to demonstrate labelling in

3,4-dihydroxyphenylacetic acid could well be due to the sharp dependence of this reaction on the rate of administration. Detailed discussion of this rate dependence would be premature as it has so far been demonstrated only in one subject. Even at this stage, however, it appears that we cannot distinguish between a hepatic or an extrahepatic site of origin for these metabolites simply by comparing the results from the oral and intravenous routes.

Kinetic aspects

The excretion of the acidic metabolites of tyramine do not follow identical time-courses. Comparison of the excretion courses of labelled *p*-hydroxyphenylacetic acid and *p*-hydroxymandelic acid, both after intravenous administration (Jones and Pollitt, 1976) and, to a lesser extent, after oral administration (Figure 22.2) of labelled *p*-tyramine, reveals that the production of *p*-hydroxymandelic acid is significantly delayed. This is most easily explained in terms of storage of octopamine. The similar, though more marked, delay in the excretion of homovanillic acid (Figure 22.2) points to storage of its precursor, presumably dopamine, although it could result in part from slow methylation of 3,4-dihydroxyphenylacetic acid. It will be of interest to see if this phenomenon is also seen with intravenous administration.

An attempt to analyse the results of our earlier experiments by analogue computer simulation revealed a complex situation. A good fit to the observed *p*-hydroxyphenylacetic acid excretion could be obtained using a simple two-compartment system but it has not yet been possible to fit a model to the time course of the *p*-hydroxymandelic acid labelling. Presumably, allowance must be made for the release of newly synthesised octopamine by the incoming tyramine as the infusion proceeds but also it appears that at least two octopamine pools of differing turnover times may be necessary to account for the observed results.

Practical aspects

The urinary acid extract used in this work is a very complex mixture of exogenous and endogenous metabolites. Even with the dietary restrictions imposed in these experiments, interference from other components was common and varied from individual to individual. In some cases the interfering substance emerged from the gas chromatograph only a few seconds away from the compound whose isotopic composition was being determined and either concurrent scanning of the swept portion of the mass range using the ultraviolet recorder or very careful alignment of the multiple ion monitor traces was necessary to detect it. The interfering substance could usually be separated using a different gas chromatographic stationary phase or a high-resolution capillary column. The use of capillary columns can create its own problems, however, as the isotopically labelled compound is often completely separated from the corresponding natural compound which has a longer retention time. Under these circumstances the natural compound can no longer act as carrier for a smaller amount of deuterated analogue so that unless losses in the gas chromatograph and at the GC–MS interface are kept to a minimum, selective destruction of the labelled material occurs. It is also important to bear in mind that with high gas chromatographic resolution, a fragment containing ^{30}Si will not necessarily appear concurrently with a fragment of the same nominal mass containing two ^{13}C atoms or two ^{2}H atoms so that at low levels of deuterium labelling rather complex peak shapes, which need careful analysis, will be seen on the single ion recording.

Notwithstanding the difficulties just discussed, it still appears that for this type of work stable isotopes possess considerable practical advantages over radioactive tracers. In most instances in the present study it was relatively easy to obtain quantitative results for labelling. The methods are far less tedious than those required for radioactive tracers where purification of each urinary acid can be very time-consuming and specific activities are particularly difficult to determine accurately with the limited doses of tracer that can be administered to human subjects (see, for example, the study by Coppen *et al.*, 1974). Against this must be set the theoretical disadvantage of having to use large doses of precursor to achieve reasonable labelling levels in the metabolites. Obviously, for the present experiments where it is required to measure pressor response and where ring hydroxylation only becomes prominent at higher input rates, this drawback is not so significant.

ACKNOWLEDGEMENTS

We would like to thank Dr C. Paschalis for clinical supervision of the tyramine administration. R. A. D. Jones is an MRC Scholar.

REFERENCES

Axelrod, J., Inscoe, J. K. and Daly, J. (1965). *J. Pharm. Exp. Ther.*, **149**, 16
Blackwell, B. (1963). *Lancet*, **2**, 849
Boulton, A. A. and Marjerrison, G. L. (1972). *Nature*, **236**, 76
Coppen, A., Brooksbank, B. W. L., Eccleston, E., Peet, M. and White, S. G. (1974). *Psychol. Med.*, **4**, 164
Creveling, C. R., Levitt, M. and Udenfriend, S. (1962). *Life Sci.*, **10**, 523
Faraj, B. A., Bowen, P. A., Isaacs, J. W. and Rudman, D. (1976). *New Engl. J. Med.*, **294**, 1360
Ghose, K., Turner, P. and Coppen, A. (1975). *Lancet*, **1**, 1317
Hanington, E. (1967). *Brit. Med. J.*, **2**, 550
Horowitz, D., Lovenberg, W., Engleman, K. and Sjoerdsma, A. (1964). *J. Amer. Med. Assoc.*, **188**, 1108
Imaizumi, R., Yoshida, H., Hiramatsu, H. and Omori, K. (1958). *Jap. J. Pharmacol.*, **8**, 22
Jones, R. A. D. and Pollitt, R. J. (1976). *J. Pharm. Pharmacol.*, **28**, 461
Lee, C. R. (1976). *Biomed. Mass Spectrom.*, **3**, 48
Lemberger, L., Klutch, A. and Kuntzman, R. (1966). *J. Pharm. Exp. Ther.*, **153**, 183
Lemberger, L., Kuntzman, R., Conney, A. H. and Burns, J. J. (1965). *J. Pharm. Exp. Ther.*, **150**, 292
Smith, I., March, S. E. and Gordon, A. J. (1972). *Clin. Chim. Acta*, **40**, 415
Smith, I. and Mitchell, P. D. (1974). *Biochem. J.*, **142**, 189
Tacker, M., Creaven, P. J. and McIsaac, W. M. (1972). *J. Pharm. Pharmacol.*, **24**, 247
Youdim, M. B. H., Bonham-Carter, S., Sandler, M., Hanington, E. and Wilkinson, M. (1971). *Nature*, **230**, 127

23

Studies in man using semi-tracer doses of deuterated phenylalanine and tyrosine: implications for the investigation of phenylketonuria using the deuterated phenylalanine load test

J. A. Hoskins and R. J. Pollitt (MRC Unit for Metabolic Studies in Psychiatry, University Department of Psychiatry, Middlewood Hospital, P.O. Box 134, Sheffield, S6 1TP, UK)

INTRODUCTION

In classical phenylketonuria there is a virtual absence of phenylalanine hydroxylase activity in the liver, the organ usually responsible for catabolism of excessive dietary phenylalanine. The accumulation of phenylalanine, and possibly some of its metabolites, results in severe mental retardation. Consequently, great effort is put into detecting the disorder in the first few weeks of life so that it can be treated by restriction of phenylalanine intake. Unfortunately, it is sometimes difficult to classify cases detected by screening in the newborn period and there appears to be an almost continuous gradation from the classical phenylketonuric to the normal individual. Attempts to differentiate sub-groups of the disease by estimating phenylalanine hydroxylase in liver biopsy specimens have been made at various centres but the situation still remains confused. The normal phenylalanine loading tests using oral doses of 100–200 mg/kg body weight are difficult to interpret and, although the results of intravenous loading tests are more amenable to rigorous mathematical treatment (Woolf, Cranston and Goodwin, 1967), this has not proved to be an acceptable procedure for routine use with the newborn child.

Deuterium-labelled phenylalanine has been employed in an attempt to obtain more information from loading tests. This approach was pioneered by Curtius, Völlmin and Baerlocher (1972a,b) and is being followed in other centres. Curtius and co-workers found that after a load of deuterium-labelled phenylalanine, deuterium was present in urinary acids derived from tyrosine in a normal subject but not in a case of typical phenylketonuria or a variant form with only a modest elevation of plasma phenylalanine. This result was interpreted as showing that the atypical patient had a complete block in the conversion of phenylalanine to tyrosine and that possibly another pathway of phenylalanine degradation was available to this patient but not to the classical phenylketonuric. This original study had several unsatisfactory features. The phenylalanine used was labelled rather non-specifically (27% $^2H_2$) making labelling patterns difficult to interpret. Deuterated D,L-phenylalanine (200 mg/kg) was used and this is open to the objection that the unnatural D-isomer could be metabolised in an unphysiological manner. These objections were recognised and discussed at the time (Curtius *et al.*, 1972b). Another, potentially more serious, objection is that this approach makes no allowance for possible metabolic compartmentation. It assumes that the labelling of *p*-hydroxylated metabolites in urine accurately reflects the degree of labelling of tyrosine in the liver. The results presented in this communication show that this assumption is not valid and that the readily measured urinary *p*-hydroxy acids (Figure 23.1) are derived from several anatomically distinct compartments.

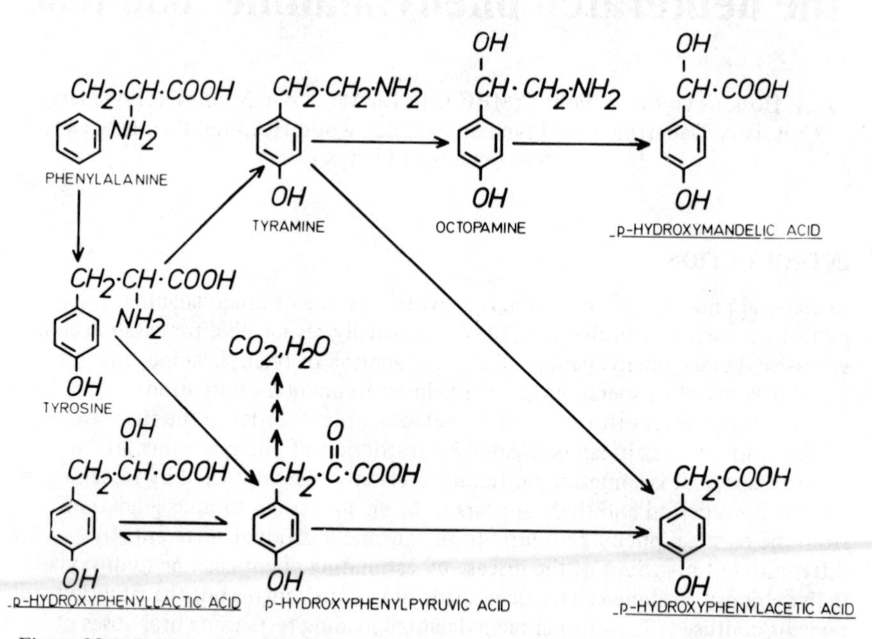

Figure 23.1 Origins of the *p*-hydroxylated metabolites studied. The compounds examined in urine are underlined

EXPERIMENTAL

Phenylalanine was deuterated in the ring by the method of Bu'Lock and Ryles (1970) using two cycles of deuterium exchange. The material used was at least 95 per cent $^2H_5$-labelled and contained less than 0.1 per cent of the D-isomer.

L-[$^2H_2$]-Tyrosine was synthesised by the method of Kirby and Ogunkoya (1965). The deuterium-labelled compounds were administered orally to normal subjects who were maintained on a milk and water diet throughout the experimental period. Venous blood samples were taken from the arm and the amino acids were extracted from the heparinised plasma by ion-exchange chromatography and converted to their trimethylsilyl derivatives. The extraction of the urinary organic acids and their analysis by gas chromatography–mass spectrometry was performed as described previously (Jones, Lee and Pollitt, 1977). For the determination of the isotope content of the amino acids phenylalanine and tyrosine and of the organic acids derived from them, ring-containing fragment ions of the hydrogen form and the corresponding deuterium form were compared. The actual ions measured and their origins are shown in Table 23.1.

Table 23.1 Fragments used to determine degree of labelling

	Ion measured (*m/e*) H form	Origin
Tyrosine	179	M–CH(NHTMS)COOTMS
Phenylalanine	192	M–COOTMS
p-Hydroxymandelic acid	267	M–COOTMS
p-Hydroxyphenyllactic acid	179	M–CH_2CH(OTMS)COOTMS
	308	M–TMSOH
p-Hydroxyphenylacetic acid	296	Molecular ion
m-Hydroxyphenylacetic acid	296	Molecular ion
o-Hydroxyphenylacetic acid	296	Molecular ion
	253	M–(CH_3 + CO)

RESULTS

The typical time-course of labelling of plasma tyrosine and of urinary *p*-hydroxyphenyllactic, *p*-hydroxymandelic and *p*-hydroxyphenylacetic acids after an oral load of L-[$^2H_2$]-tyrosine (12.5 mg/kg) is shown in Figure 23.2. Results from another subject are shown in Figure 23.3. Figure 23.4 shows the results from a subject following an oral load of 50 mg L-[$^2H_2$]-tyrosine/kg body weight. Figure 23.5 demonstrates the labelling pattern of selected metabolites in an individual taking an oral load of L-[$^2H_5$]-phenylalanine (25 mg/kg) and L-[$^2H_2$]-tyrosine (12.5 mg/kg) simultaneously. Details of the labelling of *o*- and *m*-hydroxyphenylacetic acid in this experiment are shown in Figure 23.6.

DISCUSSION

The labelling curves obtained after oral [$^2H_2$]-tyrosine loads (12.5 mg/kg) are rather varied. In some cases the time courses of labelling of plasma tyrosine and of the urinary acids approximate to those expected for a precursor–product relationship, with the descending portion of the tyrosine curve cutting the acid

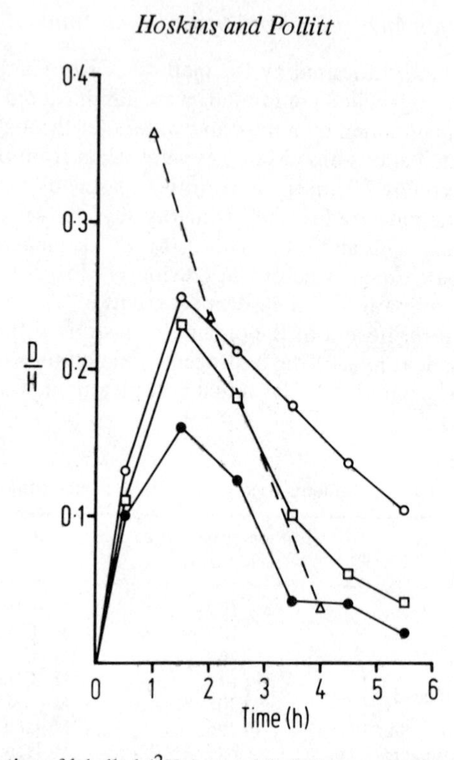

Figure 23.2 Molar ratios of labelled ($^2H_2$) to unlabelled compounds after ingestion of L-[$^2H_2$]-tyrosine (12.5 mg/kg) by a normal subject (subject A). Plasma tyrosine, △; urinary *p*-hydroxyphenyllactic acid, ○; urinary *p*-hydroxymandelic acid, □; urinary *p*-hydroxyphenylacetic acid, ●

metabolites curves at their highest points. However, it is usually necessary to invoke two other sources from the urinary acids to explain the results (Figure 23.2). One of these sources is in the enterohepatic system and shows a very strong 'first pass' effect, producing a high degree of labelling soon after the load. This is particularly prominent for the *p*-hydroxymandelic acid labelling shown in Figure 23.3. The other source consists of pools which are not readily accessible to plasma tyrosine and which consequently reduce the degree of labelling observed. This is shown especially by the labelling of *p*-hydroxyphenylacetic acid and is most easily explained by bacterial production of acidic metabolites in the bowel. The increased labelling of this acid found several hours after the ingestion of a higher (50 mg/kg) dose of deuterated tyrosine (Figure 23.4) can be ascribed to the action of gut bacteria on unabsorbed amino acid.

The large contribution of peripheral (non-enterohepatic) tyrosine pools to the urinary *p*-hydroxylated acids is to be expected. For *p*-hydroxyphenyllactic acid it may be predicted from the distribution of tyrosine aminotransferase and *p*-hydroxyphenylpyruvate oxidase (Fellman *et al.*, 1972). The general distribution of aromatic amino acid decarboxylase is reflected in the peripheral contribution to *p*-hydroxyphenylacetic acid labelling. Similarly, *p*-hydroxymandelic acid is the major end-metabolite of octopamine, which is formed by the action of dopamine-β-hydroxylase on tyramine in the

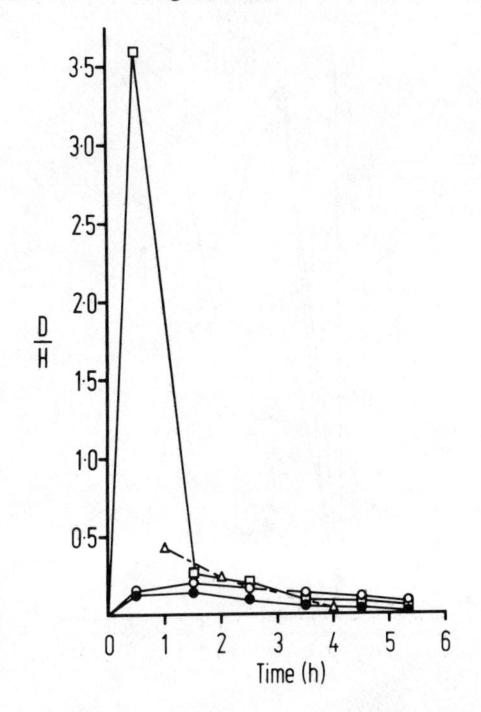

Figure 23.3 Molar ratios of labelled ($^2H_2$) to unlabelled compounds after ingestion of L-[$^2H_2$]-tyrosine (12.5 mg/kg) by an apparently normal subject. Plasma tyrosine, △; urinary *p*-hydroxyphenyllactic acid, ○; urinary *p*-hydroxymandelic acid, □; urinary *p*-hydroxyphenylacetic acid, ●

noradrenergic system. The very high degree of *p*-hydroxymandelic acid labelling shown in Figure 23.3 demonstrates that on occasion the enterohepatic system can produce large amounts of this metabolite; the excretion of unlabelled *p*-hydroxymandelic acid was essentially unchanged. It is of interest that this very high degree of labelling was achieved without much affecting the labelling pattern of *p*-hydroxyphenylacetic acid – a major metabolite of the postulated intermediate, tyramine. Somewhat analogous results have been obtained by perfusing rat liver with phenylalanine, when at high phenylalanine levels mandelic acid production outstrips that of phenylacetic acid (Blau *et al.*, 1976). The high level of *p*-hydroxymandelic acid labelling is also achieved without excessive labelling of urinary 3,4-dihydroxyphenylacetic acid, a somewhat unexpected finding in view of the labelling patterns obtained with oral tyramine (Jones *et al.*, 1977).

The easily measured urinary acids derived from tyrosine originate from several sites, the liver often representing only a minor source. This fact, and the variability of the metabolite labelling patterns found, may make it impossible to derive information about the labelling of the hepatic tyrosine pool(s) from these patterns. In an experiment where L-[$^2H_5$]-phenylalanine (25 mg/kg) was taken simultaneously with an 'internal standard' of L-[$^2H_2$]-tyrosine (12.5 mg/kg), the ratio of $^2H_2$ to $^2H_4$ labelling was different for different acids (Figure 23.5).

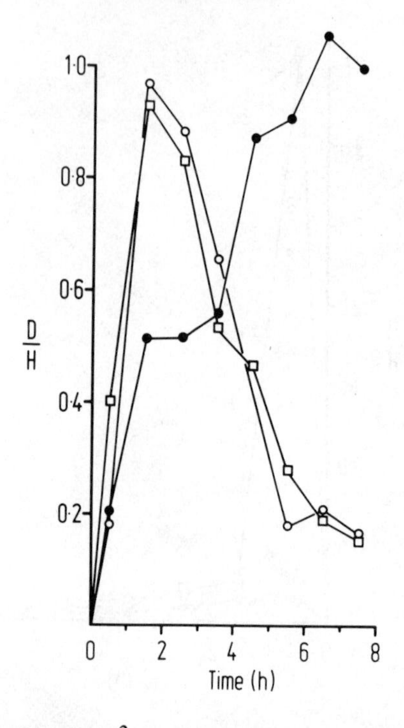

Figure 23.4 Molar ratios of labelled ($^2H_2$) to unlabelled urinary metabolites after ingestion of L-[$^2H_2$]-tyrosine (50 mg/kg) by a normal subject. *p*-Hydroxyphenyllactic acid, ○; *p*-hydroxymandelic acid, □; *p*-hydroxyphenylacetic acid, ●

The labelling of plasma tyrosine itself may be a more reliable guide; we do not have sufficient results to generalise. However, it appears unlikely that liver and plasma tyrosine are in complete equilibrium and it is possible to envisage that under some circumstances only a small proportion of the tyrosine being produced by the action of phenylalanine hydroxylase may find its way into the general circulation, the majority being transaminated and oxidised via homogentisate. Such a phenomenon might explain the unexpected results obtained by Curtius and co-workers (1972a and b) in their atypical case of phenylketonuria, results since confirmed by other workers (S. K. Wadman, personal communication, 1976). This explanation would involve a more extreme degree of metabolic compartmentation than is seen in normals though the relatively slow production of tyrosine by the residual phenylalanine hydroxylase believed to be active in atypical (mild) phenylketonuria variants may be a factor. However, the situation is still further complicated by the increased excretion of *p*-hydroxyphenyllactic and *p*-hydroxyphenylpyruvic acids which is seen in patients with high plasma phenylalanine levels (Chalmers and Watts, 1974) and is thought to be due to inhibition of *p*-hydroxyphenylpyruvate oxidase by phenylpyruvic acid. If, as seems likely, these excessive metabolites originate in the liver then it is hard to ascribe the absence of labelling under load conditions to the metabolic compartmentation we have observed using semi-tracer doses.

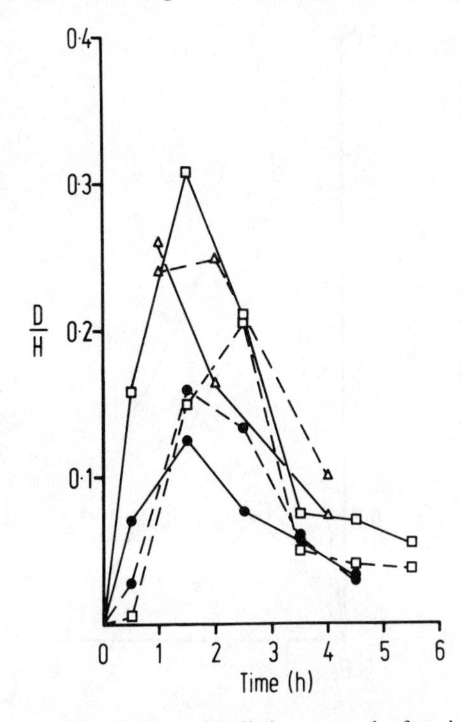

Figure 23.5 Molar ratios of labelled to unlabelled compounds after simultaneous ingestion of L-[$^2H_2$]-tyrosine (12.5 mg/kg) and L-[$^2H_5$]-phenylalanine (25 mg/kg) by subject A. Plasma [$^2H_2$]-tyrosine, △– – – –△; plasma [$^2H_4$]-tyrosine, △———△; urinary [$^2H_2$]-*p*-hydroxymandelic acid, □– – – –□; urinary [$^2H_4$]-*p*-hydroxymandelic acid, □———□; urinary [$^2H_2$]-*p*-hydroxyphenylacetic acid, ●– – – –●; urinary [$^2H_4$]-*p*-hydroxyphenylacetic acid, ●———●

A metabolite which merits further investigation in the context of phenylketonuria is *m*-hydroxyphenylacetic acid. Curtius and co-workers noted that this metabolite was labelled in a normal but not in a phenylketonuric subject after a deuterated phenylalanine load. The *m*-hydroxyphenylacetic acid can arise in two ways, either by dehydroxylation of catechol derivatives such as dopamine (Boulton and Dyck, 1974) or from *m*-tyrosine formed as a minor product of the direct action of phenylalanine hydroxylase (Coulson, Henson and Jepson, 1968) or tyrosine hydroxylase (Tong, D'Iorio and Benoiton, 1971). In the former case the product from L-[$^2H_5$]-phenylalanine would contain three deuterium atoms, in the latter, four. Our results from oral loads of L-[$^2H_5$]-phenylalanine (25 mg/kg) show that the major product is $^2H_4$-labelled and the time-course (Figure 23.6), which is similar to that for the *ortho* isomer, is consistent with production in the liver (Fellman *et al.*, 1972). We have confirmed Curtius' result that detectable labelling does not occur in a case of typical phenylketonuria. Little is known of the metabolism of *m*-tyrosine in man but if *meta*-hydroxylation should lead to a dead-end pathway these metabolites may provide a better indication of phenylalanine hydroxylase activity than the *p*-hydroxylated ones.

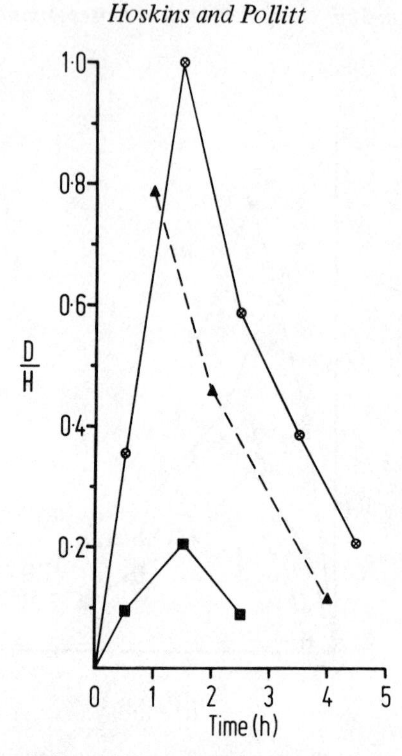

Figure 23.6 Molar ratios of phenylalanine and metabolites in the experiment illustrated in Figure 23.5. Plasma [$^2H_5$]-phenylalanine, ▲; urinary [$^2H_4$]-*o*-hydroxyphenylacetic acid, ⊕; urinary [$^2H_4$]-*m*-hydroxyphenylacetic acid, ■

ACKNOWLEDGEMENTS

We would like to thank Miss V. Fell for the synthesis of the labelled compounds and for other help and Drs C. Paschalis and V. H. Lyons for clinical assistance.

REFERENCES

Blau, K., Goodwin, B., Woods, H. F. and Youdim, M. B. H. (1976). *Brit. J. Pharmacol.*, **58**, 474P

Boulton, A. A. and Dyck, L. E. (1974). *Life Sci.*, **14**, 2497

Bu'Lock, J. D. and Ryles, A. P. (1970). *Chem. Commun.*, 1404

Chalmers, R. A. and Watts, R. W. E. (1974). *Clin. Chim. Acta*, **55**, 281

Coulson, W. F., Henson, G. and Jepson, J. B. (1968). *Biochim. Biophys. Acta*, **156**, 135

Curtius, H.-Ch., Völlmin, J. A. and Baerlocher, K. (1972a). *Clin. Chim. Acta*, **37**, 277

Curtius, H.-Ch., Völlmin, J. A. and Baerlocher, K. (1972b). *Organic Acidurias, S.S.I.E.M. Symposium 9* (ed. J. Stern and C. Toothill), Churchill Livingstone, Edinburgh, p. 146

Fellman, J. H., Buist, N. R. M., Kennaway, N. G. and Swanson, R. E. (1972). *Clin. Chim. Acta*, **39**, 243

Jones, R. A. D., Lee, C. R. and Pollitt, R. J. (1977). This symposium.

Kirby, G. W. and Ogunkoya, J. (1965). *J. Chem. Soc.*, 6914

Tong, J. H., D'Iorio, A. and Benoiton, N. L. (1971). *Biochem. Biophys. Res. Commun.*, **44**, 229

Woolf, L. I., Cranston, W. I. and Goodwin, B. L. (1967). *Nature*, **213**, 882

24
Application of chemical ionisation mass spectrometry and stable isotopes in studies of α-methyldopa metabolism

N. Castagnoli, K. L. Melmon, C. R. Freed, M. M. Ames, A. Kalir and R. Weinkam
(Department of Pharmaceutical Chemistry and Division of Clinical Pharmacology, University of California, San Francisco, California, 94143, USA)

Following the definitive clinical studies of Oates and co-workers (Oates *et al.*, 1960) on the antihypertensive properties of α-methyldopa (I), a number of investigators have attempted to establish the mechanistic features underlying the pharmacological properties of this clinically useful drug. The most widely accepted theory was advanced a few years ago by Henning (1968), who proposed that α-methyldopa is transported into the brain, where it undergoes metabolism to α-methyldopamine (II) which is subsequently converted to α-methylnorepinephrine (III). The α-methylnorepinephrine and/or α-methyldopamine displace the endogenous central catecholamines dopamine (IV) and norepinephrine (V) and act as 'false neurotransmitters', leading to the attenuation of sympathetic outflow from the central nervous system and to a decrease in blood pressure. A number of key experiments have contributed to the development of this theory and include the observations that the antihypertensive properties of α-methyldopa are associated exclusively with the S-enantiomer (S-I) (Gillespie *et al.*, 1962) and that peripheral decarboxylase inhibitors do not block the antihypertensive effects of the drug whereas central decarboxylase inhibitors do (Jaju, Tangri and Bhargava, 1966).

The analytical techniques employed in assessing the various parameters associated with the interactions of α-methyldopa and its metabolites with the brain and peripheral endogenous catecholamines have relied extensively on spectrophotofluorometric assays (Bogdanski *et al.*, 1956). While these approaches have proved extremely useful, it is likely that limitations in specificity would hamper attempts to quantitatively estimate all four amines (II–V) in complex biological matrices such as the brain. The sensitivity and specificity which can be achieved by mass spectrometric analyses employing stable-isotope-labelled internal standards have provided useful information on

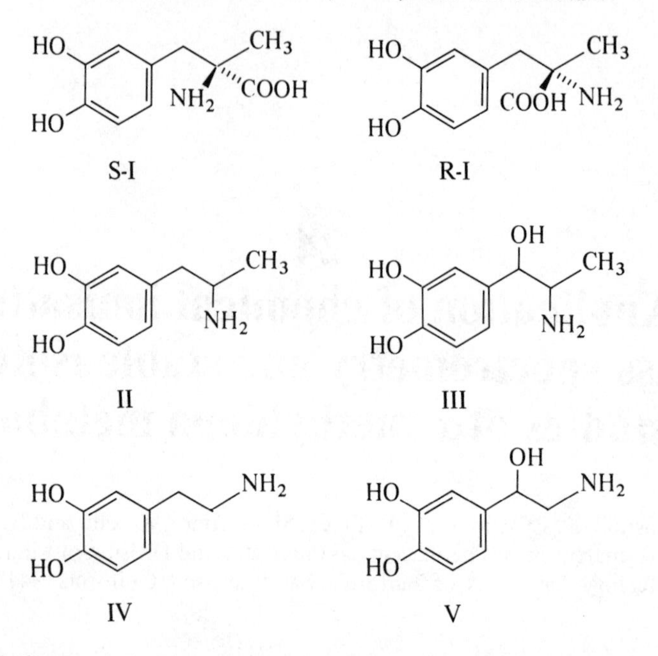

the metabolism and disposition of small molecules in biological systems (Koslow, Cattabeni and Costa, 1962). We have been interested in developing techniques utilising stable isotopes and chemical ionisation mass spectrometry in the area of α-methyldopa metabolism with the hope of obtaining definitive structural and quantitative information relating to the mode of action of this drug in the brain. This presentation summarises our studies in this area and includes descriptions of (1) the syntheses of (S)-α-methyldopa-^{13}C and deuterium-labelled II–V; (2) the stereochemical specificities associated with the brain metabolism of α-methyldopa; and (3) the interactions of metabolically generated α-methylated catecholamines with endogenous catecholamines in the presence and absence of the peripheral decarboxylase inhibitor MK 486 (Henning, 1969). The discussion which follows is divided into four areas: (1) synthetic studies, (2) analytical techniques, (3) stereochemical studies and (4) dopa/α-methyldopa interaction studies.

SYNTHETIC STUDIES

Synthesis of (S)-α-methyldopa-^{13}C (S-I-^{13}C)

The synthetic sequence leading to the ^{13}C-enriched α-methyldopa is summarised in Figure 24.1. The basic requirement was to introduce the ^{13}C-label at the metabolically stable benzylic position. This was readily accomplished by lithiation of the bromo compound VI, followed by carbonation of the lithio derivative VII to form 3,4-dibenzyloxybenzoic acid-^{13}C (VIII-^{13}C). Attempted two-electron reduction of the carboxylic acid group under various conditions was unsuccessful. However, the required aldehyde X-^{13}C could be obtained in high yield by first reducing the carboxyl function with $LiAlH_4$ followed by the

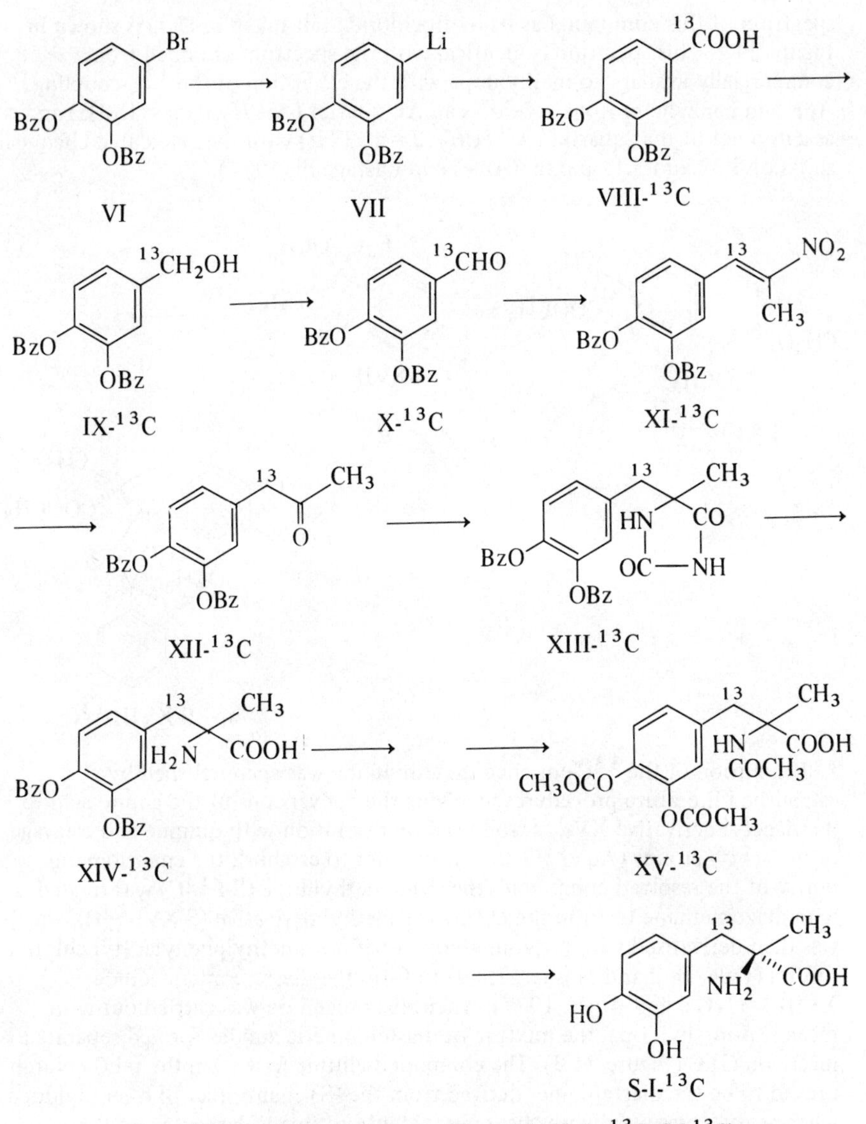

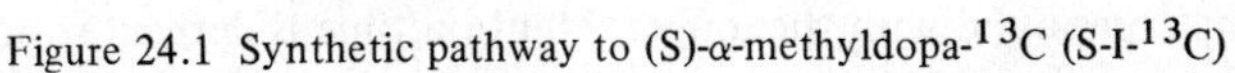

Figure 24.1 Synthetic pathway to (S)-α-methyldopa-^{13}C (S-I-^{13}C)

selective two-electron oxidation of the resulting carbinol with chromium trioxide immobilised on graphite (Lehachette, Rollin and Dumas, 1972). Condensation of 3,4-dibenzyloxybenzaldehyde-^{13}C with nitroethane gave the 1-phenyl-2-nitropropene XI-^{13}C, which was converted to the phenyl-2-propanone XII-^{13}C with Fe/acetic acid (Sadeh, Pelah and Kalir, 1967). Formation of the hydantoin XIII-^{13}C, followed by base hydrolysis to a protected amino acid XIV-^{13}C and acid cleavage of the benzyl ether groups, gave the desired ^{13}C-enriched α-methyldopa (89 per cent ^{13}C). The 60 MHz PMR spectrum of this compound as its hydrochloride salt taken in D_2O is shown in Figure 24.2. This spectrum is identical with the spectrum obtained from commercially available α-methyldopa with the exception of the ^{13}C-coupling. The two benzylic protons appear as an AB quartet ($ArCH_2, J_{AB} = 15$ Hz) and as a doublet of that quartet ($Ar^{13}CH_2, J = 133$ Hz) with the calculated chemical shifts of 3.32 and 3.12 p.p.m. (Ames and Castagnoli, 1974).

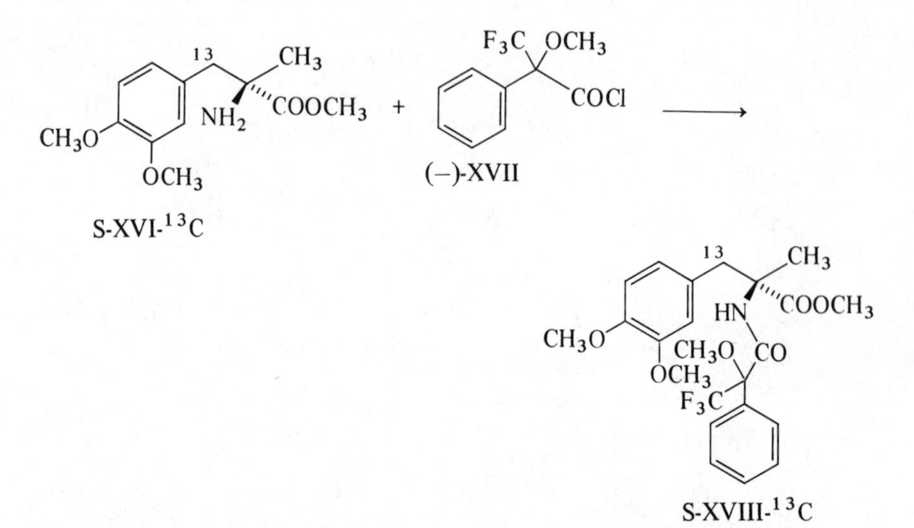

Resolution of the ^{13}C-enriched α-methyldopa was accomplished by established literature procedures involving the conversions of the amino acid to its triacetyl derivative XV-^{13}C followed by resolution with quinine and cleavage of the acetyl groups (Au *et al.*, 1972). In order to establish the enantiomeric purity of the resolved compound, the (S)-α-methyldopa (S-I-^{13}C) was treated with diazomethane to form the *O,O,O*-tris-methyl derivative (S-XVI-^{13}C), which was then derivatised with (–)-α-methoxy-α-trifluoromethylphenylacetyl chloride (XVII) (Dale, Dull and Mosher, 1969) to form the diastereomeric amide XVIII-^{13}C (Gal and Ames, 1977). When this procedure was carried out with racemic α-methyldopa, the mixture of diastereomeric amides formed separated nicely on GLC (Figure 24.3). The compound eluting first from the GLC column proved to be the diastereomer derived from the (S)-enantiomer of α-methyldopa whereas the compound with the longer retention time is derived from the (R)-enantiomer. The enantiomeric purity of the ^{13}C-enriched α-methyldopa is illustrated in Figure 24.4, which shows that the compound is essentially free of any of the (R)-isomer.

Synthesis of deuterium-labelled dopamine and α-methyldopamine

Side-chain labelled dopamine-d_2 (IV-d_2) and α-methyldopamine-d_2 (II-d_2) were readily obtained by $LiAlD_4$ reduction of the corresponding phenylnitroalkene dimethyl ethers XIX and XX followed by HBr cleavage of the methyl ethers

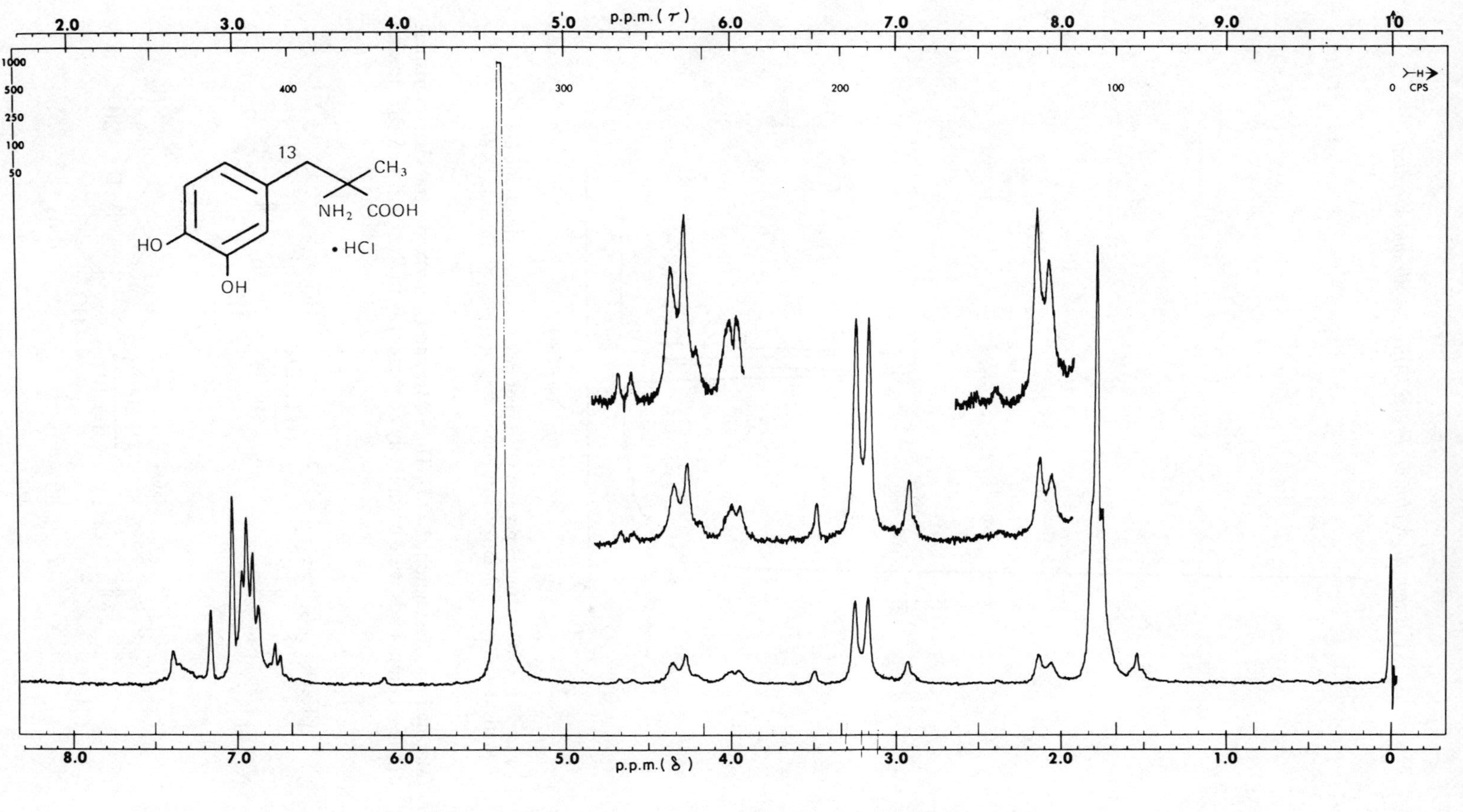

Figure 24.2 NMR spectrum (D_2O) of racemic α-methyldopa-^{13}C hydrochloride

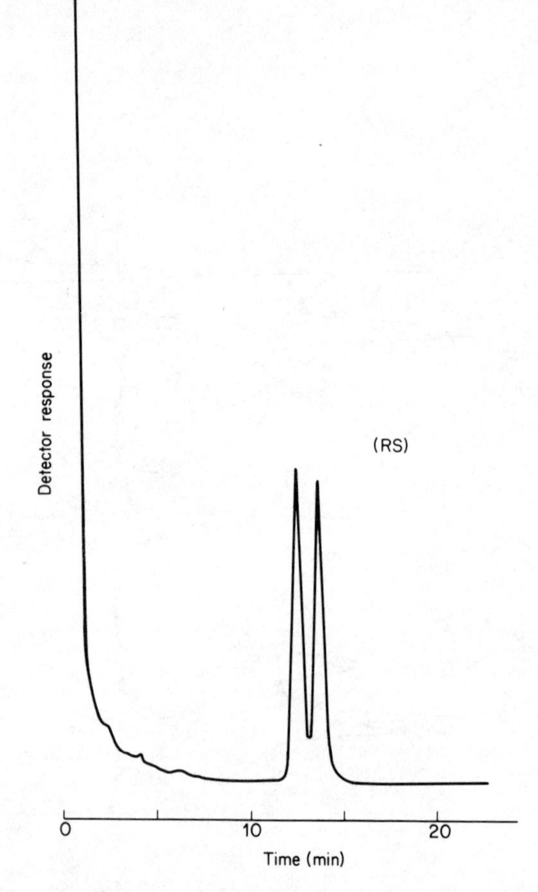

Figure 24.3 GLC tracing (3 per cent OV-17) of racemic α-methyldopa dimethylether derivatised with (–)-α-methoxy-α-trifluoromethylphenylacetyl chloride (XVII)

(Marshall and Castagnoli, 1973). The NMR and CI mass spectral characteristics of these compounds were completely consistent with their assigned structures.

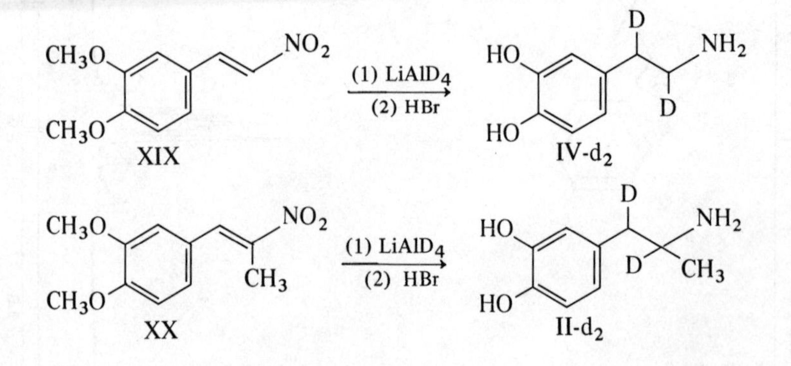

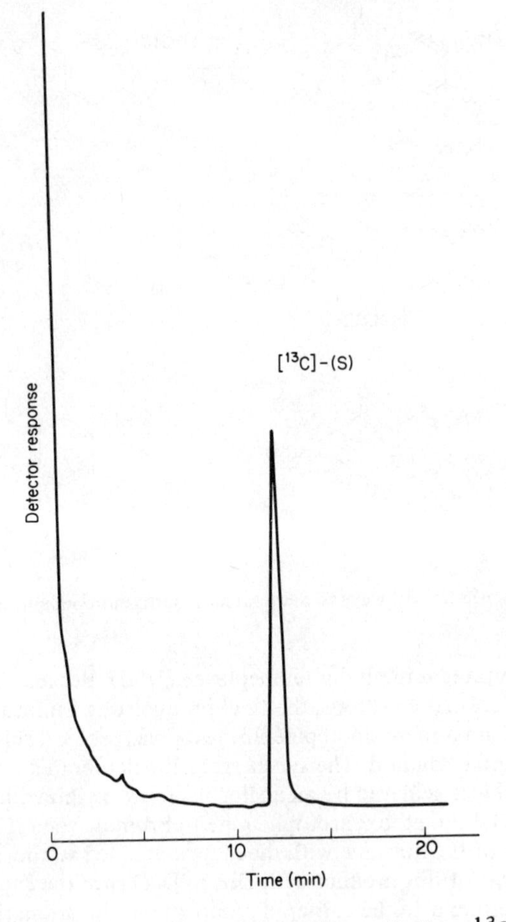

Figure 24.4 GLC tracing (3 per cent OV-17) of the (S)-α-methyldopa-^{13}C amide derivative (XVIII-^{13}C)

Syntheses of deuterium-labelled *erythro*-α-methylnorepinephrine and norepinephrine

The basic synthetic approach to the preparation of the deuterated alkanolamines started from 3,4-dimethoxyacetophenone (XXI). In the case of α-methylnorepinephrine, the acetophenone was selectively methylated with methyl-d_3 iodide to form the corresponding propiophenone-d_3 (XXII-d_3) as shown in Figure 24.5. Subsequent manipulations of the XXII-d_3 involved bromination to the α-bromoketone XXIII-d_3, followed by displacement of the bromo group with dibenzylamine to give XXIV-d_3. Treatment of this product with DBr in D_2O led to the cleavage of the methyl ether groups and introduction of deuterium both α to the carbonyl group and also at the aromatic positions of the catecholaminoketone XXV-d. Catalytic hydrogenation/hydrogenolysis gave

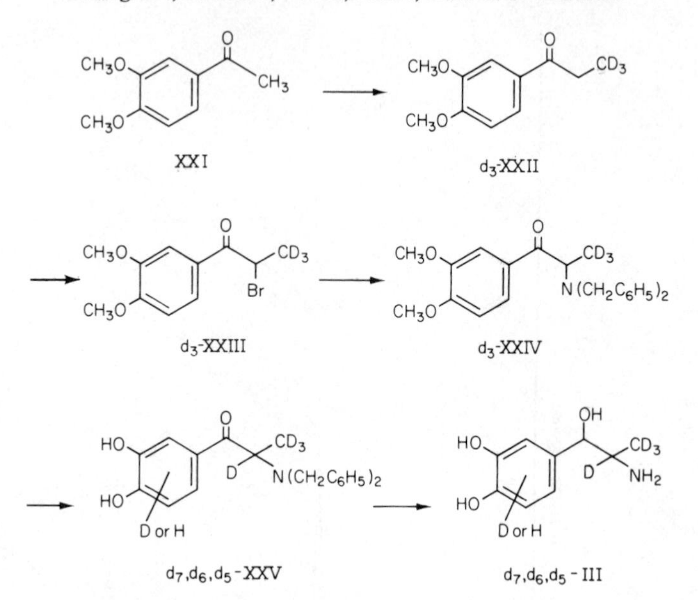

Figure 24.5 Synthetic pathways to deuterated α-methylnorepinephrine (III-d_7,d_6,d_5)

the desired deuterated α-methylnorepinephrine (III-d). Because of incomplete exchange of the aromatic protons, the final product was a mixture of d_7, d_6 and d_5 species. Since no α-methylnorepinephrine-d_0 was present, this product was a satisfactory internal standard. The synthesis of the deuterated norepinephrine internal standard was achieved by an analogous route as shown in Figure 24.6. Bromination of 3,4-dimethoxyacetophenone to bromoketone XXVI, followed by displacement of the bormine with dibenzylamine, led to the aminoketone XXVII. Treatment of this product with DBr in D_2O gave the catechol XXVIII-d containing deuterium α to the carbonyl group and in the aromatic ring (partial exchange). Reduction of XXVIII-d with deuterium gas provided the deuterated norepinephrine as a mixture of d_6 and d_5 species (V-d_6,d_5). The NMR spectra of these products confirmed the partial deuterium incorporation in the aromatic ring and the essentially complete deuterium content of the side chain positions (Kalir *et al.*, 1977).

ANALYTICAL PROCEDURES

Our studies on the identification and quantitative estimation of the various catecholamines in rat brain utilised chemical ionisation mass spectrometric analyses. The direct insertion probe of a modified AEI MS902 mass spectrometer was employed with isobutane (0.5–1.0 Torr) as the reagent gas and at probe temperatures ranging between 170 and 290°.

Our initial efforts to examine the underivatised catecholamines and amino acids confirmed our suspicions that derivatisation of these substances would be

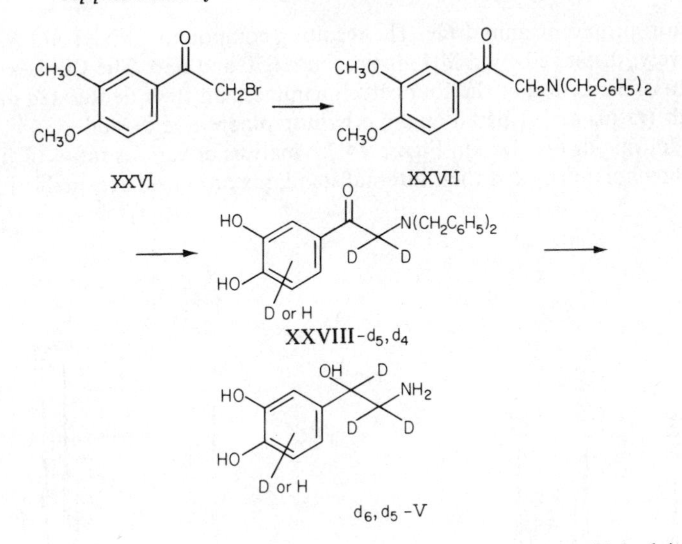

Figure 24.6 Synthetic pathway to deuterated norepinephrine (V-d_6,d_5)

necessary if strong protonated molecular ions (MH^+) were to be obtained. Derivatisation of α-methyldopamine and dopamine with pentafluoropropionic anhydride provided the tris-pentafluoroproponyl (PFP) derivatives XXIX and XXX, respectively, which gave abundant protonated molecular (MH^+) ions at *m/e* 606 and 592. α-Methylnorepinephrine and norepinephrine were converted to their corresponding β-ethyl ethers, prior to derivatisation with

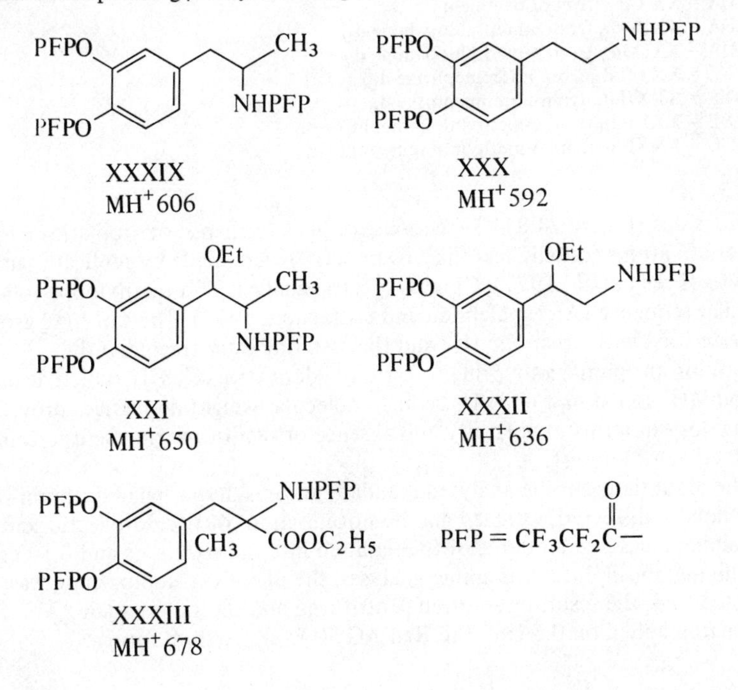

pentafluoropropionic anhydride. The resulting compounds, XXXI and XXXII respectively, displayed good MH^+ ions at *m/e* 650 and 636. The CI mass spectrum of a mixture of the four catecholamines and their deuterated internal standards (ratios of 1:1 based on α-methylnorepinephrine-d_7 and norepinephrine-d_6) is given in Figure 24.7. Analysis of various ratios of the native catecholamines and their internal standards produced acceptable linear

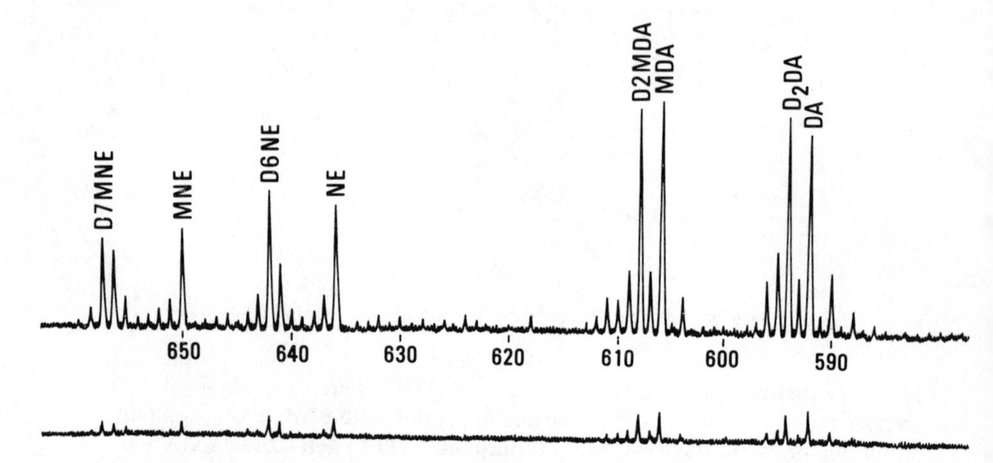

Figure 24.7 Chemical ionisation mass spectrum of derivatised catecholamines. See accompanying structures.
DA = XXX-d_0 from dopamine-d_0.
D_2DA = XXX-d_2 from dopamine-d_2.
MDA = XXIX-d_0 from α-methyldopamine-d_0.
D2MDA = XXIX-d_2 from α-methyldopamine-d_2.
NE = XXXII-d_0 from norepinephrine-d_0.
D6NE = XXXII-d_6 from norepinephrine-d_6.
MNE = XXXI-d_0 from α-methylnorepinephrine-d_0.
D7MNE = XXXI-d_7 from α-methylnorepinephrine-d_7

relationships (Figure 24.8). Corrections for back-exchange of aromatic deuterium atoms (usually less than 10 per cent) were made by replicate sample analysis (Freed *et al.*, 1977). CI mass spectral analysis of α-methyldopa followed a similar sequence (Ames, Melmon and Castagnoli, 1977). The carboxyl group was esterified with ethanolic HCl and the resulting ester treated with pentafluoropropionic anhydride to form the derivative XXXIII, which displayed a good MH^+ ion at *m/e* 678. These high molecular weight derivatives proved satisfactory in terms of volatility and absence of significant fragment ions in the mass region of interest.

The brain tissues to be analysed (caudate nucleus, hypothalamus, brain stem) were rapidly dissected, weighed and homogenised in 5% trichloroacetic acid containing measured amounts of the required internal standards and 0.5 per cent sodium metabisulphite. For amine analyses, the pH of the homogenate was adjusted to 4, the resulting solution centrifuged and the supernatant chromatographed on 0.5 cm^3 Bio Rad AG 50 W-x8 (sodium form).

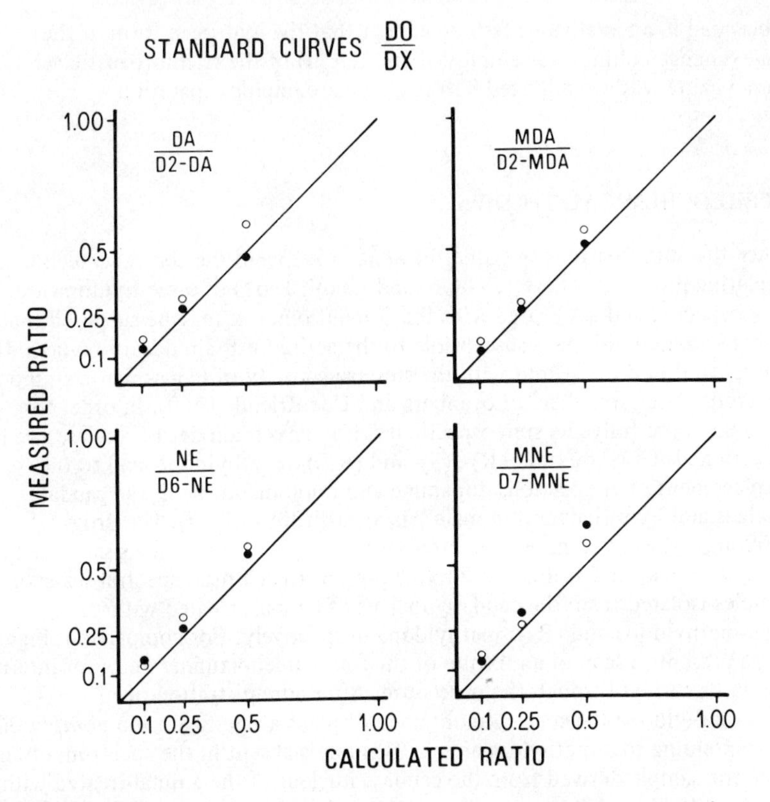

Figure 24.8 Standard curves of native to deuterated catecholamines. Two nmol of deuterium-labelled compound was combined with 0.2, 0.5 and 1 nmol of native catecholamine. Duplicates are shown.

DA = dopamine-d_0.
D2-DA = dopamine-d_2.
MDA = α-methyldopamine-d_0.
D2-MDA = α-methyldopamine-d_2.
NE = norepinephrine-d_0.
D6-NE = norepinephrine-d_6.
MNE = α-methylnorepinephrine-d_0.
D7-MNE = α-methylnorepinephrine-d_7

The column was eluted sequentially with 0.5 M acetic acid–sodium acetate (pH 5, 5 ml), 0.1 N HCl (10 ml), anhydrous ethanol (10 ml) and 2 N anhydrous ethanolic HCl (8 ml). The last fraction, which contained the amines of interest, was evaporated to dryness and the residue warmed for 60 min at 60° in 1 N ethanolic HCl (0.25 ml). The solvent was removed under nitrogen and the residue treated with pentafluoropropionic anhydride (0.25 ml, 60°, 20 min) and the reaction mixture, after concentrating to about 10 μl, was analysed by direct insertion probe CI mass spectrometry. Three to five scans over the relevant mass region were obtained for each sample and the appropriate ion current intensity lines were averaged. Those samples which did not provide sequential ion intensity ratios within 5 per cent were discarded. The analysis for α-methyldopa proceeded in an analogous fashion except that the hydrogen form of the ion-exchange column was employed and the pH of the trichloroacetic acid homogenate was not adjusted before chromatographic separation.

STEREOCHEMICAL STUDIES

Since the antihypertensive (Gillespie *et al.*, 1972) and the central catecholamine depleting properties (Porter, Totaro and Leiby, 1961) of α-methyldopa are stereospecific and associated with the S-enantiomer, it may be surmised that only (S)-α-methyldopa is susceptible to the action of brain decarboxylase. This interpretation is consistent with the stereospecificity of (S)-aromatic amino acid decarboxylase (Weissbach, Lovenberg and Udenfriend, 1960). In order to examine more fully the stereospecificity of *in vivo* brain decarboxylase, we have determined to what extent (R)-, (S)- and (R,S)-α-methyldopa lead to the displacement of endogenous dopamine and norepinephrine in the caudate nucleus and hypothalamus in male Wistar rats (300–400 g). The drugs (200 mg/kg) were administered interperitoneally and the animals sacrificed by decapitation after 5 h. Figures 24.9b,c present the CI mass spectral scans of samples isolated from the caudate nucleus of animals treated with (S)-α-methyldopa and (R)-α-methyldopa, respectively. For comparison, Figure 24.9a presents a scan of a mixture of the four catecholamines and four internal standards carried through the procedure. After administration of (S)-α-methyldopa the spectrum obtained displays a significant ion at *m/e* 606 corresponding to α-methyldopamine. This ion is absent in the spectrum obtained from the sample derived from the caudate nucleus of the animal treated with (R)-α-methyldopa. The level of endogenous dopamine in the caudate nucleus after (S)-α-methyldopa calculated from a series of experiments was approximately 15 nmol/g of tissue whereas the averaged dopamine level after (R)-α-methyldopa was approximately 65 nmol/g of tissue. Thus it was clear the (S)-α-methyldopa had led to the displacement of endogenous dopamine. When ^{13}C-labelled (S)-α-methyldopa was administered, a similar spectrum to that shown in Figure 24.9b was obtained except that the ion at *m/e* 606 was replaced by an ion at *m/e* 607. Consistent with previous reports (Blaschko, 1972), levels of norepinephrine and α-methylnorepinephrine in the caudate nucleus were very low or undetectable even though the internal standards were readily observed.

A similar series of experiments was carried out in which the hypothalamus was examined. Figure 24.10a shows the mass spectral scan of a hypothalamic sample obtained from an animal treated with (S)-α-methyldopa. A strong ion at *m/e* 606 corresponds to α-methyldopamine and at *m/e* 650 to α-methylnorepinephrine. Apparently, under these conditions, all of the endogenous norepinephrine was displaced since the ion intensity of *m/e* 636 corresponding to norepinephrine is not above background. Figure 24.10b shows the corresponding spectrum of the hypothalamic sample obtained after treatment with (S)-α-methyldopa-^{13}C. The ion at *m/e* 606 has shifted to *m/e* 607 and the ion at *m/e* 650 to *m/e* 651, fully confirming the above structural assignments. We also administered a ‘pseudo-racemic’ mixture consisting of (S)-α-methyldopa-^{13}C and (R)-α-methyldopa. The α-methyldopamine ion was

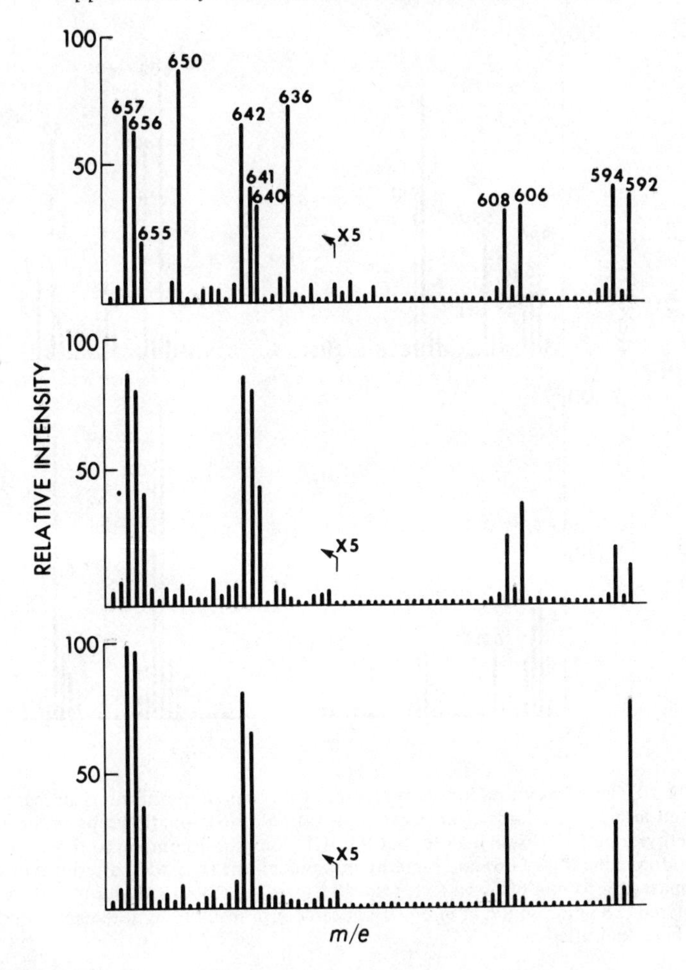

Figure 24.9 Chemical ionisation mass spectra of derivatised catecholamines, α-methylcatecholamines and their deuterated internal standards: upper, mixture of derivatised standard amines; centre, from rat brain caudate nucleus after i.p. (S)-α-methyldopa; lower, from rat brain caudate nucleus after i.p. (R)-α-methyldopa. The mass numbers corresponding to the MH^+ ions of the derivatised catecholamines are as follows: *m/e* 657 (α-methylnorepinephrine-d_7), 650 (α-methylnorepinephrine-d_0), 642 (norepinephrine-d_6), 636 (norepinephrine-d_0), 608 (α-methyldopamine-d_2), 606 (α-methyldopamine-d_0) 594 (dopamine-d_2) and 592 (dopamine-d_0)

again observed at *m/e* 607 while the ion at *m/e* 606 was not above background. Similarly, the ion at *m/e* 651 corresponding to the ^{13}C derived α-methylnorepinephrine was abundant whereas there was no significant ion current at *m/e* 650. Thus it was apparent from these studies that administration of (S)-α-methyldopa but not (R)-α-methyldopa led to the formation of the α-methylated catecholamines in the caudate nucleus and hypothalamus. The structure assignment based on mass spectrometry were unambiguously defined by the shifts of the anticipated ions by one mass unit

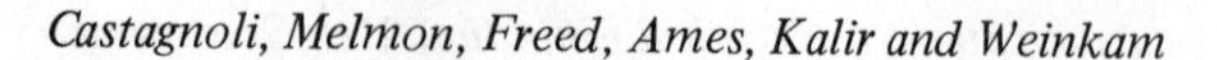

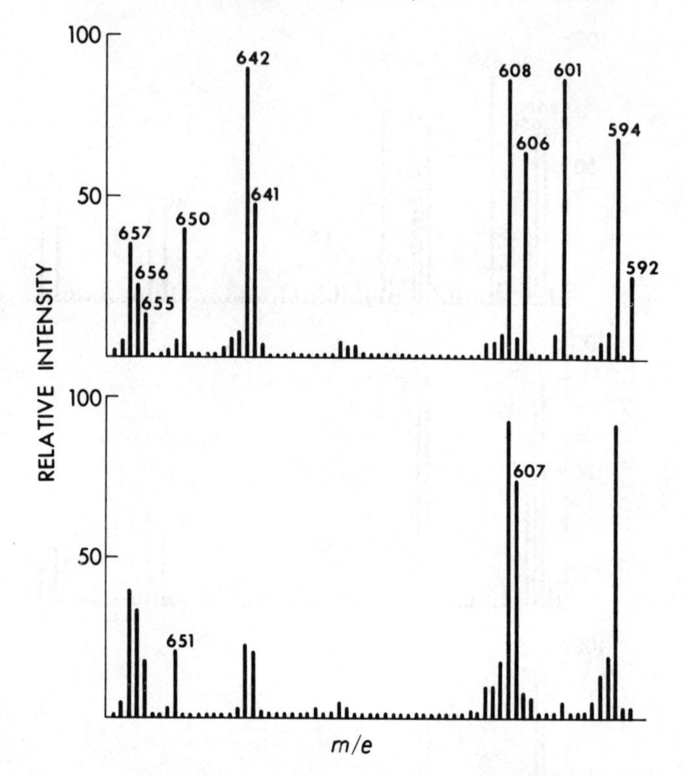

Figure 24.10 Chemical ionisation mass spectra of rat brain hypothalamus amines and deuterated internal standards after i.p. administration of (S)-α-methyldopa (upper) and (S)-α-methyldopa-^{13}C (lower). Note that the MH^+ ions corresponding to the derivatised α-methyldopamine (*m/e* 606) and α-methylnorepinephrine (*m/e* 650) are shifted by one atomic mass unit to *m/e* 607 and 651, respectively, when (S)-α-methyldopa-^{13}C is administered. The intense ion at *m/e* 601 appeared erratically in hypothalamic studies and has not been identified

when normal abundance (S)-α-methyldopa was replaced with (S)-α-methyldopa-^{13}C.

These results could be interpreted in terms of either the stereospecific central decarboxylation of (S)-α-methyldopa or of the stereospecific transport of (S)-α-methyldopa into the brain, or due to both factors. In order to investigate this question we examined the caudate nucleus of rats for α-methyldopa following treatment with (R)- and (S)-α-methyldopa. In these experiments the unlabelled amino acid was administered to the animals and the ^{13}C-labelled amino acid was used as an internal standard. Figure 24.11a shows the CI mass spectral scan of the caudate nucleus sample obtained 1 h after administration of (S)-α-methyldopa. The ion appearing at *m/e* 678 corresponds to the unlabelled amino acid and at *m/e* 679 to the ^{13}C internal standard. Under similar conditions when (R)-α-methyldopa was administered, the ion intensity at *m/e* 678, when corrected for the 11 per cent contamination of the ^{12}C compound in the ^{13}C internal standard, proved to be only about 10 per cent of the

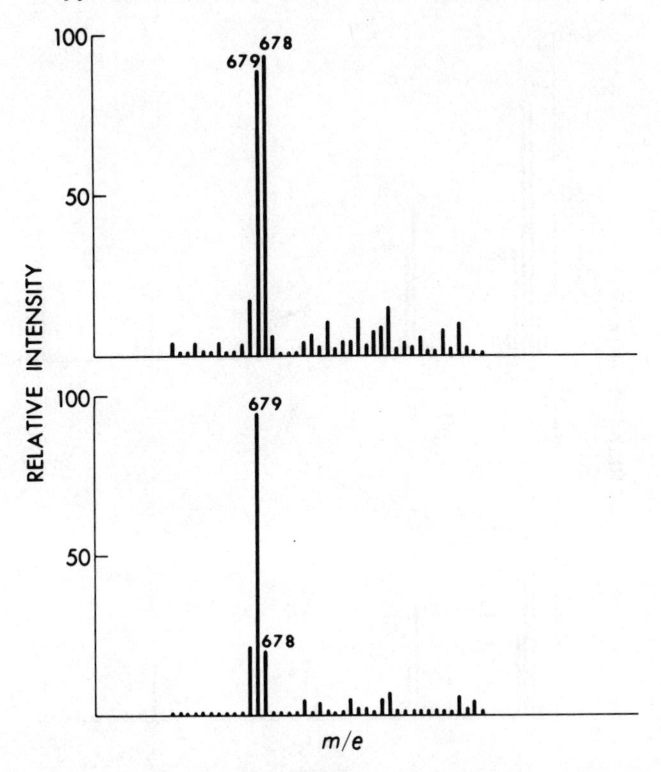

Figure 24.11 Chemical ionisation mass spectra of caudate nucleus extracts worked-up for amino acids: upper, obtained after i.p. administration of (S)-α-methyldopa; lower, obtained after i.p. administration of (R)-α-methyldopa. The ions observed at *m/e* 679 in both spectra are derived from the α-methyldopa-^{13}C internal standard. The ion at *m/e* 678 is derived from the administered (S)-α-methyldopa-^{13}C. Both tracings are corrected for the 11% ^{12}C-contaminant in the ^{13}C-internal standard

corresponding ion intensity found after administration of (S)-α-methyldopa (Figure 24.11b). We have concluded from these results that the transport of α-methyldopa into the rat brain is stereoselective. In an attempt to by-pass this stereochemical barrier, we administered both (S)- and (R)-α-methyldopa directly into the fourth ventricle of the rat. Figure 24.12a shows the CI mass spectral scan of the caudate nucleus sample obtained after intraventricular administration of (S)-α-methyldopa and Figure 24.12b shows the corresponding scan after administration of (R)-α-methyldopa. It is clear from these data that only (S)-α-methyldopa leads to measurable amounts of α-methyldopamine in the caudate nucleus.

On the basis of these experiments it may be concluded that in mammals α-methyldopa interacts with macromolecules in a stereospecific fashion analogous to naturally occurring substrates both in terms of active transport mechanisms and enzymatic conversions. The results are consistent with the false neurotransmitter hypothesis involving the stereoselective transport and central metabolism of the (S) -enantiomer.

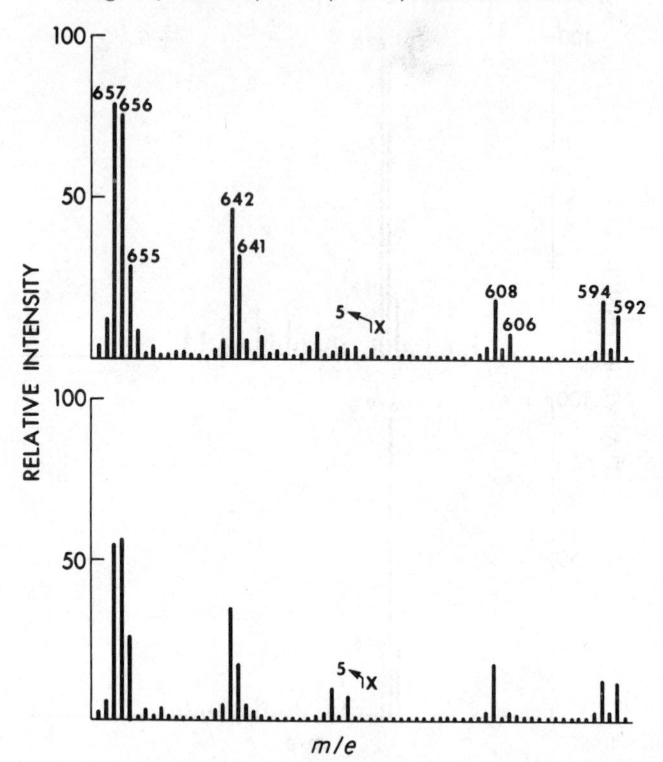

Figure 24.12 Chemical ionisation mass spectra of derivatised catecholamines, α-methylcatecholamines and their deuterated internal standard found in the rat caudate nucleus: upper, after intraventricular administration of (S)-α-methyldopa; lower, after intraventricular administration of (R)-α-methyldopa

STUDIES ON INTERACTIONS OF (S)-α-METHYLDOPA AND (S)-DOPA

An intriguing feature of α-methyldopa and its biotransformation products is their close structural relationship to dopa and its metabolites. In order to gain some understanding of the *in vivo* interactions of these compounds, we examined the effects of simultaneous administration of (S)-α-methyldopa and (S)-dopa on brain catecholamine levels (Freed *et al.*, 1977) and (under steady state infusion conditions) on blood pressure. The analytical procedures followed were essentially as described for the stereochemical studies. For the steady state infusion studies, catheters were placed in the external jugular vein and left carotid artery. Twenty-four h after surgery, base-line blood pressures were recorded and 24 h drug infusions were started. A second recording of the blood pressure was obtained following termination of the infusion and then the animals were sacrificed for brain catecholamine determinations.

Table 24.1 summarises the caudate nucleus results obtained 6 h following single injections of the indicated drug(s). Following (S)-α-methyldopa alone, essentially all of the dopamine of the caudate nucleus was replaced by

Table 24.1 Concentrations of dopamine and α-methyldopamine (nmol/g) found in the rat caudate nucleus 6 h following indicated drug treatment[a]

Amine[b]	Control	(S)-dopa	(S)-α-methyldopa	MK 486	(S)-dopa + (S)-α-methyldopa	(S)-dopa + (S)-α-methyldopa + MK 486
Dopamine	59–65	48–57	0	69–77	18–28	90–98
α-Methyl-dopamine	0	0	67–75	0	70–80	28–35

[a] Doses of (S)-α-methyldopa and (S)-dopa were 200 mg/kg and MK 486 was 50 mg/kg. All drugs were administered as i.p. bolus doses.
[b] The levels of norepinephrine and α-methylnorepinephrine were below detectable limits. In general, assays were performed on pairs of animals.

α-methyldopamine. Dopamine levels were partially maintained when (S)-dopa was co-administered with (S)-α-methyldopa (both at 200 mg/kg) even though the level of α-methyldopamine remained high. When the peripheral decarboxylase inhibitor MK 486 was also given, then supranormal levels of dopamine were achieved (94 nmol/g *vs* 62 nmol/g control) and the levels of α-methyldopamine were somewhat depressed. These data are consistent with reports that (S)-dopa suffers extensive peripheral decarboxylation, a fate not totally shared by (S)-α-methyldopa (Henning, 1969; Buhs *et al.*, 1964). It would appear, however, that (S)-dopa is a better substrate than (S)-α-methyldopa for brain decarboxylase since in the presence of MK 486 dopamine levels in the caudate nucleus were three times those of α-methyldopamine, even though the drugs were co-administered at the same dose. Until we can compare the turnover rates of dopamine and α-methyldopamine under these experimental conditions, it will not be possible to approximate enzyme specificities. It is likely that α-methyldopamine, a phenylisopropylamine, is not metabolised by monoamine oxidase (MAO), a factor which could prolong vesicular storate time. As was previously noted, the levels of norepinephrine and α-methylnorepinephrine in the rat caudate nucleus were below detectable limits in all experiments.

Analogous results were obtained with the hypothalamus (Table 24.2).

Table 24.2 Concentration of amines (nmol/g) found in the hypothalamus 6 h following indicated drug treatment[a]

	Control	(S)-dopa	(R,S)-α-methyldopa	(S)-dopa + (S)-α-methyldopa	MK 486	(S)-dopa + (S)-α-methyldopa + MK 486
Dopamine	2.2–3.0	1.7–2.7	1.8–3.5	2.0–3.0	4.3–5.5	12.5–15.9
α-Methyl-dopamine	0	0	8.8–38.9	9–15	0.7–1.3	5.4–10.4
Norepinephrine	5.7–9.0	3.9–6.3	0	0.2–1.2	5.3–9.5	4.6–7.8
α-Methylnor-epinephrine	0	0	12.6–18.5	5.5–7.1	0	4.7–6.5

[a] See footnote, Table 24.1.

Treatment with (S)-α-methyldopa alone or with co-administration of (S)-dopa lowered hypothalamic norepinephrine levels to essentially zero, while both α-methyldopamine and α-methylnorepinephrine reached levels comparable to or greater than the corresponding control levels of the native amines. The peripheral decarboxylase inhibitor MK 486, when co-administered with the combined drugs, appeared to protect (S)-dopa from peripheral decarboxylation since hypothalamic norepinephrine levels were maintained and dopamine levels were supranormal. The need for turnover rate studies is obvious if the significance of these inter-relationships is to be understood on a quantitative basis.

In a separate series of experiments, we examined the hypothalamic levels of all four amines following a 24 h infusion of (S)-α-methyldopa alone and in the presence of MK 486 and/or (S)-dopa. In general, these preliminary steady-state infusion results are consistent with the single injection results. Norepinephrine levels were essentially zero, α-methylnorepinephrine levels ranged between 8 and 15 nmol/g, dopamine levels ranged between 2 and 10 nmol/g and α-methyldopamine levels ranged between 1 and 20 nmol/g. Pre- and post-infusion blood pressure measurements were obtained and with the exception of α-methyldopamine no correlations were observed between amine levels and changes in blood pressure. On the other hand, a reasonably linear relationship for lowering of blood pressure and α-methyldopamine levels was observed. A similar correlation has been previously reported (Maître, Hedwall and Waldmeier, 1974). Although further study is required, these data indicate that dopaminergic pathways in the brain may be associated with the blood pressure regulating properties of (S)-α-methyldopa.

REFERENCES

Ames, M. M. and Castagnoli, Jr., N. (1974). *J. Lab. Comp.*, **10**, 195

Ames, M. M., Melmon, K. L. and Castagnoli, Jr., N. (1977). *Biochem. Pharmacol.* (in press)

Au, W. Y. W., Dring, L. G., Grahame-Smith, D. G., Isaac, P. and Williams, R. T. (1972). *Biochem. J.*, **129**, 1

Blaschko, H. (1972). *The Distribution of Catecholamines in Vertebrates* (ed. H. Blaschko and E. Muschall), Springer-Verlag, Berlin, p. 148

Bogdanski, D. F., Pletscher, A., Brodie, B. and Udenfriend, S. (1956). *J. Pharmacol. Exp. Ther.*, **117**, 82

Buhs, R. P., Beck, J. L., Speth, O. C., Smith, J. L., Trenner, N. R., Cannon, P. J. and Laragh, J. H. (1964). *J. Pharmacol. Exp. Ther.*, **143**, 205

Dale, J. A., Dull, D. L. and Mosher, H. S. (1969). *J. Org. Chem.*, **34**, 2543

Freed, C. R., Weinkam, R. J., Melmon, K. L. and Castagnoli, Jr., N. (1977). *Anal. Biochem.* (in press)

Gal, J. and Ames, M. M. (1977). Submitted for publication to *Anal. Biochem.*

Gillespie, L., Oates, J. A., Crout, J. R. and Sjoerdsma, A. (1962). *Circulation*, **25**, 281

Henning, M. (1968). *Acta Physiol. Scand., Suppl.*, **322**, 1

Henning, M. (1969). *Acta Pharmacol. Toxicol.*, **27**, 135

Jaju, B. P., Tangri, K. K. and Bhargava, K. P. (1966). *Can. J. Physiol. Pharmacol.*, **44**, 687

Kalir, A., Freed, C., Melmon, K. L. and Castagnoli, Jr., N. (1977). *J. Lab. Comp. Radiopharm.*, **18**, 41

Koslow, S. H., Cattabeni, F. and Costa, E. (1972). *Science, N.Y.*, **176**, 177

Lehacette, J.-M., Rollin, G. and Dumas, P. (1972). *Can. J. Chem.*, **50**, 3058

Maître, L., Hedwall, P. R. and Waldmeier, P. C. (1974). *Ciba Foundation Symp.*, Vol. 22, *Aromatic Amino Acids in the Brain*, p. 335
Marshall, K. S. and Castagnoli, Jr., N. (1973). *J. Med. Chem.*, **16**, 266
Oates, J. A., Gillespie, L., Udenfriend, S. and Sjoerdsma, A. (1960). *Science, N.Y.*, **131**, 1890
Porter, C. C., Totaro, J. A. and Leiby, C. M. (1961). *J. Pharmacol. Exp. Ther.*, **134**, 139
Sadeh, S., Pelah, Z. and Kalir, A. (1967). *Israel J. Chem.*, **5**, 44p
Weissbach, H., Lovenberg, W. and Udenfriend, S. (1960). *Biochem. Biophys. Res. Commun.*, **3**, 225

25
The role of prostaglandins in the pathogenesis of human disease: elucidation with stable isotopic methods

J. A. Oates, H. W. Seyberth, J. C. Frölich, B. J. Sweetman and J. T. Watson
(Departments of Medicine and Pharmacology, The Vanderbilt University School of Medicine, Nashville, Tennessee 37232, USA)

Prostaglandins are extremely potent biogenic compounds and as such are formed in quite small amounts *in vivo*. The total body production of prostaglandin E_2 (PGE_2), for example, is less than one μmol per day. Accordingly, the prostaglandins have provided a number of analytical challenges that have been soluble only by the use of selected ion monitoring with mass spectrometry. This is accomplished ultimately by employing stable-isotope-labelled compounds both as internal standards and as carriers.

In addition to the analytical problem posed by the biosynthesis of only trace quantities, the prostaglandins are quite rapidly metabolised. Therefore, the assessment of *in vivo* biosynthesis of prostaglandins through measurement of their metabolites has been a valuable approach in investigations of clinical disorders of prostaglandin biosynthesis. The major initial step in the metabolism of PGE_2 is 15-dehydrogenation followed closely by 13,14-reduction, yielding 15-keto-13,14-dihydro-PGE_2 which is the principal metabolite present in the circulation (Änggård, Gréen and Samuelson, 1965). As a large fraction of PGE_2 is metabolised to 15-keto-13,14-dihydro-PGE_2 in a single pass through the lung, little reaches the systemic circulation unchanged. Accordingly, it is not possible for PGE_2 to accumulate appreciably in the blood. Because of the extremely low levels in blood and the variable release of prostaglandin E_2 from platelets in the process of blood withdrawal, it has not been easy to develop specific and reliable methods for quantifying this prostaglandin in the venous circulation. On the other hand, 15-keto-13,14-dihydro-PGE_2 does accumulate and can be measured by a mass spectrometric method (Gréen *et al.*, 1973). 15-Keto-13,14-dihydro-PGE_2 is further metabolised by β- and ω-oxidation, yielding the major urinary metabolite 7α-hydroxy-5,11-diketo-tetranorprostane dioic acid (PGE-M) (Hamberg and Samuelsson, 1971). Measurement of levels of this major urinary metabolite may be employed as an indicator of total body PGE synthesis (Hamberg, 1972; Seyberth *et al.*, 1976b).

There are two clinical disorders in which prostaglandin E_2 overproduction has been clearly identified and quantified by mass spectrometric methods. One of these is the hypercalcaemia associated with certain solid tumours. Although some neoplastic disorders produce hypercalcaemia in association with metastasis to bone, a subset of tumour-associated hypercalcaemia is humoral in origin. Thus resection of the primary tumours causes remission of the hypercalcaemia in such patients. A number of bone-resorbing substances have been proposed as the humoral mediators of this hypercalcaemia. The finding of Klein and Raisz (1970) that PGE_2 is a potent stimulator of bone-resorption *in vitro* led to the consideration of prostaglandins among the postulated mediators. The studies of Tashjian *et al.* (1972) demonstrating that the hypercalcaemia caused by fibrosarcomas in mice can be reversed with indomethacin further raised the possibility that prostaglandins might mediate tumour-associated hypercalcaemia in man.

We initially investigated the possible participation in tumour-associated hypercalcaemia by determining whether any of these patients excreted increased amounts of the major urinary metabolite of PGE_2. The mass spectrometric assay for the major urinary metabolites employs a stable-isotope-labelled standard. Initially, the standard for PGE-M was obtained by the chemical synthesis of 15-keto-dihydro-PGE_1, and infusion of this intermediate metabolite into the rabbit in order to convert it to the major urinary metabolite, which is then isolated from rabbit urine after preparation of the methyl ester (ME). Deuterium is incorporated by derivatising the methyl ester of PGE-M with $^2H_3$-methoxyamine. This deuterated methoxime (MO) derivative is added following initial isolation of PGE-M from urine on XAD-2 and its derivatisation to the ME-MO-PGE-M. The urine extract containing both deuterated standard and a tritium-labelled marker is further isolated on reversed-phase partition chromatography and thin-layer chromatography, then converted to the trimethylsilyl derivative. The ratio of standard to endogenous PGE-M is determined by monitoring the ion at *m/e* 365 [M – (90 + 31)] as well as at *m/e* 368, the corresponding ion for the standard. Selected ion monitoring is performed with a computer-controlled display-oriented data system (Watson *et al.*, 1973), yielding a gas chromatographic elution profile of these ion currents that reveals both compounds to be present in a high level of purity (Figure 25.1). Recently a heptadeutero-PGE-M, obtained entirely by chemical synthesis, has been employed as the internal standard; this of course may be added to the urine immediately after its collection (Rosegay and Taub, 1976).

In patients with hypercalcaemia associated with solid tumours, the excretion of the major urinary metabolite of the E prostaglandins was considerably increased (median for the group equals 59 ng/mg of creatinine) compared with normals (7.1 ng/mg of creatinine) (Seyberth *et al.*, 1975). Of the 14 patients with tumour-associated hypercalcaemia, the excretion of PGE-M was normal in only two. By contrast, only slightly elevated values for PGE-M were found in normocalcaemic patients with solid tumours. The levels of PGE-M were normal

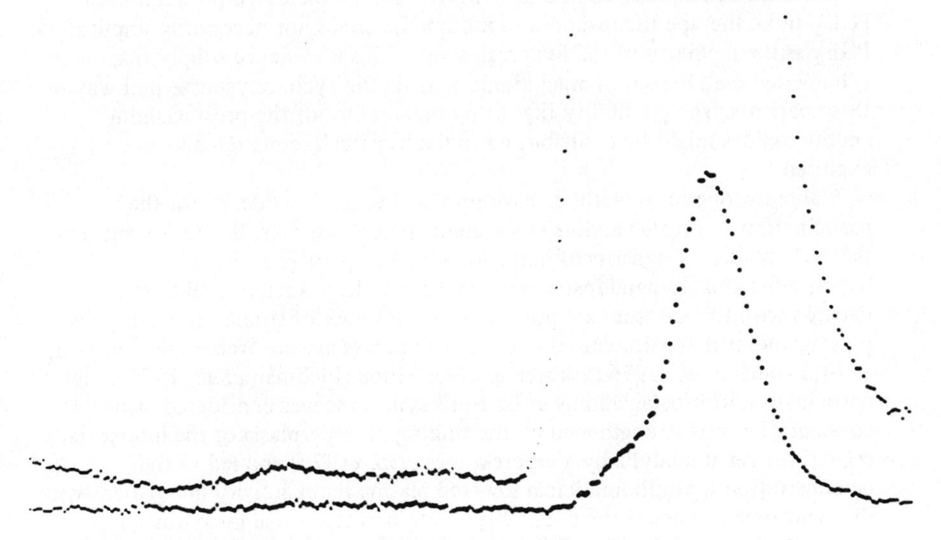

Figure 25.1 Ion profiles obtained during the analysis of PGE-M in human urine; the upper trace represents m/e 368 (internal standard-carrier) and the lower m/e 365

in hypercalcaemic patients with primary hyperparathyroidism and with haematological malignancies such as multiple myeloma and lymphosarcoma.

Following this initial investigation demonstrating that many of the patients with hypercalcaemia associated with solid tumours have increased biosynthesis of E prostaglandins, the therapeutic effects of indomethacin and aspirin on serum calcium were evaluated in relation to the levels of PGE-M (Seyberth *et al.*, 1976a). In those patients whose hypercalcaemia was associated with elevated PGE-M, either indomethacin or aspirin reduced levels of serum calcium into the normal range when there were no clinically detectable bony metastases, and reduced serum calcium significantly, but only slightly, and not into the normal range, when bony metastases were evident. When the hypercalcaemia associated with solid tumours was not associated with an elevation in PGE-M, no reduction in serum calcium resulted from the inhibition of prostaglandin synthesis. Thus assessment of total body prostaglandin E_2 synthesis by a stable isotope method for measuring its principal urinary metabolite permits the identification of patients in whom the tumour-associated hypercalcaemia is prostaglandin dependent. In such patients, aspirin or indomethacin will produce a clinically important reduction in serum calcium if bony metastases are not clinically apparent. In those patients in whom the hypercalcaemia is not dependent on prostaglandins, there was a tendency for the renal clearance of cyclic AMP to be higher and for serum chlorides to be higher than in the prostaglandin-dependent group. The identification of a prostaglandin-independent group by this method should enable any searches for other tumour-derived humoral substances to be more effective by focusing them on the subset of hypercalcaemic patients that are not dependent on the prostaglandin system.

It should be emphasised that identification of an increased production of PGE_2 and a therapeutic response to indomethacin do not necessarily implicate PGE_2 as the mediator of the hypercalcaemia. This is evidence simply that there is increased metabolism of arachidonic acid via the cyclo-oxygenase pathway in these patients. The possibility that other metabolites of the prostaglandin endoperoxides might be contributing to the hypercalcaemia remains to be explored.

Stable-isotope-ratio methods have provided valuable evidence for the participation of prostaglandins in a second clinical disorder, Bartter's syndrome. Bartter's syndrome consists of juxtaglomerular hyperplasia with hyperreninaemia, hyperaldosteronism and severe hypokalaemic alkalosis. Patients with this disorder are normotensive. Because of studies indicating that prostaglandins may stimulate the release of renin (Larsson, Weber and Änggård, 1974; Frolich *et al.*, 1976a) as well as aldosterone (Fichman *et al.*, 1972), the participation of prostaglandins in Bartter's syndrome was considered. This consideration was strengthened by the finding of hyperplasia of the interstitial cells of the renal medulla by Verberckmoes *et al.* (1976) and led to the demonstration that indomethacin lowered plasma renin activity in a patient with the syndrome. To assess the possibility of an increased renal biosynthesis of prostaglandins in patients with Bartter's syndrome, the excretion of PGE_2 in the urine was measured. Several lines of evidence indicate that PGE_2 in the urine derives primarily from the kidney and it is therefore an indicator of renal PGE_2 synthesis (Frölich *et al.*, 1975). Stop-flow studies in the dog have demonstrated that the loop of Henle is the major site of entry of prostaglandins into the tubular fluid (Frölich *et al.*, 1976b).

Urinary PGE_2 was measured by a mass spectrometric method utilising selected ion monitoring. 3,3,4,4-Tetradeutero-PGE_2, generously provided by the Upjohn Company, was employed as an internal standard and the ions at m/e 419 and 423 from the methyl ester, methoxime, bis-acetate derivative were monitored. It should be emphasised that not even selected ion monitoring is a completely specific method and extensive prior purification of the urine extract is required. Such purification was achieved with high-performance liquid chromatography, which separates the prostaglandins with a high degree of resolution and purifies PGE_2 sufficiently to yield a discrete peak on the selected ion monitor. The extraordinary resolution achieved by high-performance liquid chromatography not unexpectedly yielded some separation based on isotope effect. If small fractions are taken from the high-performance liquid chromatograph, it can be seen that there is a slight separation of the tetradeutero-PGE_2 (measured absorbance at 278 nm after conversion to PGB_2) from the heptatritio-PGE_2 that is employed as a tracer in the isolation procedures (Figure 25.2). Furthermore, a slight separation of the deuterated and non-deuterated PGE_2 can be detected when 0.2 ml aliquots of the column eluate are collected and the ratio of non-deuterated to deuterated PGE_2 is determined by monitoring ions at m/e 419 and 423, respectively, by mass spectrometry (Figure 25.3). If the possibility of this resolution of isotopes is not taken into account, it is possible not only to get poor recovery of the deuterated internal standard as carrier in the tritium-containing fractions of the column, but also to obtain material from small fractions of the column eluate that are not representative of the true ratios of deuterated and non-deuterated compounds

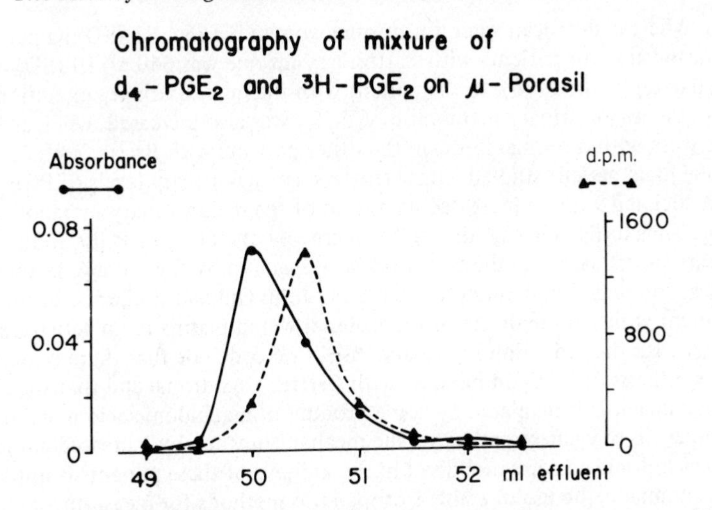

Figure 25.2 Elution profiles of d_4-PGE_2 and ^3H-PGE_2 from a silicic acid column with high-performance liquid chromatography. d_4-PGE_2 was measured by absorbance at 278 mμ after conversion to d_4-PGB_2

present in the original sample; such a separation, of course, would invalidate the principle of isotope ratio analysis employed in the quantification. Thus the extent of any isotope effect on elution from high-performance liquid chromatographs must be determined as a basis for selection of fractions from the columns.

Urinary PGE_2 was quantified by this method in five female patients with Bartter's syndrome, and the values were elevated above normal in all five of

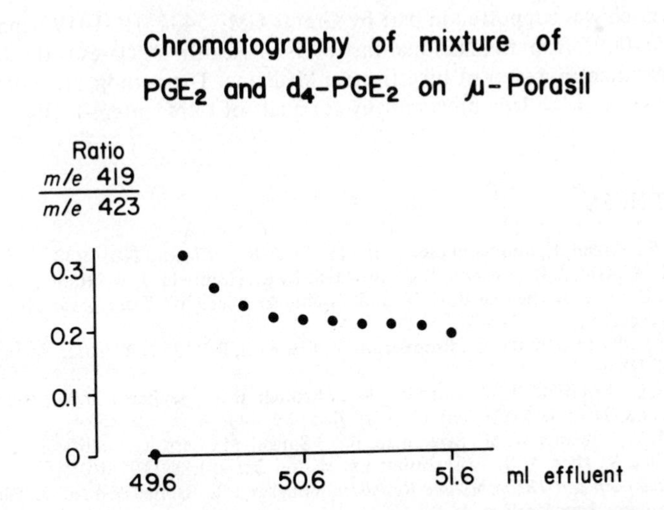

Figure 25.3 Ratios of PGE_2 and d_4-PGE_2 during elution from a silicic acid column with high-performance liquid chromatography; 0.2 ml fractions were analysed by gas chromatography–mass spectrometry in the selected-ion monitoring mode to determine the ratio of *m/e* 419 (PGE_2) to *m/e* 423 (d_4-PGE_2)

them. Whereas the mean value for normal women is 185 ± 81 (SD) ng per day, the mean value for patients with Bartter's syndrome was 640 ± 119 (SD) ng per day (Bartter *et al.*, 1976). In the patient with the highest urinary excretion of PGE_2, the major urinary metabolite of PGE_2 was also increased, whereas PGE-M levels were within normal limits in the other patients with Bartter's syndrome. In the one male patient studied, consistently very high urinary levels of PGE_2 were also associated with an increased excretion of the major urinary metabolite of PGE_2. Thus in Bartter's syndrome the increased excretion of PGE_2 in the urine suggests an increased synthesis of that prostaglandin by the kidney. In these studies, conducted in conjunction with Drs John Gill and Frederick Bartter, treatment with indomethacin or ibuprofen lowered plasma renin activity in parallel with the reduction in urinary PGE_2. We conclude that there is increased renal synthesis of PGE_2 in patients with Bartter's syndrome and that the hyperreninaemia is mediated by some product of arachidonic acid metabolism via the cyclo-oxygenase pathway. The mechanism whereby increased amounts of free arachidonic acid are mobilised in the kidneys of these patients is not known.

In summary, the use of stable-isotope-ratio methods for measuring prostaglandin E_2 and its metabolites has permitted elucidation of the participation of the prostaglandin system in the hypercalcaemia of some patients with cancer, and, at a more local level, the participation of a prostaglandin in the renal disorders in Bartter's syndrome. Whether PGE_2 itself is the mediator or merely an indicator of disordered arachidonic metabolism in these syndromes remains to be determined.

ACKNOWLEDGEMENTS

This research was supported in part by Grants GM-15431, HL-14192 and 5-MO1-55-0095 from the National Institutes of Health. Dr Oates is the Joe and Morris Werthan Professor of Investigative Medicine. The authors are indebted to Dr Udo Axen of the Upjohn Company for a gift of tetradeutero-PGE_2.

REFERENCES

Änggård, E., Gréen, K. and Samuelsson, B. (1965). *J. Biol. Chem.*, **240**, 1932

Bartter, F. C., Gill, J. R., Frölich, J. C., Bowden, R. E., Hollifield, J. W., Radfar, N., Keiser, H. R., Oates, J. A., Seyberth, H. W. and Taylor, A. A. (1976). *Trans. Assn. Am. Physicians*, **89**, 77

Fichman, M. P., Littenburg, G., Brooker, A. and Horton, R. (1972). *Circulat. Res.*, **30/31** (suppl. II), 19

Frölich, J. C., Hollifield, J. W., Dormois, J. C., Frölich, B. L., Seyberth, H. W., Michelakis, A. M. and Oates, J. A. (1976a). *Circulat. Res.*, **39**, 447

Frölich, J. C., Williams, W. M., Sweetman, B. J., Smigel, M., Carr, K., Hollifield, J. W., Fleischer, S., Nies, A. S., Frisk-Holmberg, M. and Oates, J. A. (1976b). *Adv. Prostaglandin and Thromboxane Research*, Vol. 1 (ed. B. Samuelsson and K. Paoletti), Raven Press, New York, p. 65

Frölich, J. C. Williams, W. M., Sweetman, B. J., Smigel, M., Nies, A. S., Carr, K. Watson, J. T. and Oates, J. A. (1975). *J. Clin. Invest.*, **55**, 763

Gréen, K., Granström, E., Samuelsson, B. and Axen, U. (1973). *Anal. Biochem.*, **54**, 434

Hamberg, M. (1972). *Biochem. Biophys. Res. Comm.*, **49**, 720
Hamberg, M. and Samuelsson, B. (1971). *J. Biol. Chem.*, **246**, 6713
Klein, D. C. and Raisz, L. G. (1970). *Endocrinology*, **86**, 1436
Larsson, C., Weber, P. and Änggård, E. (1974). *Europ. J. Pharmacol.*, **28**, 391
Rosegay, A. and Taub, O. (1976). *Prostaglandins*, **12**, 785
Seyberth, H. W., Segre, G. V., Hamet, P., Sweetman, B. J., Potts, J. T., Jr. and Oates, J. A. (1976a). *Trans. Ass. Am. Physicians,* **89**, 92
Seyberth, H. W., Segre, G. V., Morgan, J. L., Sweetman, B. J., Potts, J. T. and Oates, J. A. (1975). *New Engl. J. Med.*, **293**, 1278
Seyberth, H. W., Sweetman, B. J., Frölich, J. C. and Oates, J. A. (1976b). *Prostaglandins,* **11**, 381
Tashjian, A. H. Jr., Voelkel, E. F., Levine, L. and Goldhaber, P. (1972). *J. Exper. Med.*, **136**, 1329
Verberckmoes, R., Van Damme, B., Clement, J., Amery, A. and Michielsen, P. (1976). *Kidney International*, **9**, 302
Watson, J. T., Pelster, D. R., Sweetman, B. J. and Oates, J. A. (1973). *Anal. Chem.*, **45**, 2071

26

Application of deuterium labelling in studies of the biosynthesis and metabolism of prostaglandin $F_{2\alpha}$ in man

A. R. Brash, M. E. Conolly, G. H. Draffan,* P. Tippett and T. A. Baillie
(Department of Clinical Pharmacology, Royal Postgraduate Medical School, London W12 0HS, UK)

INTRODUCTION

Quantitative determination of the major urinary metabolites of the E and F series prostaglandins represents a valuable approach to the study of prostaglandin biosynthesis in man. The primary prostaglandins, such as prostaglandin $F_{2\alpha}$ ($PGF_{2\alpha}$), are first metabolised to their biologically inactive 15-keto-13,14-dihydro derivatives. These, in turn, are converted into several more polar derivatives which are excreted in the urine. The main end-product in the metabolism of $PGF_{2\alpha}$ is 5α,7α-dihydroxy-11-ketotetranor-prostane-1,16-dioic acid (referred to below as PGF-M), while an analogous C_{16} dioic acid is formed from PGE_2 (Figure 26.1).

The development of methods for the analysis of these metabolites has been severely hampered by the lack of available reference compounds. This is particularly true of methods based on gas chromatography–mass spectrometry (GC–MS), where it is desirable to employ a stable-isotope-labelled analogue of the metabolite as internal standard. Although some prostaglandin metabolites have been synthesised chemically (e.g. Nidy and Johnson, 1975), the metabolic conversion of an appropriate patent prostaglandin in a suitable animal species remains an important approach to the production of such compounds. However, the preparation of isotopically labelled metabolites by this technique requires that the heavy atoms in the parent compound occupy metabolically stable positions, a condition which is not satisfied in the case of the 3,3,4,4-tetradeutero analogues which have been synthesised for use in the analysis of the primary

* Present address: Inveresk Research International, Inveresk Gate, Musselburgh, Midlothian EH21 7UB, UK.

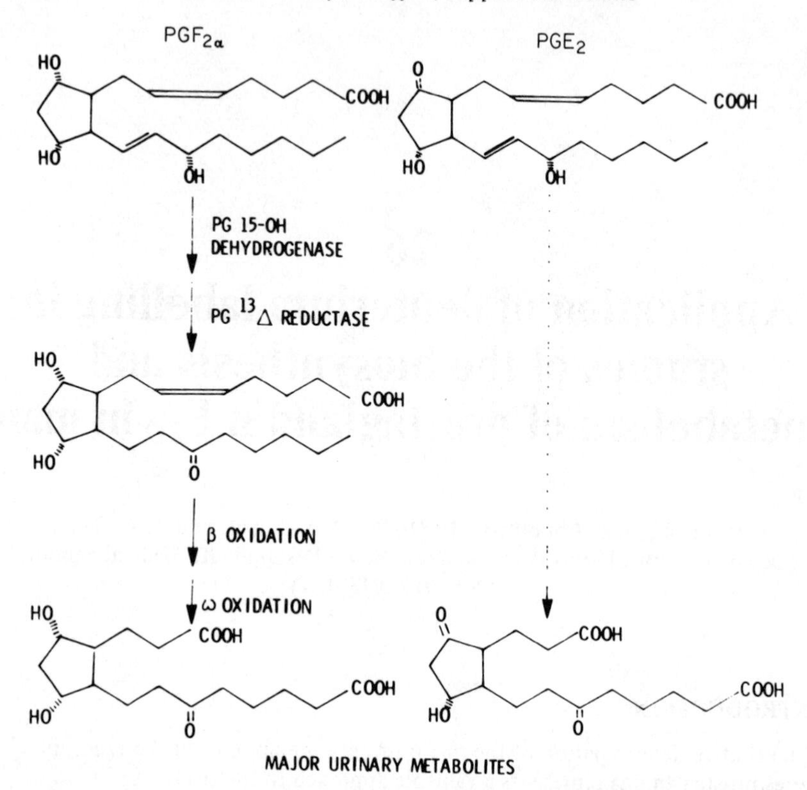

Figure 26.1 The major pathways in the metabolism of $PGF_{2\alpha}$ and PGE_2 in man

prostaglandins (Gréen *et al.*, 1973). We have provided a solution to this problem for PGF-M, and potentially for all PGF urinary metabolites, by the development of a simple procedure by which comparatively large quantities of $PGF_{2\alpha}$ and $PGF_{2\beta}$, labelled specifically with deuterium in the cyclopentane ring, may be prepared from PGE_2, a readily available starting material. The $[^2H_3]$-$PGF_{2\alpha}$ obtained in this manner was converted by *in vivo* metabolism, without loss of label, into $[^2H_3]$-PGF-M, which served as internal standard in the GC–MS assay of the corresponding endogenous metabolite.

In this paper the basis of the assay and two applications of the method are described. The first application is to a study of the possible role of $PGF_{2\alpha}$ in the pathogenesis of bronchial asthma and the second is concerned with the effect of indomethacin on the biosynthesis and metabolism of $PGF_{2\alpha}$ in man.

MATERIALS AND METHODS

Preparation of dimethyl [5β-^{3}H] 5α,7α-dihydroxy-11-ketotetranor-prostane-1,16-dioate ([5β-^{3}H]-PGF-M methyl ester)

A reference sample of [5β-^{3}H]-PGF-M was isolated from pooled urine collections from rats which had received subcutaneous injections of [9β-^{3}H]-$PGF_{2\alpha}$. The separation sequence involved extraction on a column of

Amberlite XAD-2, methylation and purification by liquid-gel chromatography (LGC). Isolation of [5β-^{3}H]-PGF-M methyl ester and the corresponding δ-lactone derivative was carried out by reversed-phase LGC (Nyström and Sjövall, 1973) (Figure 26.2). Subsequent purification by straight-phase LGC (Brash and Jones, 1974) gave radiochemically pure [5β-^{3}H]-PGF-M which could be quantified from the known specific activity of its tritiated $PGF_{2\alpha}$ precursor.

Preparation of dimethyl [5β-^{3}H; 4,6,6-$^2H_3$]-5α,7α-dihydroxy-11-ketotetranor-prostane-1,16-dioate ([5β-^{3}H; 4,6,6-$^2H_3$]-PGF-M methyl ester

The preparation of this derivative from PGE_2 is outlined in Figure 26.3. Introduction of deuterium at the C-8 and C-10 positions is achieved under conditions of base-catalysed enolisation, a reaction which was originally described as a means of preparing 8-iso-PGE_2 (Pike, Lincoln and Schneider, 1969). At equilibrium, the mixture of PGE_2 and 8-iso-PGE_2 is about 9 : 1 in favour of the naturally occurring isomer. Following sodium borohydride reduction of the reaction mixture, four PGF_2 isomers were obtained with the deuterium atoms at C-8 and C-10 now in chemically non-exchangeable and metabolically stable positions. Deuterated $PGF_{2\alpha}$ was isolated by reversed-phase LGC and was administered to a rhesus monkey, together with a small amount of [9β-^{3}H]-$PGF_{2\alpha}$ which served as a tracer. A mixture of deuterated $PGF_{2\alpha}$ metabolites was excreted in the urine and the required deuterated PGF-M was isolated by LGC.

Details of the procedures described above have been published elsewhere (Brash *et al.*, 1976). The mass spectra of the methyl ester *tert*-butyldimethylsilyl (*tert*-BDMS) ether derivative of unlabelled and deuterium-labelled PGF-M are illustrated in Figure 26.4.

GAS CHROMATOGRAPHY–MASS SPECTROMETRY (GC–MS)

Recording of complete mass spectra and development of the GC–MS assay for PGF-M were carried out on an AEI MS12 mass spectrometer, coupled to a Varian series 1400 gas chromatograph via a silicone membrane separator (Brash *et al.*, 1976).

Routine quantitative assays of urinary PGF-M were based on selected ion monitoring (SIM) analyses using a Finnigan 3200 GC–MS instrument. A 5 ft x 2 mm i.d. glass column packed with 3 per cent OV-1 on Gas Chrom Q (100–200 mesh) was used at a temperature of 260° and with a helium flow rate of 30 ml/min. Under these conditions the retention time of the PGF-M methyl ester *tert*-BDMS ether derivative was approximately 3.5 min. The mass spectrometer was operated in the electron impact mode with an electron energy of 25 eV and an emission current of 400 μA. The ions recorded were *m/e* 397 for the derivative of endogenous PGF-M and *m/e* 400 for the internal standard. Data acquisition and reduction were performed on-line by a Finnigan Model 6000 computer system.

RESULTS AND DISCUSSION

Quantitative determination of PGF-M in urine

The analysis of PGF-M in urine was based on a stable isotope dilution assay which employed the deuterium-labelled metabolite as the internal standard.

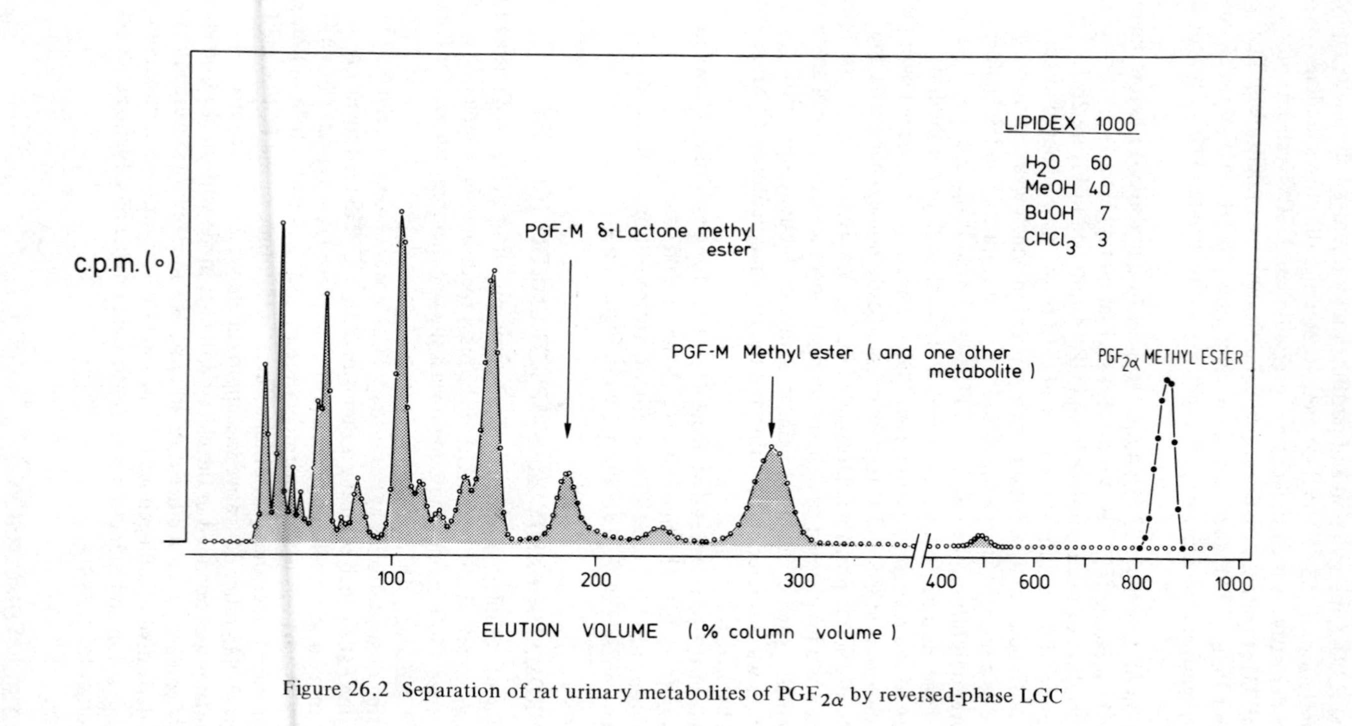

Figure 26.2 Separation of rat urinary metabolites of $PGF_{2\alpha}$ by reversed-phase LGC

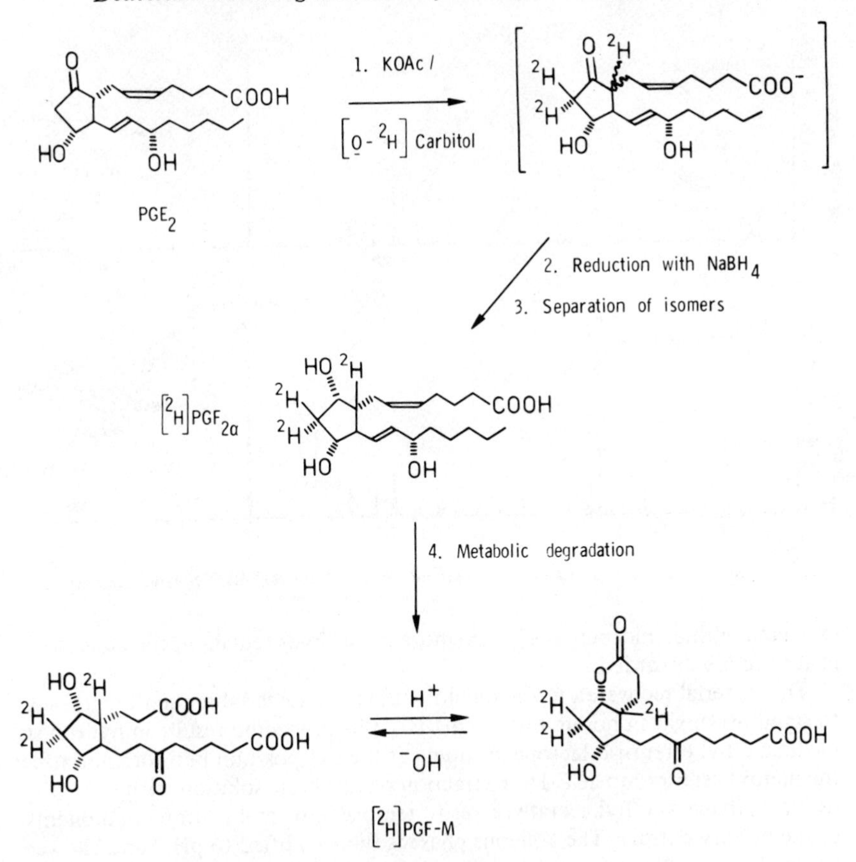

Figure 26.3 Preparation of deuterium-labelled PGF-M

Experimental details of the method have been published (Brash *et al.*, 1976) and therefore the emphasis in the account given below has been placed on the rationale behind the analytical procedure. Figure 26.5 illustrates the formation and interconversion of PGF-M derivatives throughout the various steps in the method.

Ten ml of the urine sample is 'spiked' with 200 ng of the methyl ester of the deuterated metabolite and the mixture is allowed to stand overnight under alkaline conditions (pH $>$ 12) at room temperature. This treatment results in quantitative hydrolysis of the deuterated PGF-M methyl ester and in ring-opening of the δ-lactone form of the endogenous metabolite; labelled and unlabelled species of PGF-M are thus equilibrated. The urine is then acidified and the metabolite is extracted using a small column of Amberlite XAD-2. The acidic conditions employed during this procedure cause partial conversion of the dioic acid to its δ-lactone derivative. The corresponding methyl esters are subsequently prepared by treatment of the urinary extract with tetramethylammonium hydroxide/methyl iodine in dimethylacetamide (Greeley, 1974) and, following addition of water to the reaction mixture, the derivatives are

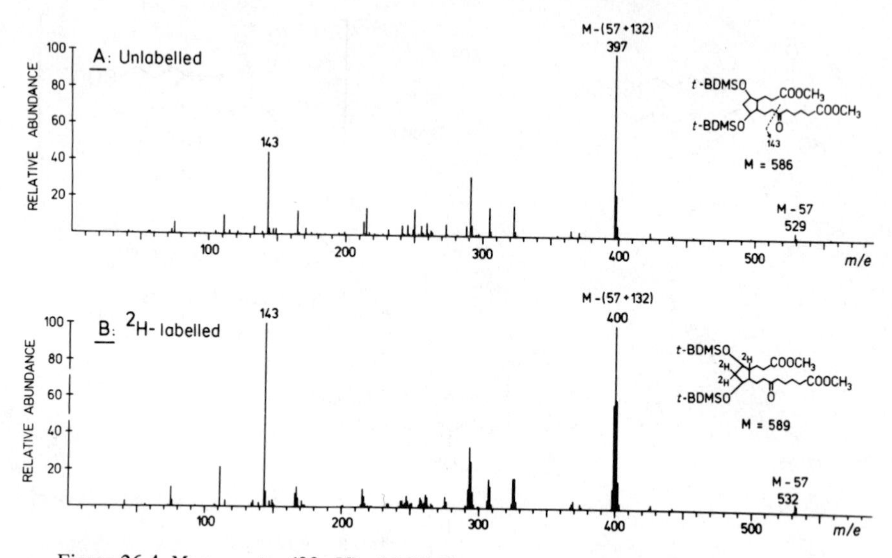

Figure 26.4 Mass spectra (23 eV) of PGF-M methyl ester *tert*-BDMS ether derivatives

extracted with dichloroethane, while more polar lipids remain in the aqueous phase and are discarded.

The material recovered in the dichloroethane phase is subsequently allowed to stand overnight in borate buffer (pH 10); this procedure results in hydrolysis of the methyl ester or δ-lactone grouping at the C-1 position but does not affect the methyl ester group at C-16. Extraction of this basic solution with dichloroethane or ethyl acetate serves to remove basic and neutral components of the urinary extract. The aqueous phase is then acidified to pH 2 and allowed to stand at room temperature for 1 h, resulting in dehydration of the 1-monoic acid 16-methyl ester to the δ-lactone 16-methyl ester derivative. This neutral compound is then extracted into dichloroethane after the aqueous phase has been re-adjusted to pH 8. The net result of this selective hydrolysis/back-extraction sequence is effectively to remove all basic and neutral urinary metabolites, together with most of the urinary acids, and to give a 50- to 100-fold purification of the extract.

The dimethyl ester of PGF-M is now re-formed from the δ-lactone methyl ester; advantage is taken of the basic conditions of the methylation procedure to obtain the δ-lactone ⇌ 5-hydroxy acid ⇌ 5-hydroxy ester as a favourable equilibrium. The methylated extract is then treated with a mixture of *tert*-butyldimethylchlorosilane/imidazole/dimethylformamide at room temperature overnight and the resulting *tert*-BDMS ether derivative is extracted and purified by thin-layer chromatography (R_f = 0.45 on silica gel 60; solvent system : ethyl acetate/heptane; 2 : 3, v/v). The extract is now sufficiently pure for analysis by GC–MS.

The GC–MS instrument is focused to monitor the ion currents at *m/e* 397 (base peak of the methyl ester *tert*-BDMS ether derivative of unlabelled PGF-M) and at *m/e* 400 (base peak of the deuterium-labelled derivative). Typical ion current profiles obtained from a mixture of standards and from a urine extract

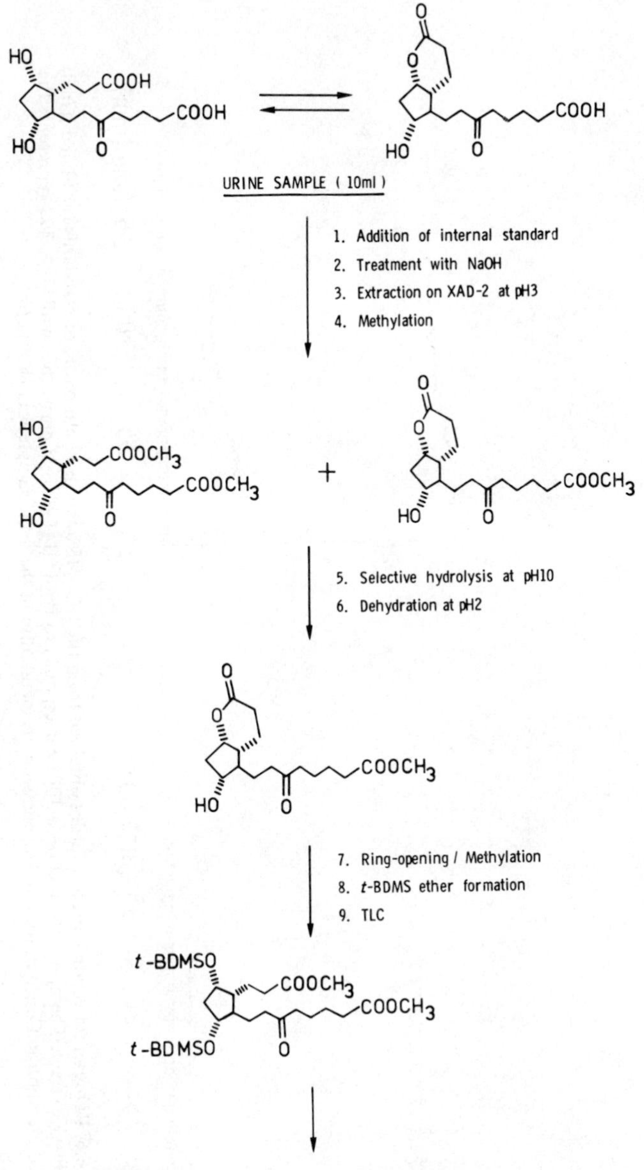

Figure 26.5 Analytical procedure for the determination of PGF-M in urine

are shown in Figure 26.6. The ratio of peak heights is measured and the amount of metabolite in the original urine specimen is determined from a standard curve. Measurements of urinary PGF-M in the normal range of 5–40 ng/ml may be made with 2 per cent precision from a 10 ml sample of urine, while the lower limit of detection is in the order of 1 ng/ml.

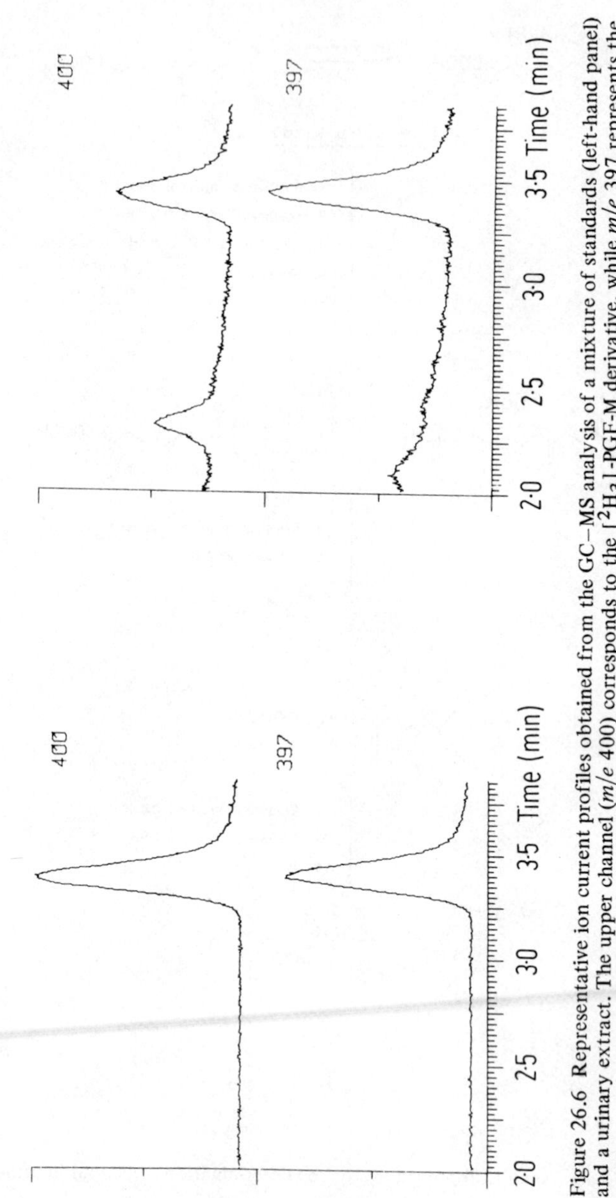

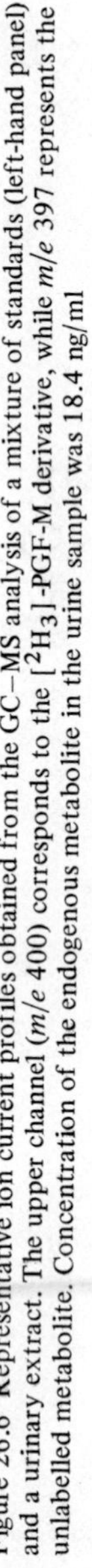

Figure 26.6 Representative ion current profiles obtained from the GC–MS analysis of a mixture of standards (left-hand panel) and a urinary extract. The upper channel (*m/e* 400) corresponds to the [$^{2}H_{3}$]-PGF-M derivative, while *m/e* 397 represents the unlabelled metabolite. Concentration of the endogenous metabolite in the urine sample was 18.4 ng/ml

The values obtained for daily excretion rates of PGF-M in healthy male and female subjects are in good agreement with those obtained by radioimmunoassay (Ohki *et al.*, 1976; Granström and Kindahl, 1976) and by an earlier mass spectrometric method (Hamberg, 1973a). In 18 males the values ranged from 10.6 to 66.9 μg/24 h (mean value 22.2 ± 13.2 (S.D.) μg/24 h) and in 17 females from 6.5 to 18.8 μg/24 h (mean value 11.6 ± 3.7 (S.D.) μg/24 h).

Excretion of PGF-M during an asthmatic attack

Pharmacological studies on the effect of prostaglandins on bronchial smooth muscle and respiratory function have indicated that endogenously synthesised prostaglandins could be important mediators in the pathogenesis of human bronchial asthma (Horton, 1969). Several prostaglandins are bronchoconstrictors and of these $PGF_{2\alpha}$ has been the most extensively studied. Using isolated human lung tissue it has been demonstrated that $PGF_{2\alpha}$ can be synthesised and released in response to mechanical or immunological provocation (Piper and Walker, 1973) and recently it has been shown that immunological challenge to sensitive human subjects causes an increase in the circulating levels of 15-keto-13,14-dihydro-$PGF_{2\alpha}$, the main metabolite of $PGF_{2\alpha}$ in plasma (Gréen, Hedqvist and Svanborg, 1974). In the latter experiment the increase in metabolite levels was in proportion to the severity of the induced bronchospasm and up to tenfold changes in metabolite levels were recorded. If these changes were maintained over the course of a more prolonged attack of bronchial asthma, then one would expect to find a corresponding increase in the levels of PGF-M in urine.

We have measured urinary PGF-M output in patients admitted to hospital for exacerbation of intrinsic bronchial asthma (i.e. asthma which is not of immunological origin). The object of the investigation was to determine whether, in this type of asthmatic patient, the rate of production of $PGF_{2\alpha}$ altered with changes in respiratory function and whether it was modified by drug treatment. Respiratory function was monitored throughout the period of hospitalisation and the results are expressed as the peak expiratory flow rate (PEFR).

Seven male and two female asthmatics have been studied; the PEFR values and corresponding urinary PGF-M excretions are given in Table 26.1. The patients each presented with severe respiratory obstruction; on the day of admission to hospital the PEFR averaged 131 l/min in the male subjects and

Table 26.1 Excretion of PGF-M (ng/mg creatinine) and PEFR during an asthmatic attack

	Male *n* = 5		Female *n* = 2	
Day	Excretion rate (normal = 6.0–38.7)	PEFR (l/min)	Excretion rate (normal = 4.6–19.0)	PEFR (l/min)
1	15.9 ± 6.3	131 ± 19.0	7.3 ± 1.3	98 ± 27.5
2	13.7 ± 4.3	165 ± 23.3	10.6 ± 4.5	133 ± 47.5
3	15.0 ± 4.5	195 ± 39.4	14.8 ± 6.3	158 ± 52.5
4	15.3 ± 5.2	249 ± 44.9	10.2 ± 0.1	238 ± 62.5
5	14.0 ± 4.4	338 ± 58.9	12.4 ± 1.1	360 ± 5.0

98 l/min in the females (Table 26.1). Treatment with beta-adrenergic agonists and steroids was initiated and continued with appropriate modifications until the PEFR reached more normal values. During this period of recovery the urinary excretion of PGF-M remained essentially unchanged. In each of the asthmatics the metabolite levels were within the range found in normal subjects at all times, and as is evident from the data in Table 26.1, there was no apparent trend towards a fall in PEFR concomitant with improvement in lung function.

It can be concluded that there was no abnormality in $PGF_{2\alpha}$ production in the patients we have studied and that there was no demonstrable effect of drug therapy on prostaglandin biosynthesis. The results suggest that $PGF_{2\alpha}$ was not an important mediator of the asthma in these patients. It is apparent that the increased production of $PGF_{2\alpha}$ which is seen on immunological challenge of sensitive subjects is not maintained throughout a prolonged attack of the intrinsic type of bronchial asthma.

The effect of indomethacin on the biosynthesis and metabolism of prostaglandin $F_{2\alpha}$ in man

It is well established that indomethacin and other non-steroidal anti-inflammatory drugs are inhibitors of prostaglandin biosynthesis and there is now a great deal of evidence to suggest that this may represent the mechanism of action of these drugs (Vane, 1974). It has been taken as strong support for the latter hypothesis that the administration of indomethacin or aspirin to animals or man results in a marked reduction in the excretion of prostaglandin urinary metabolites. This was first demonstrated in the guinea-pig, when it was also shown that doses of indomethacin which suppressed endogenous metabolite output had no effect on the bioconversion of tritium-labelled PGE_2 into labelled urinary metabolites (Hamberg and Samuelsson, 1972). However, only two animals were used in these preliminary studies.

Recently it has been shown that in rat kidney homogenates indomethacin can act as an inhibitor of the prostaglandin-metabolising enzymes 9-hydroxy dehydrogenase, Δ^{13} reductase and 15-hydroxy dehydrogenase (Pace-Asciak and Cole, 1975). It has also been demonstrated that the metabolism of $PGF_{2\alpha}$ in the perfused rabbit kidney is inhibited by indomethacin (Bito, 1976). The concentrations of indomethacin required for inhibition of prostaglandin metabolism in the above experiments were higher than those required for inhibition of prostaglandin biosynthesis, but they were of the same order of magnitude as would be obtained from therapeutic doses of the drug in man (Palmér *et al.*, 1974). Substantial doses of non-steroidal anti-inflammatory drugs have been used to demonstrate decreased excretion of PGE-M or PGF-M in man – in the case of indomethacin, the dose was 200 mg daily for 3 days (Hamberg, 1973b; Granström and Kindahl, 1976) – and therefore the question arises whether a significant component of this reduction in metabolite output could be due to an inhibition of prostaglandin metabolism. Since the only direct evidence against this possibility derives from the aforementioned experiments in the guinea-pig, we decided to further investigate the problem in human subjects.

Essentially, our experiment was designed to measure the effect of indomethacin on the conversion of exogenously administered $PGF_{2\alpha}$ into PGF-M and also to determine its effect on the daily output of endogenous

PGF-M. This presented a problem of how to measure the PGF-M derived from the administered $PGF_{2\alpha}$ in the presence of endogenous metabolite. The use of radio-labelled $PGF_{2\alpha}$ was ruled out because the accurate quantitative determination of radioactive PGF-M in the complex pattern of urinary $PGF_{2\alpha}$ metabolites would have presented a severe analytical problem. The use of unlabelled $PGF_{2\alpha}$ would require a dose large enough to swamp out a naturally occurring background of unlabelled metabolite, which itself is markedly altered by indomethacin. It can be shown by simple arithmetic that, for any given dose, a more accurate analysis can be performed using a deuterium-labelled tracer because the effect of variations in the biological background is minimised. In this case the deuterium-labelled PGF-M which is formed from $[^2H_3]$-$PGF_{2\alpha}$ can be assayed by selected ion monitoring GC–MS using a large excess of the unlabelled material (25–50 times the biological background) as internal standard.

The minimum amount of labelled $PGF_{2\alpha}$ which was required for the infusion in order to obtain good analytical precision in the measurement of the deuterated metabolite in urine was calculated as follows.

(1) The rate of excretion of endogenous PGF-M in healthy subjects is usually 1 μg/h or less. It was therefore necessary to add at least 20 μg of unlabelled PGF-M as internal standard to a 1 h urine sample. This amount would render fluctuations in the endogenous metabolite of little consequence to the measurement of deuterated metabolite.

(2) Since the ratio of the ions at *m/e* 400/397 in the unlabelled PGF-M methyl ester *tert*-BDMS ether derivative was known to be 0.02, the 'blank' in the *m/e* 400 channel would be equivalent to (0.02 x 20) μg/h = 0.4 μg/h in the absence of deuterated metabolite.

(3) It was assumed that an increment in the ratio *m/e* 400/397 of 0.01 was the minimum which could be measured as being significantly different from the 'blank' value of 0.02. It followed that the minimum detectable amount of deuterated metabolite would be (0.01/0.02 x 0.4) μg/h = 0.2 μg/h.

(4) It was known that only about 20 per cent of infused $PGF_{2\alpha}$ is excreted as PGF-M (Granström and Samuelsson, 1971) and it was decided that 5 per cent of this material should represent the limit of sensitivity of the assay (i.e. sensitivity in terms of a significant signal above the 'blank' rather than as absolute signal-to-noise ratio). It followed that the minimum dose of deuterated $PGF_{2\alpha}$ which could be infused was (0.2 x 100/20 x 100/5) μg = 20 μg.

(5) Finally, since the isotopic purity of the labelled $PGF_{2\alpha}$ was only 60 per cent $^2H_3$, the response of the corresponding deuterated PGF-M derivative at *m/e* 400 would be approximately one-half of the value obtained in a pure trideuterated species. To allow for this factor the minimum dose had to be doubled, to 40 μg. To provide a suitable margin for error in the above calculations, it was decided to increase the infused dose to 60 μg of deuterated $PGF_{2\alpha}$.

The experiment was carried out in four healthy male volunteers. Endogenous PGF-M levels were measured in a control 24 h urine and in a 24 h collection made on the fourth day of treatment with 200 mg of indomethacin (50 mg qds). On a different control day the subjects were given an intravenous infusion of 60 μg [8,10,10-$^2H_3$]-$PGF_{2\alpha}$ over a period of 10 min and this infusion was repeated on the fifth day of treatment with indomethacin. On the second occasion a 10 ml blood sample was taken immediately before the infusion; the plasma was stored at −20° and later analysed for indomethacin by high-pressure liquid chromatography (Skellern and Salole, 1975). After the infusion the subjects were permitted to continue their normal daily routine. Hourly urine

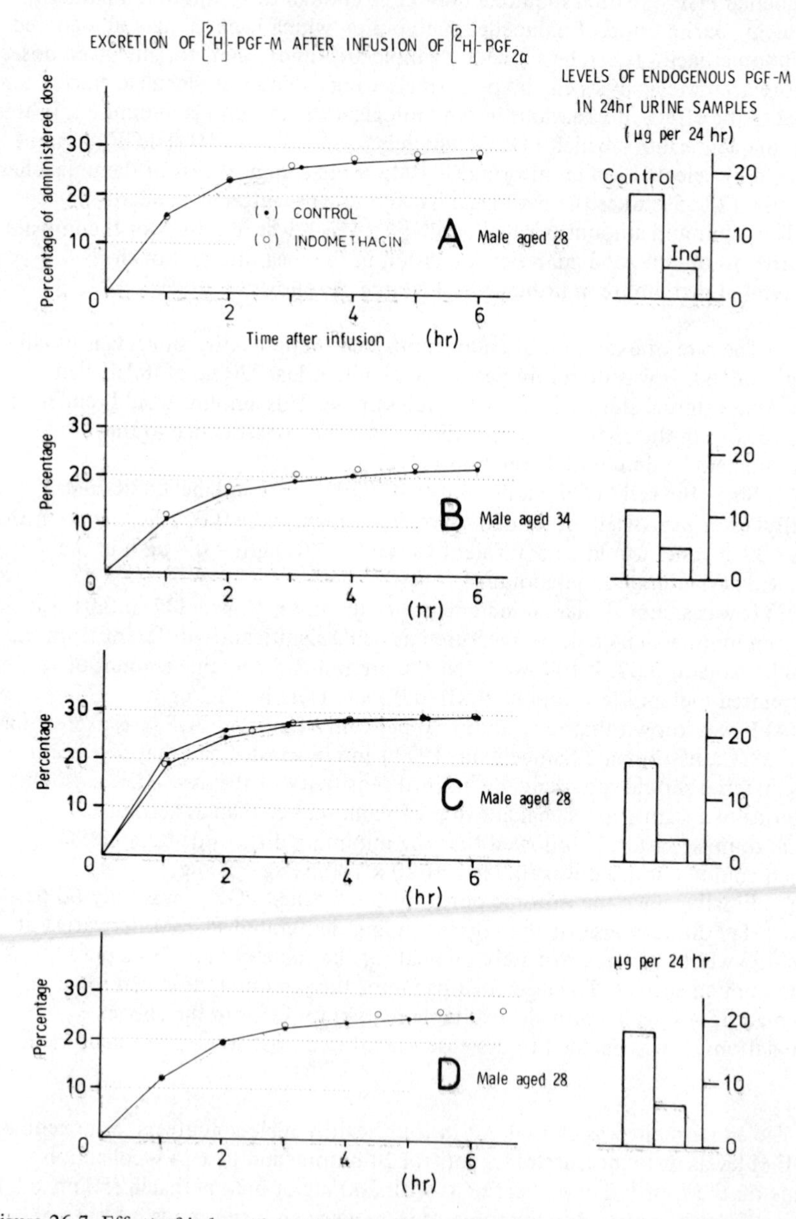

Figure 26.7 Effect of indomethacin (200 mg/day) on the biosynthesis and metabolism of $PGF_{2\alpha}$

collections were made and the first six samples were analysed for deuterium-labelled PGF-M; 10 per cent of each hourly collection was 'spiked' with 2 μg of unlabelled PGF-M, and the samples were then carried through the assay procedure described above and analysed by GC–MS.

The results of the experiment are summarised in Figure 26.7. Looking down the right-hand columns it can be seen that indomethacin (50 mg qds) caused a reduction in endogenous metabolite levels of between 55 and 75 per cent in the four individuals. From the data on the left-hand side of the figure, it is apparent that indomethacin had no effect on the bioconversion of infused deuterium-labelled $PGF_{2\alpha}$ to deuterated PGF-M; both the rate of excretion of the deuterated metabolite and the total amount formed were unaltered. The plasma concentrations of indomethacin at the time of infusion were 1.8, 1.2, 1.4 and 1.0 μg/ml in subjects A, B, C and D, respectively. It can be concluded from these results that indomethacin, at a dose of 200 mg daily, causes a reduction in the excretion of PGF-M through its effect on the biosynthesis, rather than on the metabolism, of $PGF_{2\alpha}$.

ACKNOWLEDGEMENTS

The authors would like to thank Dr J. E. Pike of the Upjohn Company and Dr K. Crowshaw of May and Baker Ltd for gifts of prostaglandins and Dr V. Tarkkanen (Packard-Becker) for samples of Lipidex gels. This work was supported by a grant from the Medical Research Council.

REFERENCES

Bito, L. Z. (1976). *Prostaglandins*, **12**, 639
Brash, A. R., Baillie, T. A., Clare, R. A. and Draffan, G. H. (1976). *Biochem. Med.*, **16**, 77
Brash, A. R. and Jones, R. L. (1974). *Prostaglandins*, **5**, 441
Granström, E. and Kindahl, H. (1976). *Advances in Prostaglandin and Thromboxane Research*, Vol. 1 (ed. B. Samuelsson and R. Paoletti), Raven Press, New York, p. 81
Granström, E. and Samuelsson, B. (1971). *J. Biol. Chem.*, **246**, 5254
Greeley, R. H. (1974). *J. Chromatogr.*, **88**, 229
Gréen, K., Granström, E., Samuelsson, B. and Axén, U. (1973). *Anal. Biochem.*, **54**, 434
Gréen, K., Hedqvist, P. and Svanborg, N. (1974). *Lancet*, **2**, 1419
Hamberg, M. and Samuelsson, B. (1972). *J. Biol. Chem.*, **247**, 3495
Hamberg, M. (1973a). *Anal. Biochem.*, **55**, 368
Hamberg, M. (1973b). *Biochem. Biophys. Res. Commun.*, **49**, 720
Horton, E. W. (1969). *Physiol. Rev.*, **49**, 122
Nidy, E. G. and Johnson, R. A. (1975). *J. Org. Chem.*, **40**, 1415
Nyström, E. and Sjövall, J. (1973). *Anal Lett.*, **6**, 155
Ohki, S., Nishigaki, Y., Imaki, K., Kurono, M., Hirata, F., Hanyu, T. and Nakazawa, N. (1976). *Prostaglandins*, **12**, 181
Pace-Asciak, C. and Cole, S. (1975). *Experientia*, **31**, 143
Palmér, L., Bertilsson, L., Alvan, G., Orme, M., Sjöqvist, F. and Holmstedt, B. (1974). *Prostaglandin Synthetase Inhibitors* (ed. H. J. Robinson and J. R. Vane), Raven Press, New York, p. 91
Pike, J. E., Lincoln, F. H. and Schneider, W. P. (1969). *J. Org. Chem.*, **34**, 3552
Piper, P. J. and Walker, J. L. (1973). *Brit. J. Pharmacol.*, **47**, 291
Skellern, G. C. and Salole, E. G. (1975). *J. Chromatogr.*, **114**, 483
Vane, J. R. (1974). *Prostaglandin Synthetase Inhibitors* (ed. H. J. Robinson and J. R. Vane), Raven Press, New York, p. 155

27

Balance of water, urea and creatinine in peritoneal dialysis

R. Medina, H.-L. Schmidt and W. Hoppe (Institut fur Chemie Weihenstephan der Technischen Universität München, D-8050 Freising-Weihenstephan, Federal Republic of Germany) and G. Traut (II. Medizinische Klinik der Universität des Saarlandes, D-6650 Homburg/Saar, Federal Republic of Germany)

INTRODUCTION

Following kidney failure, metabolites normally excreted in urine accumulate in the body. They can, however, be eliminated by dialysis procedures, most commonly by haemodialysis (artificial kidney), or in some cases by peritoneal dialysis (Traut, 1969; Jutzler and Traut, 1973); in the latter method, the patient's peritoneum is the dialysis membrane.

In general, the transport of metabolites through the peritoneum is mediated by diffusion forces, but the absolute velocity of this transport may depend on several factors. The kinetics and balance of the diffusion process can be derived from the total pool of the metabolites in the body and from its variations during the treatment. Parameters of a kinetic model for the process can be determined by isotope dilution analyses and from the decrease of concentrations and labelling patterns of the metabolites as a function of time.

The aim of the present investigations was to determine the balance of certain metabolites during peritoneal dialysis and to develop a method for the rapid estimation of individual diffusion parameters in order to establish the optimal conditions for the treatment of a patient.

METHODS

Two litres of the dialysis liquid was introduced into the patient's abdomen, 1 litre of which was replaced every 15 min by fresh liquid. After 6–8 cycles, the total liquid was drained off. The treatment was continued for 24 h, after which time about 120 litres of dialysis liquid had been in contact with the patient's peritoneum.

Normally, patients were given 10 ml of D_2O orally and/or 150 mg urea-[$^{15}N_2$] or 35 mg creatinine-[^{15}N] i.v. in saline solution before treatment; after a 3 h equilibration period, the dialysis was started. In some cases, administration of D_2O was repeated after the treatment was complete. Urea and creatinine were determined in samples of blood (10 ml) and of dialysate (200 ml) (Ceriotti, 1974), and metabolites were isolated for isotope analysis according to the scheme shown in Figure 27.1.

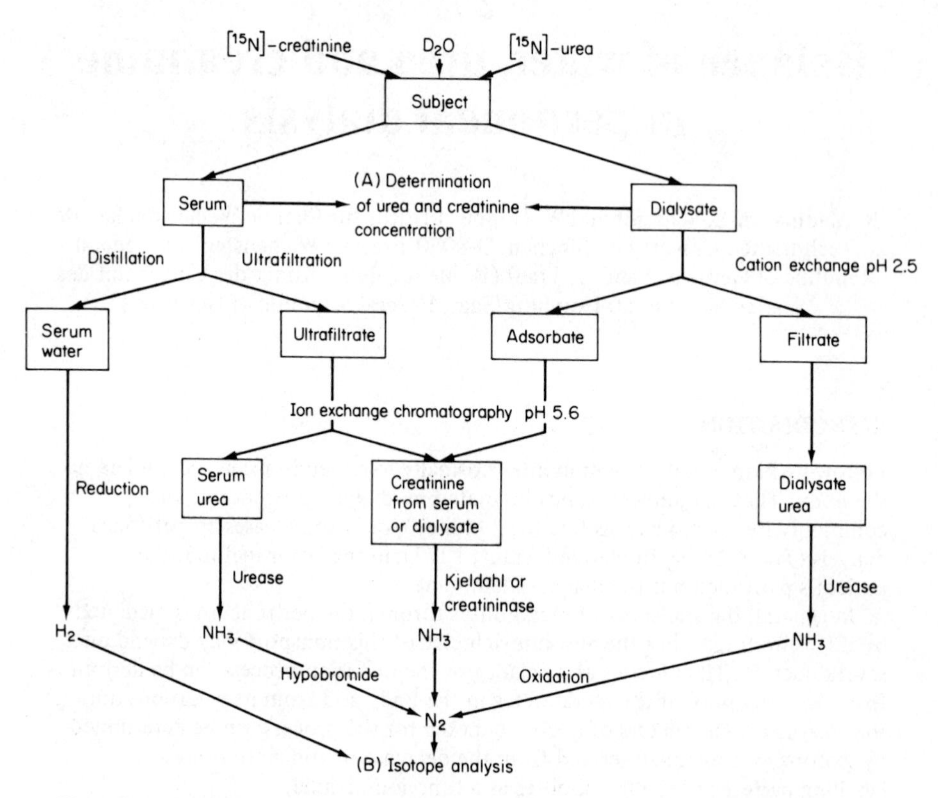

Figure 27.1 Procedure of isolation of urea and creatinine from serum and dialysis liquid for isotope analysis

The first step in the isolation of creatinine from serum was ultrafiltration (centriflo membrane cones Type CF 25, Amicon Corporation, Massachusetts, USA), while for the measurement of the metabolite in the dialysate, it was found to be necessary to concentrate the fluid by filtration on Dowex 50 W x 80 (200–400 mesh, column volume 8 ml) after adjustment to pH 2.5. Further purification was performed by chromatography on M 21 ion-exchange resin (Beckmann Instruments International) with 2 N citrate buffer, pH 5.6 (buffer flux 68 ml/h, temperature 56°); the fractions containing creatinine were identified by reaction with sodium picrate (Ceriotti, 1974). Creatinine was combusted by Kjeldahl oxidation or by incubation with creatininase (EC 3.5.4.21, Beckmann Instruments, Intern. Bioproducts/Microbics Depts.

Geneve, 2.9 U/ml, 1 h, pH 8.5) (McLean, Gallas and Hendrixson, 1973); this enzyme splits off the amino group of creatinine which contains the ^{15}N label (Medina and Schmidt, 1976).

The fractions containing urea were treated directly with urease (20 U/ml, 30 min, pH 7), and the NH_3 formed was isolated by Conway diffusion (Fiedler and Proksch, 1975). Preparation of samples for isotope analysis is described elsewhere (Schmidt, 1974); a mass spectrometer model MM 903, VG-Isotopes Ltd, Nat Lane, Winsford, Cheshire, UK) was used for deuterium and ^{15}N determinations.

RESULTS AND DISCUSSION

Total body water and volumes of distribution for urea and creatinine

Total body water (TBW) is determined from isotope dilution with D_2O (Heinz *et al.*, 1963) and is normally expressed as a percentage of body weight. The values found for three patients were in the normal range, while one was rather low (Table 27.1). Differences in TBW before and after dialysis corresponded to the patients' loss of body weight during the treatment.

Table 27.1 Total body water (TBW), urea and creatinine spaces. All values in % of body weight

Subject	TBW	Urea space	Creatinine space
Z	53.1	53.5	53.7
S	62.4	62.1	–
P	57.9	58.0	–
M	44.6	44.6	–

The total amounts of urea and creatinine in the patients were determined by isotope dilution analyses with urea-[$^{15}N_2$] and creatinine-[^{15}N]; total urea was in the range of 70 g, while creatinine was in the range of 4 g per patient. From these data and from the serum concentrations of the metabolites, their volumes of distribution were calculated. As shown in Table 27.1, they were identical with the TBW; therefore urea and creatinine must be uniformly distributed over the whole body, a result which, in the case of urea, had already been obtained by others (San Pietro and Rittenberg, 1953). Thus the TBW may be determined by use of either of the labelled compounds used in this study.

Equilibration of water and metabolites between body water and dialysis liquid

Peritoneal dialysis may be described by a two-compartment model which is derived from a gradient mixing system (Schmidt *et al.*, 1975). In this model a pool of unlabelled water, the dialysis liquid, is in contact with the second (labelled) pool, namely the TBW; yet total mixing is prevented by a permeability barrier.

The isotope abundance of the serum water at a given time a_t is a function of the degree of labelling a_0 at the beginning of the dialysis, the total body water TBW and the flux f [1/hr] of the dialysis liquid (equation 27.1):

$$a_t = a_0 - (0.015e^{-\alpha f t/\text{TBW}}) + 0.015 \text{ atom } \% \qquad (27.1)$$

In this equation, 0.015 atom % is the natural abundance of deuterium in water while α is an equilibration factor, which would equal 1 in the case of total mixing of the two pools; it is an expression for a relative diffusion velocity, depending on the constitution and the surface area of the separating membrane.

Experimental data were in complete accordance with this equation; a typical curve for the decrease of the deuterium content in the serum water during the treatment is shown in Figure 27.2. Regression analysis showed that the experimental data in this case are best represented by equation (27.1) for $\alpha = 0.358$.

A corresponding model for the time course of elimination of the metabolites must take into consideration that the pool of these substances is enlarged by their *de novo* synthesis while they are being eliminated by dialysis. Thus, in the

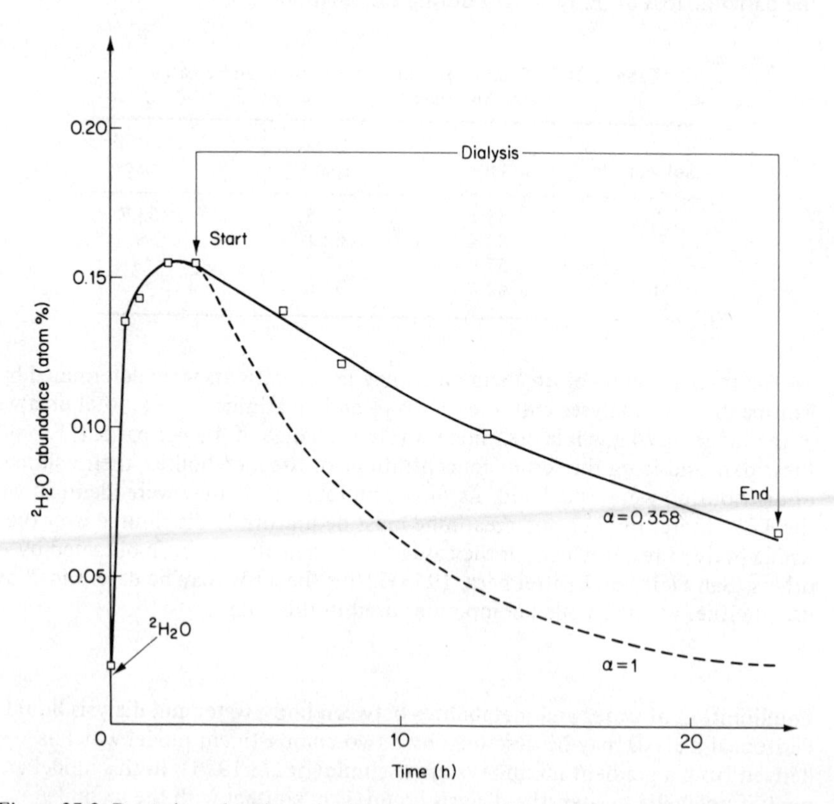

Figure 27.2 Deuterium abundance in the serum water of a patient during peritoneal dialysis (oral dose of D_2O 50 ml). Solid line calculated from equation (27.1) for $\alpha = 0.358$, dotted line for $\alpha = 1.0$

case of urea, if A_t and A_0 are the total amounts of the metabolite present in the body at time t and time zero, respectively, with c_t and c_0 *as the corresponding* concentrations, and if p_t is the total amount of urea synthesised at time t, then the course of these parameters would be described by equations (27.2):

$$A_t = (A_0 + p_t)\,\mathrm{e}^{-\beta f t/\mathrm{TBW}} \text{ g} \tag{27.2a}$$

and

$$c_t = (c_0 + p_t/\mathrm{TBW})\,\mathrm{e}^{-\beta f t/\mathrm{TBW}} \text{ g/l} \tag{27.2b}$$

where β is the equilibration factor for urea. p_t can be calculated from the overall balance by introducing B_t as the total amount of urea eliminated at time t:

$$p_t = A_t + B_t - A_0 \text{ g} \tag{27.3}$$

Figure 27.3 shows typical curves from which β and p values have been calculated. β values thus obtained varied between 0.12 and 0.30 (Table 27.2). Total amounts of urea eliminated in 20 h were found to be between 40 and 59 g while synthesis during this interval was 10–16 g (0.5–0.8 g/hr) (Table 27.3); the

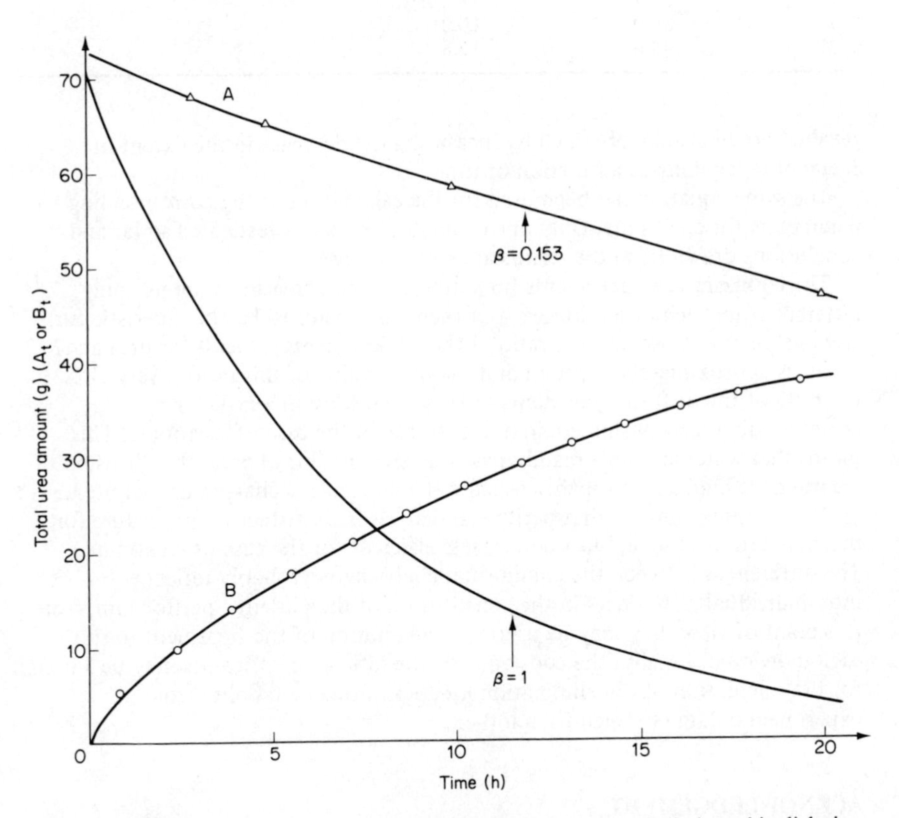

Figure 27.3 Total amount of urea in a patient's body (curve A) and eliminated in dialysis liquid (curve B) during peritoneal dialysis. β values were calculated from equations (27.2)

Table 27.2 Equilibration coefficients for water, urea and creatinine calculated from equations (27.1) and (27.2)

Subject	Equilibration coefficients for			Ratio of coefficients	
	water (α)	urea (β)	creatinine (γ)	β/α	γ/α
Z	0.212	0.117	0.046	0.55	0.22
S	0.358	0.280	–	0.78	–
P	0.203	0.153	–	0.66	–
M	0.410	0.280	–	0.68	–

Table 27.3 Data of overall balance for urea and creatinine during peritoneal dialysis. Values of B_t and p_t for a 20 h period

Subject	Urea (g/20 h)		Creatinine (g/20 h)	
	eliminated (B_t)	synthesised (p_t)	eliminated (B_t)	synthesised (p_t)
Z	54.2	15.6	1.40	0.83
S	53.2	13.2	–	–
P	59.0	10.0	–	–
M	40.0	12.8	–	–

p values could also be obtained by measuring the decrease in the extent of metabolite labelling as a function of time.

The same equation has been used for the calculation of the corresponding parameters for creatinine. Only one example has been investigated so far and conclusions drawn from the results are still tentative.

The equilibration coefficients for water, urea and creatinine are not only different from each other, but each of them also seems to be characteristic for any one person. However, the ratio of these coefficients, at least for urea and water, is approximately constant and the mean value of this ratio is very close to the ratio of the diffusion constants of these two substances (0.67); a corresponding relationship could not be found in the case of creatinine. This means that water and urea readily pass the peritoneum, whereas the diffusion of creatinine is hindered, probably because the molecule is charged or hydrated.

These results confirm that peritoneal dialysis is a satisfactory procedure for the elimination of urea, but that it is less efficient for the case of creatinine. The differences between the equilibrium coefficients probably reflect inter-individual differences in the constitution of the patients' peritoneum; from this point of view they may be useful for adaptation of the treatment to suit each individual. Finally, the constancy of the ratio of coefficients may be a basis for the calculation of the elimination kinetics of one metabolite from experimental data obtained for another.

ACKNOWLEDGEMENT

We thank the Deutsche Forschungsgemeinschaft for support given to R.M.

REFERENCES

Ceriotti, G. (1974). *Clinical Biochemistry Principles and Methods*, Vol. II (ed. H. Ch. Curtius and M. Roth), Walter de Gruyter, Berlin–New York, pp. 1124, 1135

Fiedler, R. and Proksch, G. (1975). *Anal. Chim. Acta,* **78**, 53

Heinz, R., Brass, H., Baumann, F. and Paul, U. (1963). *Klin. Wochschr.*, **41**, 359

Jutzler, G. A. and Traut, G. (1973). *Praxis der Dialysebehandlung* (ed. H. L. Franz), Georg Thieme Verlag, Stuttgart, p. 93

McLean, M. H., Gallas, J. and Hendrixson, M. (1973). *Clin. Chem.,* **19**, 623

Medina, R. and Schmidt, H.-L. (1976). *J. Labelled Compd. Radiopharm.*, **4**, 565

San Pietro, A. and Rittenberg, D. (1953). *J. Biol. Chem.*, **201**, 445

Schmidt, H.-L. (1974). *Messung von radioaktiven und stabilen Isotopen* (ed. H. Simon), Springer-Verlag, Berlin–Heidelberg–New York, p. 296

Schmidt, H.-L., Kirch, P., Traut, G. and Keller, H.-E. (1975). *Proceedings of the Second International Conference on Stable Isotopes* (ed. E. R. Klein and P. D. Klein), Oak Brook, Illinois, USA, p. 411. National Technical Information Service Document CONF-751027

Traut, G. (1969). Kolloquium: *Niereninsuffizienz und Dialyseverfahren*, July 1969, Homburg, Saar, p. 14

Index